KNOWLEDGE
MANAGEMENT
HANDBOOK

Edited by

Jay Liebowitz

Robert W. Deutsch Distinguished Professor
in Information Systems
Department of Information Systems
University of Maryland—Baltimore County

CRC Press

Boca Raton London New York Washington, D.C.

Acquiring Editor:	Ron Powers
Project Editor:	Carol Whitehead
Marketing Manager:	Jane Stark
Cover design:	Jon Pennell
PrePress:	Carlos Esser
Manufacturing:	Carol Slatter

Library of Congress Cataloging-in-Publication Data

Knowledge management handbook / edited by Jay Liebowitz.
      p.   cm.
    Includes bibliographical references and index.
    ISBN 0-8493-0238-2 (alk. paper)
    1. Knowledge management.   I. Liebowitz, Jay.
HD30.2.K637   1999
658.4'038--dc21

          98-45326
          CIP

No claim to original U.S. Government works
International Standard Book Number 0-8493-0238-2
Library of Congress Card Number 98-45326
Printed in the United States of America   2 3 4 5 6 7 8 9 0
Printed on acid-free paper

Preface

Knowledge Management: Fact or Fiction?

The knowledge management (KM) craze has commenced! Companies such as Buckman Labs, Dow Chemical, Skandia, Hewlett Packard, Monsanto, FedEx, the Big 5, RWD Technologies, and others have jumped on the knowledge management bandwagon to harness the intellectual capital, especially the human capital, in their organizations. Many CEOs will agree that their most competitive advantage is their "brainware" or their "human capital." It's refreshing to see companies like RWD invest in their human capital in a myriad of ways, including certification programs, training and education courses, Friday forums, knowledge sharing sessions, and the like. However, is knowledge management really a new concept?

Not really! Knowledge management deals with the process of creating value from an organization's intangible assets. It's an amalgamation of concepts borrowed from the artificial intelligence/knowledge-based systems, software engineering, BPR, human resource management, and organizational behavior fields. We have been concentrating on the collection of information and perhaps knowledge bases, but now we have the "connectivity" in Web-based and intranet technologies to connect to and between these knowledge bases to promote knowledge sharing. Instead of having isolated islands of knowledge, we can now build bridges between these islands.

Some companies, like Caterpillar, have a graying employee/management base where about one third of their workforce will retire in the next five years. Caterpillar is looking at knowledge management strategies to best acquire/collect their knowledge and make it available in an interactive mode to those at Caterpillar who could benefit from this knowledge. Expert systems can be a viable technology to assist in these knowledge management endeavors.

Johnson and Johnson, for example, has Knowledge Fairs and Knowledge Exchanges within their organization to promote a culture that encourages knowledge sharing among its employees. Other companies, like AMS (American Management Systems), have created Corporate Knowledge Centers within their organization. Some people, like Larry Prusak of IBM, believe that 70 to 80% of what's learned is through informal means versus formal methods like reading books, brochures, documents, etc. In this manner, these Knowledge Fairs and Corporate Knowledge Centers may be a worthwhile approach to learning through observations and discussions (although learning by doing would probably be a better approach).

RWD is moving in the right direction of seeing how knowledge management can best be used within their organization. A central theme is how to best leverage knowledge down to the customers and point of sale. RWD has recently formed a Knowledge Management Council and has started Knowledge Management Forums within the company. This is an excellent first step. Some thought should also be given to how best to encourage the sharing of knowledge in the organization and build a supportive culture to do this. Additionally, knowledge repositories/corporate memories should also be created to include lessons learned, best and worst practices, and other pertinent information. A knowledge transfer department or some other infrastructure should also be considered to oversee the development and maintenance of these knowledge management systems, as well as considering having a Chief Knowledge Officer (CKO) or equivalent position created to keep the knowledge management strategies and implementations alive within RWD.

Companies are realizing that their competitive edge is mostly the brainpower or intellectual capital of their employees and management. Many organizations are drowning in information but starving for knowledge. In order to stay ahead of the pack, companies must leverage their knowledge internally and externally to survive. With Web-based and intranet technologies, the connectivity and possible sharing of knowledge are greatly enabled to build the knowledge infrastructure of the firm. Knowledge management is believed to be the current savior of organizations, but it is much more than developing Lotus Notes lessons learned databases. Knowledge management deals with the conceptualization, review, consolidation, and action phases of creating, securing, combining, coordinating, and retrieving knowledge.

So what makes knowledge management so hard? First, the organization must create a knowledge sharing environment. Some firms provide incentives (monetary rewards, frequent flyer air miles to the first "x" number of employees to use the knowledge management system, etc.) to promote this climate until it becomes the norm. Other organizations require their employees to actively contribute and use knowledge in the organization's knowledge repositories as part of their annual job performance review. From a recent benchmarking of 150 companies, most people were not concerned about keeping their knowledge close to heart to maintain their own competitive edge. Rather, it was people who didn't want to use other people's knowledge because they couldn't put their own thumbprint on the knowledge used.

The second enigma is how can senior executives value the knowledge in their organization to show some tangible benefits. A number of individuals like Leif Edvinsson, Michael Malone, Karl Sveiby, Tom Stewart, Annie Brooking, Rob van der Spek, Robert de Hoog, myself, and others have developed methodologies to value knowledge. Some of these techniques value knowledge at the "global" (firm-wide) level, and others value knowledge at the "knowledge item" (lower level). It's not easy to value the intellectual capital (especially the "human capital") in the organization. Unless we develop ways to do this, top management may not place much emphasis on knowledge management, and may not stress the importance of intellectual capital in the organization.

A third obstacle is the belief that knowledge management is the same as information management. Knowledge is information with a process applied to it that may eventually become wisdom or expertise. Many organizations are having their IT (information technology) directors become Chief Knowledge Officers, as top management often feels they are comparable positions. This is a mistake, because knowledge management draws from many disciplines, including IT, and is broader in scope than the technology functions that an IT director often oversees. A new breed of knowledge officers or knowledge analysts is needed to fill the roles of knowledge managers in organizations.

A fourth fallacy that organizations haven't fully realized is knowledge management works best when the CEO and all others on down are actively using the knowledge management systems designed for their organizations. Without senior management's commitment and involvement, the knowledge management systems and infrastructure may be pushed aside and not be integrated within the mainstream of the organization. Buckman Labs' K'Netix (their knowledge management network) was successful largely due to the backing and use of it by their CEO, Bob Buckman.

The last major concern regarding the survivability of knowledge management is the misnomers being labeled on almost every tool as a knowledge management tool. This hype will kill the "good" from knowledge management principles in the same way that the previous fad, BPR (business process reengineering), died out. Many consulting firms are proclaiming their expertise in knowledge management. But the truth is that knowledge management, as a field, is almost too young for many experts to already exist.

Without organizations fully understanding these five major concerns, the fear is that the mystique of knowledge management will remain cloudy and shapeless. With more researchers and practitioners working together to further define and develop the KM field, the mystique of KM will begin to produce an aura of fundamentally sound principles, concepts, methodologies, techniques, and tools. Let's all work together toward this goal!

This *Handbook on Knowledge Management* will, it is hoped, serve the role as a key reference book in integrating views of researchers in and practitioners of knowledge management. Even though the knowledge management field is evolving and still maturing, this *Handbook* is a first step in helping to formulate methodologies, techniques, and practices for making knowledge management a sound field. Many of the leading individuals and organizations in the knowledge management field have contributed to this *Handbook*. I would like to thank these individuals, as well as the reviewers and the CRC Press staff. I especially want to express my gratitude to Ron Powers, the publisher at CRC, for his insight and continued support throughout the years. I would also like to express my gratitude to Jenny Preece and my colleagues and students at the University of Maryland–Baltimore County, George Washington University, RWD Technologies, and the Robert W. Deutsch Foundation. Last, this *Handbook* would not have been possible without the continued love and support of my family — Janet, Jason, Kenny, and my parents.

Jay Liebowitz
Editor

Contributors

Angela Abell
TFPL Ltd.
London
United Kingdom

Seung Baek
Saint Joseph's University
Philadelphia, Pennsylvania

Glenn Becker
Thomson Technology Labs
Rockville, Maryland

Thomas J. Beckman
The George Washington University
Washington, D. C.

David Coleman
Collaborative Strategies
San Francisco, California

Thomas H. Davenport
Boston University School of Management
Boston, Massachusetts

John S. Edwards
Aston University
Birmingham
United Kingdom

Patricia S. Foy
PricewaterhouseCoopers
Stamford, Connecticut

Mary Granger
The George Washington University
Washington, D. C.

Clyde W. Holsapple
Carol M. Gatton College of Business and Economics
University of Kentucky
Lexington, Kentucky

Robert de Hoog
Department of Social Science Informatics
University of Amsterdam
The Netherlands

Dustin Huntington
MultiLogic
Albuquerque, New Mexico

K. D. Joshi
Washington State University
Pullman, Washington

Jay Liebowitz
University of Maryland — Baltimore County
Baltimore, Maryland

Ron Mallis
AED Inc.
Woburn, Massachusetts

Carl R. Moore
Science Applications International
    Corporation (SAIC)
McLean, Virginia

Nigel Oxbrow
TFPL Ltd.
London
United Kingdom

Srinivas Y. Prasad
The George Washington University
Washington, D. C.

James A. Sena
California Polytechnic State University
San Luis Obispo, California

A. B. (Rami) Shani
California Polytechnic State University
San Luis Obispo, California

Ed Swanstrom
Chairman, Knowledge Management
   Consortium
Gaithersburg, Maryland

Robert M. Taylor
KPMG Management Consulting
London
United Kingdom

Bart van der Meij
Reekx
Groningen
The Netherlands

Rob van der Spek
CIBIT
Utrecht
The Netherlands

Gertjan van Heijst
CIBIT
Utrecht
The Netherlands

Karl M. Wiig
Knowledge Research Institute, Inc.
Arlington, Texas

Kathie Wright
The George Washington University
Washington, D. C.

Kim Ann Zimmermann
Consultant
East Windsor, New Jersey

Contents

SECTION IV   Knowledge Management: Knowledge Technologies

SECTION V   Knowledge Management: Applications

Section I
Knowledge Management
and Strategy

1

The Current State of Knowledge Management

Thomas J. Beckman
The George Washington University

1.1 Introduction

Knowledge management (KM) is an emerging discipline with many ideas yet to be tested, many issues yet to be resolved, and much learning yet to be discovered. This chapter surveys the existing literature and presents an overview of the current state of progress in knowledge management (KM) by examining it from a variety of perspectives:

- Conceptual
- Process
- Technology
- Organizational
- Management
- Implementation

On the **conceptual** side, the following KM topics are discussed:

- Definitions of knowledge and knowledge management
- Knowledge dimensions
- Principles of knowledge and knowledge management
- KM framework

Several **process** models will be described that explain how KM works.

The **technology** section of the chapter reviews how information technology (IT) — from Lotus Notes to expert systems — can be used to enable and augment KM:

- IT infrastructure: computing, communication, storage, presentation
- Knowledge representation schema
- Knowledge elicitation
- Knowledge repository
- Integrated performance support systems
- Knowledge transformation

In the next section, several **organizational** concepts are considered:

- Knowledge organization characteristics
- Core competencies, knowledge domains, and skills assessment
- Organizational forms: networks and centers of expertise
- Roles and responsibilities
- Organizational learning
- Corporate culture

In the **management** section, a variety of practices may be needed to make KM a reality:

- Management practices
- Measuring and valuing intellectual capital
- Reward, compensation, and motivational systems

Finally, in the **implementation** section a variety of methods, techniques, and advice are reviewed that facilitate making knowledge projects successful:

- Success factors
- Challenges and prerequisites
- IT infrastructure implementation
- KM strategies

Before discussing the six proposed perspectives and their implications, a brief chronology of important KM events provides a seventh, a **historical** perspective:

Year	Entity	Event
1980	Digital Equipment Corporation Carnegie Mellon University	One of the first commercially successful Expert Systems XCON: Configures computer components
1986	Dr. Karl Wiig	Coined KM concept at keynote address for United Nation's International Labor Organization
1989	Large management consulting firms	Start internal efforts to formally manage knowledge
1989	Price Waterhouse	One of the first to integrate KM into its business strategy
1991	*Harvard Business Review* (Nonaka and Takeuchi)	One of the first journal articles on KM published
1993	Dr. Karl Wiig	One of the first books dedicated to KM published (*Knowledge Management Foundations*)
1994	Knowledge Management Network	First KM conference held
1994	Large consulting firms	First to offer KM services to clients
1996+	Various firms and practitioners	Explosion of interest and activities

1.2   Conceptual Perspective

The conceptual perspective is concerned with defining and describing the foundations and frameworks of KM. Because the KM discipline is so young, the author believes that presenting a variety of views is better than trying to describe the subject from just one or two perspectives. This section begins with the definition of basic terms. Then the characteristics and relationships between knowledge concepts are described. Finally, several architectures, frameworks, and perspectives are proposed.

Definition of Knowledge

Definitions of knowledge range from the practical to the conceptual to the philosophical, and from narrow to broad in scope. The following definitions are relevant to the topic of KM:

- Knowledge is organized information applicable to problem solving. — Woolf [1]
- Knowledge is information that has been organized and analyzed to make it understandable and applicable to problem solving or decision making. — Turban [2]
- Knowledge encompasses the implicit and explicit restrictions placed upon objects (entities), operations, and relationships along with general and specific heuristics and inference procedures involved in the situation being modeled. — Sowa [3]
- Knowledge consists of truths and beliefs, perspectives and concepts, judgments and expectations, methodologies and know-how. — Wiig [4]
- Knowledge is the whole set of insights, experiences, and procedures that are considered correct and true and that therefore guide the thoughts, behaviors, and communications of people. — van der Spek and Spijkervet [5]
- Knowledge is reasoning about information and data to actively enable performance, problem-solving, decision-making, learning, and teaching. — Beckman [6]

There are also definitions of organizational knowledge centering around intellectual capital:

- Organizational knowledge is the collective sum of human-centered assets, intellectual property assets, infrastructure assets, and market assets. — Brooking [26]
- Organizational knowledge is processed information embedded in routines and processes that enable action. It is also knowledge captured by the organization's systems, processes, products, rules, and culture. — Myers [27]

Knowledge Dimensions

There are many dimensions around which knowledge can be characterized. In this chapter, several characteristics of knowledge will be examined: storage media, accessibility, typology, and hierarchy. In the typology section, the relationships between dimensions are explored. In addition, some principles about knowledge are offered for consideration.

Knowledge Storage Media

First, there are several **media** in which knowledge can reside: human mind, organization, document, and computer. Knowledge in the human mind is often difficult to access; organizational knowledge is often diffuse and distributed; document knowledge can range from free text to well-structured charts and tables; computer knowledge is formalized, sharable, and often well-structured and well-organized.

Knowledge Accessibility

Next, there is the dimension of knowledge **accessibility**. Nonaka and Takeuchi [7] have divided accessibility into two categories: tacit and explicit. In this author's view, there may be three stages of accessibility: tacit, implicit, and explicit [8]. Accessibility can be mapped to storage media. Knowledge gains in value as it becomes more accessible and formal:

- *Tacit* (human mind, organization) — accessible indirectly only with difficulty through knowledge elicitation and observation of behavior
- *Implicit* (human mind, organization) — accessible through querying and discussion, but informal knowledge must first be located and then communicated
- *Explicit* (document, computer) — readily accessible, as well as documented into formal knowledge sources that are often well-organized

Knowledge Typologies

Another dimension involves the **typology** of knowledge. Typologies are defined, categorized, and described in terms of knowledge type-conversion, structural features, elementary properties, purpose and use, and conceptual levels. Nonaka and Takeuchi [7] suggest the following types of knowledge:

Tacit Knowledge	Explicit Knowledge
Knowledge of experience (body skills)	Knowledge of rationality (mind)
Simultaneous knowledge (here and now)	Sequential knowledge (there and then)
Analog knowledge (practice)	Digital knowledge (theory)

Nonaka [7] has also developed a matrix for knowledge conversion based on accessibility:

Knowledge Conversion	Tacit Knowledge	Explicit Knowledge
To: From:		
Tacit knowledge	Socialization (Sympathized K)	Externalization (Conceptual K)
Explicit knowledge	Internalization (Operational K)	Combination (Systemic K)

Collins [9] also relates knowledge types to their accessibility:

- Symbol-type knowledge   (explicit)
- Embodied knowledge   (implicit)
- Embrained knowledge   (implicit/tacit)
- Encultured knowledge   (tacit)

Van der Spek and Spijkervet [5] discuss the following structural features of knowledge:

- *Availability*: form, time, location
- *Content*: structure, application

Parsaye and Chignell [10] describe five elementary properties of knowledge that can be used to define and represent objects and their interactions:

- Naming   (proper nouns)
- Describing   (adjectives)
- Organizing   (categorization and possession)
- Relating   (transitive verbs and relationship nouns)
- Constraining   (conditions)

Quinn [25] suggests the following typology based on purpose and use:

- Know-what
- Know-how
- Know-where
- Know-why
- Care-why

Finally, according to Brooking [26], there are four conceptual levels of knowledge:

1. Goal-setting or idealistic knowledge
2. Systematic knowledge
3. Pragmatic knowledge
4. Automatic knowledge

Knowledge Hierarchy

A further dimension considers the premise that knowledge can be organized into a **hierarchy**. Several authors, including Alter [11], Tobin [12], van der Spek and Spijkervet [5], and Beckman [6] draw distinctions between data, information, and knowledge:

1. *Data*:   Facts, images, or sounds (+ interpretation + meaning =)
2. *Information*:   Formatted, filtered, and summarized data (+ action + application =)
3. *Knowledge*:   Instincts, ideas, rules, and procedures that guide actions and decisions

Next, Tobin [12] adds an additional higher level, "wisdom," in his hierarchy:

1. *Data*:   (+ relevance + purpose =)
2. *Information*:   (+ application =)
3. *Knowledge*:   (+ intuition + experience =)
4. *Wisdom*

Finally, Beckman [6] proposes a five-level knowledge hierarchy in which knowledge can often be transformed from a lower level to a more valuable higher level:

1. *Data*:   Text, fact, code, image, sound
                 <attribute value>
                 (+ meaning + structure =)
2. *Information*:   Organized, structured, interpreted, summarized data
                 <object attribute value>
                 (+ reasoning + abstraction + relationships + application =)
3. *Knowledge*:   Case, rule, process, model
                 <relation object attribute value>
                 (+ selection + experience + principles + constraints + learning =)
4. *Expertise*:   Fast & accurate advice, explanation & justification of result & reasoning
                 <relation object attribute value certainty importance>
                 (+ integration + distribution + navigation =)
5. *Capability*:   Organizational expertise: knowledge repository, integrated performance support
                 system, core competence

Knowledge Principles

In the final knowledge dimension, Beckman [6] [33] has developed a set of tenets or principles regarding selected knowledge dimensions:

- Shared, formal knowledge and expertise are the key to superior organizational performance, agility, and success.
- *Explicit and tacit knowledge*:
  - Knowledge must be formalized, or made explicit, to have significant value to an organization.
  - Only formalized knowledge can be represented electronically, and be stored, shared, and effectively applied.
- *Practical and theoretical knowledge*:
  - Possessing both experiential and methodological knowledge is more valuable than either alone.
  - Practice should be integrated with methods and models.

- Learning from experience is more vivid, but not very efficient. There is also a human tendency to overgeneralize from one or several experiences. When available, it may be preferable to learn from experts, books, and training. Learning from the experience and mistakes of others is often more effective.
- "There is nothing more practical than a good theory." — Albert Einstein
- *Knowledge and learning:*
  - Balance collecting and organizing available knowledge with learning and creating new knowledge.
  - Integrate KM and organizational learning to create more value to the organization than either one separately.
- *Knowledge and expertise:*
  - Knowledge is applying information and data to make valid inferences.
  - Expertise is superior performance in reasoning using knowledge to perform tasks, solve problems, make decisions, and learn new knowledge.

Definition of Knowledge Management

The discipline of knowledge management (KM) is little more than 10 years old. Karl Wiig, management consultant and AI practitioner, is one of the field's most prominent advocates as well as its likely founder. He coined the term at a 1986 Swiss conference sponsored by the United Nations — International Labor Organization [13]. Here are some definitions of KM:

- KM is the systematic, explicit, and deliberate building, renewal, and application of knowledge to maximize an enterprise's knowledge-related effectiveness and returns from its knowledge assets.

  — Wiig [13]
- KM is the process of capturing a company's collective expertise wherever it resides — in databases, on paper, or in people's heads — and distributing it to wherever it can help produce the biggest payoff.   — Hibbard [14]
- KM is getting the right knowledge to the right people at the right time so they can make the best decision.   — Petrash [15]
- KM involves the identification and analysis of available and required knowledge, and the subsequent planning and control of actions to develop knowledge assets so as to fulfil organization objectives.   — Macintosh [16]
- KM applies systematic approaches to find, understand, and use knowledge to create value.

  — O'Dell [17]
- KM is the explicit control and management of knowledge within an organization aimed at achieving the company's objectives.   — van der Spek [5]
- KM is the formalization of and access to experience, knowledge, and expertise that create new capabilities, enable superior performance, encourage innovation, and enhance customer value.

  — Beckman [6]

KM Principles

Davenport [35] has developed ten general principles of KM:

1. Knowledge management is expensive (but so is stupidity!).
2. Effective management of knowledge requires hybrid solutions involving both people and technology.
3. Knowledge management is highly political.
4. Knowledge management requires knowledge managers.
5. Knowledge management benefits more from maps than models, more from markets than hierarchies.
6. Sharing and using knowledge are often unnatural acts.

7. Knowledge management means improving knowledge work processes.
8. Access to knowledge is only the beginning.
9. Knowledge management never ends.
10. Knowledge management requires a knowledge contract (i.e., intellectual property issues).

KM Framework

Several authors, including Holsapple and Joshi [19], Wiig [4] [13], Beckman [6] [28], and van der Spek and Spijkervet [5], are concerned with creating a KM framework and methodology. Also, several authors suggest multiple perspectives to help in further understanding KM issues.

Holsapple and Joshi [19] have proposed a descriptive KM architecture with these components:

- *KM resources*: Employee/computer, culture, artifact, infrastructure, strategy, purpose
- *KM activities*: See KM processes below
- *KM influences*: Managerial, resource, environmental

Recently they revised this framework based on expert contributions from a Delphi survey [45]:

- Recognition of knowledge need
- Configuration of knowledge manipulation activities
- KM influences
- Knowledge resources
- Learning
- Projection

DiBella and Nevis [36] suggest three perspectives for studying organizational learning:

- Normative
- Developmental
- Capability

Beckman [28] proposes the following perspectives on KM:

- Conceptual
- Process
- Technology
- Organizational
- Management
- Implementation

1.3 KM Process Perspective

In order to transform knowledge into a valuable organizational asset, knowledge, experience, and expertise must be formalized, distributed, shared, and applied. Knowledge management (KM) is considered a key part of the strategy to use expertise to create a sustainable competitive advantage in today's business environment. Many authors, most notably Holsapple and Joshi [19], Wiig [4], Marquardt [20], O'Dell [17], and Beckman [6], have proposed models for the KM process. In all the models presented below, it is assumed that steps and activities are often concurrent, sometimes repeated, and not always in linear sequence.

DiBella and Nevis [36] propose an organizational learning cycle of three phases:

- Acquire
- Disseminate
- Utilize

Marquardt's KM process [20] consists of four steps:

1. Acquisition
2. Creation
3. Transfer and utilization
4. Storage

Wiig [4] also proposes a four-step process:

1. Creation and sourcing
2. Compilation and transformation
3. Dissemination
4. Application and value realization

Van der Spek and Spijkervet [5] present the following model:

1. Developing new knowledge
2. Securing new and existing knowledge
3. Distributing knowledge
4. Combining available knowledge

Ruggles [18] suggests a three-step KM process with supporting activities:

1. Generation:
     Creation
     Acquisition
     Synthesis
     Fusion
     Adaptation
2. Codification:
     Capture
     Representation
3. Transfer

O'Dell's KM model [17] has seven steps:

1. Identify
2. Collect
3. Adapt
4. Organize
5. Apply
6. Share
7. Create

Beckman [6] proposes a similar eight-stage process for KM:

Stage 1	Identify	Determine core competencies, sourcing strategy, and knowledge domains.
Stage 2	Capture	Formalize existing knowledge.
Stage 3	Select	Assess knowledge relevance, value, and accuracy. Resolve conflicting knowledge.
Stage 4	Store	Represent corporate memory in knowledge repository with various knowledge schema.
Stage 5	Share	Distribute knowledge automatically to users based on interest and work. Collaborate on knowledge work through virtual teams.
Stage 6	Apply	Retrieve and use knowledge in making decisions, solving problems, automating or supporting work, job aids, and training.
Stage 7	Create	Discover new knowledge through research, experimenting, and creative thinking.
Stage 8	Sell	Develop and market new knowledge-based products and services.

Finally, Holsapple and Joshi [19] present an extensive framework consisting of six steps with supporting activities:

1. *Acquiring knowledge:*	Extracting	Interpreting	Transferring
2. *Selecting knowledge:*	Locating	Retrieving	Transferring
3. *Internalizing knowledge:*	Assessing	Targeting	Depositing
4. *Using knowledge*			
5. *Generating knowledge:*	Monitoring	Evaluating	Producing Transferring
6. *Externalizing knowledge:*	Targeting	Producing	Transferring

1.4　Technology Perspective

In this section, the use of information technology (IT) in support of KM objectives will be discussed. There is significant difference of opinion regarding the value of IT, and especially expert systems, to enable KM work. Beckman [6] [8] [21] has done considerable work in support of the value of IT through formulating and integrating the concepts of knowledge representation, knowledge repositories, and automated knowledge transformation. Independently, Jovanovic [40] has developed a software product, The Corporate Memory, that has implemented many of these concepts.

IT Infrastructure

In order to facilitate sharing of knowledge, an IT infrastructure must be in place. In order to be applied, knowledge and expertise must be readily accessible, understandable, and retrievable. Tobin [38] suggests building an IT knowledge network with the following components:

- Knowledge repository, most commonly a database
- Directory of knowledge sources
- Directory of learning resources
- Groupware

The IT component in Beckman's Business Model [21] consists of the following elements:

- IT architecture and standards
- IT platform: computing hardware
- Communications: data, voice, image, network, security
- Interfaces
- Data/Information
- Software applications:
    Office automation and groupware
    Transaction systems
    Process modeling and simulation
    Decision support systems and executive information systems
    Functional information systems: finance, marketing, manufacturing, HR, IS
    Intelligent systems: ES (expert systems), ML (machine learning), KD (knowledge discovery),
        IPSS (integrated performance support systems)
- User support: help desks and training

While there is disagreement about the role of expert systems and other intelligent systems, there is strong agreement about the value of global computer networks and groupware for knowledge sharing.

Knowledge Representation Schema

In this section, the author presents his schema for knowledge representation that encompasses knowledge from all five levels of the knowledge hierarchy [8]. These knowledge structures can be used to support both knowledge repositories and integrated performance support systems (IPSS). In addition, the organization and components of both knowledge repositories and IPSS will be briefly addressed.

Knowledge in expert systems (ES) can be represented as three general schema [21]: cases, rules, and models. Case-based reasoning (CBR) represents knowledge from experience such as events and specific case problems and solutions. Rule-based systems (RBS) use knowledge compiled into chunks called rules that represent heuristics that human experts often use to solve complex problems. Model-based reasoning (MBR) creates an overall framework through object technology for representing and organizing domain knowledge in terms of object attributes, behaviors, and relationships, as well as simulating domain processes. A fourth schema, constraint-satisfaction reasoning, can be represented as a combination of RBS and MBR.

Knowledge Repositories

Several authors have proposed definitions for knowledge repositories:

- A knowledge repository organizes, and makes available to all employees basic information on the company's organization, products, services, customers, and business processes. — Tobin [38]
- A knowledge repository is an on-line, computer-based storehouse of expertise, knowledge, experience, and documentation about a particular domain of expertise. In creating a knowledge repository, knowledge is collected, summarized, and integrated across sources. — Beckman [8] [21]

Few prototype knowledge repositories now exist that contain varied types of knowledge. However, for an exception, see work on corporate memory by Jovanovic [40]. The major types of knowledge structures that can be represented in knowledge repositories include:

- Images: pictures and video
- Sounds and signals
- Text: flat or hypertext
- Data: relational
- Document structure: forms/templates/reports/graphs/charts/interfaces
- Cases: CBR
- Rules: RBS
- Objects: hierarchies, client/server
- Processes: decomposable hierarchies, resources, performance characteristics
- Models: MBR, frameworks and simulations

Integrated Performance Support Systems (IPSS)

In addition to applying ES in narrow application areas, it is also possible to consider the totality of workforce needs by designing IPSS that are supported by multiple knowledge repositories. IPSS provide comprehensive support to meet a broad variety of employee needs. Expanding on Winslow and Bramer [24], these systems can be categorized as providing the following services:

- *Infrastructure:* organizing and structuring the work environment (MBR and knowledge repositories)
- *Controller:* monitoring, coordinating, and controlling IPSS services (RBS)
- *Navigation:* human–computer interaction (MBR)

- *Presentation*: users can tailor and customize data and other service aspects (MBR)
- *Acquisition*: capturing knowledge, cases, opinions, learnings, and sensory data in various media forms and transforming them to an internal form (smart templates)
- *Advisory*: provides advice, reminders, assistance (RBS and CBR)
- *Instruction*: help, job aids, tutoring, training (MBR, RBS, and CBR)
- *Learning*: Applying knowledge discovery and data mining techniques to existing stores
- *Evaluation*: assessing and certifying based on performance measures, automated quality assurance, and administered tests (CBR and RBS)
- *Reference*: source of workforce and organizational knowledge and expertise, as well as source of intranet, Internet, and other databases and information (CBR and AI search engines)

Knowledge Transformation

The concept of automated knowledge transformation [6], in conjunction with the use of IPSS, may greatly increase the value of knowledge repositories. Knowledge can be transformed by moving it up the knowledge hierarchy from a lower to a higher state. For example, data or cases can be transformed into rules by using techniques from machine learning induction and knowledge discovery and data mining [22]. In another type of knowledge transformation, Zarri [23] shows how unstructured text in document form can be transformed into "metadocument" form, represented and indexed in the forms of semantic nets, frames, and conceptual graphs. Finally, Feldman [32] has recently developed a software product that demonstrates the value of text mining — finding and organizing relationships among vast quantities of documents.

1.5 Organizational Perspective

Knowledge Organization Characteristics

Liebowitz and Beckman [8] have attempted to define the characteristics of the knowledge organization:

- High performance
- Customer-driven
- Improvement-driven
- Excellence-driven
- High flexibility and adaptiveness
- High levels of expertise and knowledge
- High rates of learning and innovation
- Innovative IT-enabled
- Self-directed and managed
- Proactive and futurist
- Values expertise and sharing knowledge

Organizational Structures

Many authors, including McGill and Slocum [31] and Martin [29], have written about the benefits of the network organizational form in supporting collaboration and knowledge sharing. To better meet the specific needs of the knowledge organization, Beckman [33] has conceived of a new organizational structure: the Center of Expertise. Additionally, Davenport and Prusak [30] discuss the types of KM Project Offices needed.

Centers of Expertise (COE)

Beckman [33] believes there should be a center of expertise (COE) for each knowledge domain, discipline, or subject matter speciality. Each COE has several roles:

- Create, research, improve, and manage the domain knowledge repository.
- Set and enforce standards, methods, and practices for domain discipline.
- Establish partnerships and align/coordinate interests with related COE specialties, projects, and processes, as well as negotiate conflicts between these entities.
- Assess workforce competency and performance, identify gaps, and remedy deficiencies.
- Support, develop, and enable the workforce by providing educational and consulting services, as well as coaching and tools.
- Supply competent workers to staff projects and processes through assignment, hiring, outsourcing, and developing.

KM Project Offices

Davenport and Prusak [30] have identified four types of KM projects and related activities:

- *Knowledge repository*

  Determining the technology for storing the knowledge
  Persuading employees to contribute to the repository
  Creating a structure for holding the knowledge
- *Knowledge transfer:* Identify, develop, and monitor both human and electronic channels for knowledge sharing
- *Knowledge asset management*

  Calculating knowledge valuations
  Negotiating with holders of desired intellectual capital
  Managing a knowledge asset portfolio
- *Infrastructure development*

  Analyzing financial needs
  Work with external vendors of technologies and services
  Develop human resources management approaches

Roles and Responsibilities

Davenport and Prusak [30] have done considerable research on this topic and have identified several significant roles in KM:

- Knowledge-oriented personnel
- Knowledge Management Specialist
- Knowledge Project Manager
- Chief Knowledge Officer

In addition to these roles, Beckman [28] would add a fifth role, that of Center of Expertise Director.

Knowledge-Oriented Personnel

This category includes all knowledge workers and everyone who creates, shares, scans for, or uses knowledge in their daily work. Clearly, all white collar workers and some blue collar workers are part of this group.

Knowledge Management Specialists

Liebowitz and Beckman [8] and Davenport and Prusak [30] disagree over the scope of this role. Davenport and Prusak see little value in the acquisition, elicitation, and structuring of expert knowledge, and believe in a relatively unstructured knowledge repository. Liebowitz and Beckman, on the other hand, believe that domain experts or senior practitioners are needed to evaluate, select, and organize knowledge for inclusion in the knowledge repository. Additionally, other KM professionals would include knowledge engineers with skills in knowledge representation and knowledge elicitation. Knowledge engineers are needed to determine the cognitive reasoning and knowledge structures used by experts in that domain, and to uncover tacit expert knowledge for inclusion in the knowledge repository.

Knowledge Project Managers

Davenport and Prusak [30] suggest that knowledge project managers must perform the traditional project management functions as well as possess technical, psychological, and business skills in order to succeed in this role. The typical project management functions include:

- Developing project objectives
- Assembling and managing teams
- Determining and managing customer expectations
- Monitoring project budgets and schedules
- Identifying and resolving project problems

Chief Knowledge Officer

According to Davenport [34], the Chief Knowledge Officer should embody a number of characteristics:

- Advocate or evangelist for knowledge and learning
- Designer, implementer, and overseer of an organization's knowledge infrastructure
- Primary liaison between external providers of information and knowledge
- Provider of critical input into the knowledge creation and use processes that already exist within the company, such as product development
- Leading role in the design and implementation of a company's knowledge architectures
- Deep experience in some aspect of knowledge management, such as its creation, dissemination, or application
- Familiarity with knowledge-oriented companies and technologies, such as libraries and groupware
- Ability to set a good example by displaying a high level of expertise and success

Liebowitz and Beckman [8] believe that the Chief Knowledge Officer should have expertise in the disciplines of business reengineering, innovative IT, change management, as well as knowledge management.

Corporate Culture

Having a healthy corporate culture is imperative for success in KM. Zand [39] believes that bureaucratic cultures suffer from a lack of trust and a failure to reward and promote cooperation and collaboration. Without a trusting and properly motivated workforce, knowledge is rarely shared or applied; innovation and risk-taking cease; and organizational cooperation and alignment are nonexistent. No wonder that most bureaucratic organizations suffer under marginal performance, and are incapable of agile, innovative behaviors leading to future success.

1.6   Management Perspective

Management Practices

Management must, if needed, change the existing culture and mindsets so that they are receptive, supportive, and committed to the precepts of the knowledge organization. Management must motivate everyone by providing equal opportunities and development as well as just appraisal and rewards. Management must measure and reward the performance, behaviors, and attitudes that are needed and desired. It is essential to measure what you reward, and reward what you measure.

Edvinsson [41] suggests some interesting employee measures that can be used to evaluate management performance:

- Leadership and motivation indices — based on satisfied customers, motivated and competent staff, and improvement/quality program
- Empowerment index
- Employee turnover
- Time in training (days/year)

Measuring and Valuing Intellectual Capital

Good measures provide focus, operationalize goals, and set standards. A multitiered, multidimensional approach to measurement seems most effective. Measures should be directly aligned to goals and strategies. According to Beckman [21], there are three general categories of measures:

- Result (Past)
- Process (Present)
- Resource (Future)

Thus, focusing on resource measures, such as intellectual assets, is key to positioning your organization for future success.

The two most influential authors in the measurement of intellectual assets are Leif Edvinsson and Karl Sveiby. Edvinsson [41] explains why and how he implemented a multidimensional measurement system at Skandia. Sveiby [42] focuses on how to measure intangible assets, and has developed some most interesting measures, especially about competence.

In Skandia's Navigator, their measurement framework, they have five focuses or categories:

- Financial (Past)
- Customer (Present)
- Process (Present)
- Human (Present)
- Renewal and development (Future)

Edvinsson further suggests areas where organizations can detect future trends:

- *Customers* — Expected changes in customer base; future customer expectations; future need for customer training and support for new products and services; effectiveness of customer/company communications
- *Market* — Spending for market intelligence; track new competitors; contribution of new markets
- *Products and Services* — Likelihood of innovations coming to market; contribution and life expectancy of new products; ratio of products < 2 yrs. old to entire product line; total investment in new product/service development; percentage devoted to basic research, product design, applications

- *Strategic Partners* — Investment in partnering and networking; electronic commerce; benchmarking
- *Infrastructure* — Investment in organizational tools; capital acquisitions; value of MIS, marketing IS, process control network; capacity and load of systems infrastructure
- *Employees* — Current average education; new competence profiles added annually; training per employee per year; recruiting programs; planned growth in value-added per employee;

Finally, Skandia computes an index of intellectual capital efficiency using percentage measures:

- Market share
- Satisfied customer index
- Leadership index
- Motivation index
- R&D resources/total resources
- Training hours/employee/year
- Performance/quality goal
- Employee retention
- Administrative efficiency/revenues (reciprocal of errors and expenses)

Sveiby [42] is most interested in measuring intangible assets, and he divides these into three categories:

- External structure — customer relations measured as satisfaction levels
- Internal structure — operational efficiency
- Competence of personnel — employee satisfaction and retention (and performance and expertise — Beckman)

Oddly enough, accounting conventions record tangible capital expenses as depreciable assets, but record investments in R&D or training as costs. As Sveiby [42] observes, the economic value of any investment, tangible or otherwise, is uncertain, so this is not an adequate justification for this practice. He also suggests a way of expressing profit margin is to compare profits to value-added created, rather than to revenues. In addition, he suggests that indicators for intangible assets should consist of components for growth/renewal, efficiency, and stability.

His measurement of professional competence is quite interesting. This author would expand his framework to apply to all types of employees: managers, professionals, support staff, and contractors. Sveiby considers contractors to be an important source of competence that should be transferred into the organization.

Under the growth/renewal category, Sveiby suggests the following competence measures:

- Overall — average number of years of experience in the profession
- Level of education and certifications — average years of education
- Training and education costs — competence development
- Grading — of managers and professionals
- Turnover — net change in years of experience
- Competence-enhancing customers — customer assignments that contribute to development

Under the category of efficiency, he suggests the following competence measures:

- Ratio of professionals — professionals divided by total employees
- Leverage — ability of professionals to generate revenue
- Value-added per professional — depends on the market, management efficiency, and payout in salaries and benefits

In the final category, stability, he suggests the following competence measures:

- Average employee age
- Seniority ratio and rookie ratio
- Industry pay position — relative positions of individual companies
- Professional turnover rate — some is desirable, but not too much

Quinn, Baruch, and Zien [44] believe that evaluations at multiple levels are needed to encourage innovation in an integrated framework:

- Individual performance (by peers)
- Customer performance (by customers)
- Collaborative performance (by group members and customers)
- Enterprise performance (by value-added measures)

Reward, Compensation, and Motivational Systems

In bureaucratic organizations, employees and managers are discouraged from sharing knowledge and expertise. In fact, the opposite is often the case: knowledge is considered a source of power, and thus hoarding is not only expected but is often rewarded. Another problem is how to get employees to use and apply expertise developed by someone else. The not-invented-here syndrome is still quite powerful in most organizations. A related issue is how to encourage organizational learning and its application in the workplace. In extremely bureaucratic organizations, expertise poses a threat to the powerbase of those managers who need it. Rather than rely on, share power with, and give rewards and recognition to their professional staff, managers in dysfunctional organizations would rather outsource those knowledge tasks, retain total control and power, and minimize and ridicule the expertise of their internal experts.

Quinn, Baruch, and Zien in innovation [44], Zand in leadership [39], McGill and Slocum in learning [31], and Beckman in KM [33] [28] take mutually supportive positions on how to promote knowledge sharing and use, as well as learning and innovation.

Quinn, Baruch, and Zien [44] state that "Behavior that is reinforced will be repeated or amplified. ... Yet too often corporate incentive systems reward safe, bureaucratic behavior rather than the risk-taking, individualistic behavior characteristic of innovators. ... Substantial research also suggests that innovators respond most to a mix of financial and nonfinancial incentives and that most talented people, to feel fulfilled, need a concrete sense of adventure, appreciation for their hard work, and recognition for successes achieved." In addition, unlike bureaucratic managers, high achievers want to be measured so that they can prove their accomplishments. Finally, they believe that innovators respond best to the following factors:

- Challenges
- Personal recognition
- Freedom of activity
- Financial rewards

Zand [39] believes that collaborative, integrative, win-win reward systems should be created in which one person's or organization's gain can also be a gain for their peers. He believes, as does Beckman [28], that collaborative rewards should be emphasized by linking bonuses to the overall profitability of the firm. Beckman goes one step further by suggesting that most individual compensation as well as rewards should be based primarily on organizational success and customer satisfaction. Unfortunately, most organizations operate under competitive, distributive, win-lose reward systems where one person's gain comes at another's expense and mistrust is the norm. In these bureaucratic settings, performance is often poor, and risk-taking and innovation are the exception.

Quinn, Baruch, and Zien [44] believe that strong incentives and a healthy culture are needed to encourage innovation. They see one of the biggest problems in achieving independent collaboration as making sure that all participants receive appropriate rewards. They state: "Unless the culture or incentives are very strong, those with existing power positions can subvert progress by refusing to undertake change or to provide needed expertise for a new venture."

McGill and Slocum [31] suggest that workers want to be competent, have control over and choices in their work, and thus would respond to motivators such as:

- Significance — work is valued by the organization
- Identity — connected to the work and making a contribution
- Autonomy — influence what they do and how they do it
- Feedback — direct and clear information about their performance

They also suggest that the keys to effectively rewarding learning are:

- Eliminate the disincentives to learn, including those policies and practices that reward the status quo, and pay and promotions based on seniority.
- Clarify the expectation for learning by specifying either skill levels or breadth dimensions of a job.
- Provide the opportunity to learn via shared information, task assignments, or job rotation.
- Provide the opportunity to exhibit learning — practice and performance.
- Recognize and reward learning.

Beckman [28] [33] proposes an integrated motivational system that might solve many of the design, implementation, and performance problems that have plagued improvement efforts similar to KM. This motivational system is comprised of the following steps:

Step 1 — Determine strategic intent
Step 2 — Develop measures and standards
Step 3 — Create opportunities
Step 4 — Monitor and assess measures
Step 5 — Reward results

Liebowitz and Beckman [8] believe in applying a multidimensional measurement approach that is then combined with core values of providing good value to the customer, serving the customer, achieving high performance, leading through expertise and innovation, and sharing and cooperating. Therefore, organizations should reward:

- Customer satisfaction
- High performance
- Personal knowledge and expertise
- Teamwork and sharing of expertise and knowledge
- Creating new and extending existing knowledge and expertise
- Using and applying the knowledge and expertise in the knowledge repository
- Proactive problem solving and problem prevention

Organizations should *not* reward (and could consider punishing):

- Buck passing
- Loyalty to the boss
- Conformance and compliance behavior — passive resistance
- Internal competition
- Bureaucratic, controlling behaviors
- Power grabbing and turf battles

Rewards should take many forms, including money, recognition, time off, empowerment, work selection, advancement, and development. And rewards should celebrate successes, as well as desired behaviors such as collaborating, experimenting, risk-taking, and learning. Reward early and often.

1.7   Implementation Perspective

In this perspective, factors that predict and facilitate success will be discussed. In addition, challenges and barriers to successful KM projects are listed. Finally, KM strategies will be examined. Beckman [21] suggests using business reengineering to guide the implementation of KM projects that are larger and more complex.

IT Infrastructure Implementation

Beckman [37] proposes a four-stage model for implementing the innovative IT needed to support and enable KM:

Stage 1: Establish an IS and IT infrastructure
Stage 2: Create knowledge repositories
Stage 3: Develop expert system applications
Stage 4: Develop IPSS and KDD capability

Success Factors

Several authors, including Davenport and Prusak [30] and DiBella and Nevis [36], have identified success factors in implementing KM in organization settings. Davenport and Prusak have identified nine factors leading to knowledge project success:

- A knowledge-oriented culture
- Technical and organizational infrastructure
- Senior management support
- A link to economics or industry value
- A modicum of process orientation
- Clarity of vision and language
- Nontrivial motivational aids
- Some level of knowledge structure
- Multiple channels for knowledge transfer

Similarly, DiBella and Nevis [36] have identified facilitating factors for organizational learning that share some commonalities with those of Davenport and Prusak:

- Scanning imperative
- Performance gap
- Concern for measurement
- Organizational curiosity
- Climate of openness
- Continuous education
- Operational variety
- Multiple advocates
- Involved leadership
- Systems perspective

Prerequisites and Challenges

Beckman [33] has suggested four prerequisites to consider to enhance the chances of success during the implementation of KM:

- Executive leadership and commitment
- Healthy culture
- Expertise
- IT infrastructure

This author sees several challenges in implementing KM in the typical organization. First, knowledge is often hoarded, rather than shared. Second, valuable knowledge developed by others is often ignored, rather than applied in daily work situations. Third, knowledge and expertise are often not valued by the corporate culture, by failing to measure intellectual assets. Fourth, employees who share knowledge and expertise are considered naive, rather than being rewarded for their valuable organizational behavior.

Zand [39] also lists potential obstacles to success:

- Success breeds complacency.
- Lack of familiarity blocks action.
- Corporate culture sets the tone.
- Fear of technology can block innovation.

KM Strategies

O'Dell and Wiig [43] have identified six major strategies used by leading organizations to gain value from implementing KM:

- Knowledge management as a business strategy
- Transfer of knowledge and best practices
- Customer-focused knowledge
- Personal responsibility for knowledge
- Intellectual asset management
- Innovation and knowledge creation

1.8 Conclusions

Until the past few years, most of the knowledge, experience, and learning about KM have been accessible to only a few practitioners. However, during the past three years an explosion of interest, writing, research, and applications in KM has occurred. This chapter has attempted to capture, organize, and summarize these accomplishments and those efforts that occurred longer ago. Inevitably, some work in KM and related fields is unknown to this writer. Every month, new knowledge is created, older knowledge is uncovered, and old knowledge is made obsolete by better ideas. On a microlevel, the same issues about knowledge management that plagued this author in writing this chapter also apply to large organizations in successfully completing complex knowledge projects.

There is some concern among practitioners that KM might suffer a fate similar to business reengineering, artificial intelligence, and total quality management. That is, interest in a discipline must last long enough to iron out the bugs while simultaneously delivering significant business value. The irony is that just when the discipline really works well, potential users often have lost interest in the fad, point to the inevitable early failures, and thus miss out on the very real benefits.

Future work should focus on building practical experience through extensive experimenting, proto-typing, and testing — especially in the process, technology, organizational, and implementation perspectives. In addition, the conceptual frameworks and integration across KM perspectives need more investigation and development.

Although considerable progress has been achieved in KM across a broad front, much work remains to fully deliver the business value that KM promises. Ultimately, in order to realize the enormous potential value from KM, organizations must motivate and enable the creating, organizing, and sharing of knowledge.

Suggested KM Readings

APQC International Benchmarking Clearinghouse. *Knowledge Management: Consortium Benchmarking Study Final Report.* American Productivity and Quality Center (APQC). 1996.

Beckman, T. *Designing Innovative Business Systems through Reengineering.* Tutorial. 210 pages. Presented at the 4th World Congress on Expert Systems. Mexico. March, 1998.

Birchall, D. and Lyons, L. *Creating Tomorrow's Organization: Unlocking the Benefits of Future Work.* Pitman Publishing. 1995.

Davenport, T. and Prusak, L. *Working Knowledge: How Organizations Manage What They Know.* Harvard Business School Press. 1998.

DiBella, A. and Nevis, E. *How Organizations Learn: An Integrated Strategy for Building Learning Capability.* Jossey-Bass. 1998.

Edvinsson, L. and Malone, M. *Intellectual Capital: Realizing Your Company's True Value by Finding its Hidden Brainpower.* Harper Business. 1997.

Holsapple, C. & Joshi, K. "In Search of a Descriptive Framework for Knowledge Management: Preliminary Delphi Results." Kentucky Initiative for Knowledge Management Paper No. 118. March 1998.

Leonard-Barton, D. *Wellsprings of Knowledge: Building and Sustaining the Sources of Innovation.* Harvard Business School Press. 1995.

Liebowitz, J. and Beckman, T. *Knowledge Organizations: What Every Manager Should Know.* St. Lucie Press. 1998.

Liebowitz, J. and Wilcox, L. (eds). *Knowledge Management and Its Integrative Elements.* CRC Press. 1997.

Marquardt, M. *Building the Learning Organization.* McGraw Hill. 1996.

Martin, J. *Cybercorp: The New Business Revolution.* AMACOM. 1996.

McGill, M. and Slocum, J. *The Smarter Organization: How to Build a Business That Learns and Adapts to Marketplace Needs.* John Wiley and Sons. 1994.

O'Leary, D. "Knowledge Management Systems: Converting and Connecting," and "Using AI in Knowledge Management: Knowledge Bases and Ontologies." *IEEE Intelligent Systems Journal.* IEEE. May/June 1998.

Quinn, J., Baruch, J. and Kien, K. *Innovation Explosion: Using Intellect and Software to Revolutionize Growth Strategies.* The Free Press. 1997.

Stewart, T. *Intellectual Capital: The New Wealth of Organizations.* Currency/Doubleday. 1997.

Sveiby, K. *The New Organization Wealth: Managing and Measuring Knowledge-Based Assets.* Berrett-Koehler Publishers. 1997.

Tobin, D. *The Knowledge-Enabled Organization: Moving from Training to Learning to Meet Business Needs.* AMACOM. 1998.

Wiig, K. "Knowledge Management: Where Did It Come From and Where Will It Go?" *Journal of Expert Systems with Applications.* Fall 1997.

Winslow, C. and Bramer, W. *Future Work: Putting Knowledge to Work in the Knowledge Economy.* The Free Press. 1994.

Zand, D. *The Leadership Triad: Knowledge, Trust, and Power.* Oxford University Press. 1997.

References

[1] Woolf, H., ed. *Webster's New World Dictionary of the American Language*. G. and C. Merriam. 1990.

[2] Turban, E. *Expert Systems and Applied Artificial Intelligence*. Macmillan. 1992.

[3] Sowa, J. *Conceptual Structures*. Addison-Wesley. 1984.

[4] Wiig, K. *Knowledge Management Foundation*. Schema Press. 1993.

[5] van der Spek, R. and Spijkervet, A. "Knowledge Management: Dealing Intelligently with Knowledge." *Knowledge Management and Its Integrative Elements*. Liebowitz & Wilcox, eds. CRC Press. 1997.

[6] Beckman, T. "A Methodology for Knowledge Management." International Association of Science and Technology for Development (IASTED) AI and Soft Computing Conference. Banff, Canada. 1997.

[7] Nonaka, I. and Takeuchi, H. *The Knowledge-Creating Company: How Japanese Companies Create the Dynamics of Innovation*. Oxford University Press. 1995.

[8] Liebowitz, J. and Beckman, T. *Knowledge Organizations: What Every Manager Should Know*. St. Lucie Press. 1998.

[9] Collins, H. "Human, Machines, and the Structure of Knowledge." Ruggles, R., ed. *Knowledge Management Tools*. Butterworth-Heinemann. 1997.

[10] Parsaye, K. and Chignell, M. *Expert Systems for Experts*. John Wiley and Sons. 1988.

[11] Alter, S. *Information Systems: A Management Perspective*. 2nd ed. Benjamin/Cummings Publishing. 1996.

[12] Tobin, D. *Transformational Learning: Renewing Your Company through Knowledge and Skills*. John Wiley & Sons. 1996.

[13] Wiig, K. "Knowledge Management: Where Did It Come From and Where Will It Go?." *Expert Systems with Applications*, Pergamon Press/Elsevier, Vol. 14, Fall 1997.

[14] Hibbard, J. "Knowing What We Know." *Information Week*. October 20, 1997.

[15] Petrash, G. "Managing Knowledge Assets for Value." Knowledge-Based Leadership Conference. Linkage, Inc. Boston. October 1996.

[16] Macintosh, A. "Position Paper on Knowledge Asset Management." Artificial Intelligence Applications Institute, University of Edinburgh, Scotland. May 1996.

[17] O'Dell, C. "A Current Review of Knowledge Management Best Practice." Conference on Knowledge Management and the Transfer of Best Practices. Business Intelligence. London. December 1996.

[18] Ruggles, R. "Tools for Knowledge Management: An Introduction." Ruggles, R., ed. *Knowledge Management Tools*. Butterworth-Heinemann. 1997.

[19] Holsapple, C. & Joshi, K. "Knowledge Management: A Three-Fold Framework." Kentucky Initiative for Knowledge Management Paper No. 104. July 1997.

[20] Marquardt, M. *Building the Learning Organization*. McGraw Hill. 1996.

[21] Beckman, T. *Designing Innovative Business Systems through Reengineering*. Tutorial. Fourth World Congress on Expert Systems. Mexico City. 1998.

[22] Fayyad, U., Piatetsky-Shapiro, G., Smyth, P., and Uthurusamy, R., eds. *Advances in Knowledge Discovery and Data Mining*. MIT Press, 1996.

[23] Zarri, G., Building up and Making Use of Corporate Knowledge Repositories. KAW '96.

[24] Winslow, C. and Bramer, W. *Future Work: Putting Knowledge to Work in the Knowledge Economy*. The Free Press. 1994.

[25] Quinn, J.B., Anderson, P. and Finkelstein, S. "Managing Professional Intellect: Making the Most of the Best." *Harvard Business Review*. March-April 1996.

[26] Brooking, A. "Introduction to Intellectual Capital." The Knowledge Broker Ltd. Cambridge, England. 1996.

[27] Myers, P. ed. *Knowledge Management and Organizational Design*. Butterworth-Heinemann. 1996.

[28] Beckman, T. *Knowledge Management Seminar.* ITESM. Monterrey Mexico. June 1998.

[29] Martin, J. *The Great Transition: Using the Seven Disciplines of Enterprise Engineering to Align People, Technology, and Strategy.* AMACOM. 1995.

[30] Davenport, T. and Prusak, L. *Working Knowledge: How Organizations Manage What They Know.* Harvard Business School Press. 1998.

[31] McGill, M. and Slocum, J. *The Smarter Organization: How to Build a Business that Learns and Adapts to Marketplace Needs.* John Wiley & Sons. 1994.

[32] Feldman, R. *Text Mining: Theory and Practice.* Tutorial. 4th World Congress on Expert Systems. Mexico City. 1998.

[33] Beckman, T. *Implementing the Knowledge Organization in Government.* Paper and presentation. 10th National Conference on Federal Quality. 1997.

[34] Davenport, T. "Coming Soon: The CKO." TechWeb, CMP Media Inc. Sept. 5, 1994.

[35] Davenport, T. "Some Principles of Knowledge Management." *Strategy, Management, Competition.* Winter, 1996.

[36] DiBella, A. and Nevis, E. *How Organizations Learn: An Integrated Strategy for Building Learning Capability.* Jossey-Bass. 1998.

[37] Beckman, T. "Expert System Applications: Designing Innovative Business Systems through Reengineering." *Handbook on Expert Systems.* Liebowitz, J., ed. CRC Press. 1997.

[38] Tobin, D. *The Knowledge-Enabled Organization: Moving from Training to Learning to Meet Business Goals.* AMACOM. 1998.

[39] Zand, D. *The Leadership Triad: Knowledge, Trust, and Power.* Oxford University Press. 1997.

[40] Jovanovic, A. *Building and Practical Implementation of Corporate Memory Systems: from Basic Idea to the Full-Scale System.* Tutorial. Presented at the 4th World Congress on Expert Systems. Mexico City. 1998.

[41] Edvinsson, L. and Malone, M. In *Intellectual Capital: Realizing Your Company's True Value by Finding its Hidden Brainpower.* Harper Business. 1997.

[42] Sveiby, K. *The New Organization Wealth: Managing and Measuring Knowledge-Based Assets.* Berrett-Koehler Publishers. 1997.

[43] APQC International Benchmarking Clearinghouse. *Knowledge Management: Consortium Benchmarking Study Final Report.* American Productivity and Quality Center. 1996.

[44] Quinn, J., Baruch, J., and Zien, K. *Innovation Explosion: Using Intellect and Software to Revolutionize Growth Strategies.* The Free Press. 1997.

[45] Holsapple, C. & Joshi, K. "In Search of a Descriptive Framework for Knowledge Management: Preliminary Delphi Results." Kentucky Initiative for Knowledge Management Paper No. 118. March 1998.

2

Knowledge Management and the Broader Firm: Strategy, Advantage, and Performance

Thomas H. Davenport
Boston University School of Management

2.1 Background/Historical Context

Any time a new management concept arises in organizations, it frequently begins life as a somewhat separate entity from the mainstream of the business. This separateness is necessary in order to demonstrate to members of the organization that the concept is truly new and different from the activities pursued in the past. But the separation ultimately leads to problems; the overall performance of the business is less likely to be improved when the improvement approach is viewed as separate from the work and the organization. Knowledge management projects in this environment are likely to remain on the margin of the business, addressing the creation, distribution, or application of knowledge in single — and possibly nonstrategic — business functions or processes.

There may be virtues to this unintegrated approach in the early phases of organizational knowledge management. While there is no clear or preferred pattern for the life span of a management concept, it would seem desirable for a new notion to first establish an independent identity, and then later to be incorporated into the day-to-day work of the organization. To try to be highly strategic about knowledge management from the beginning may be difficult to sell to senior management, and runs the risk of raising expectations beyond what the concept can deliver in a short time.

Ultimately, however, if a concept such as knowledge management is to have broad impact, it must be integrated with the firm's existing strategic direction or propel it in a new one. Knowledge management must affect the most important areas of the business, improve the firm's most critical objectives, and be viewed as an integral part of strategic business initiatives. Through such integration of knowledge management the ongoing strategic objectives, core competencies, and employee capabilities will be transformed, and the performance of the organization improved in a noticeable fashion.

Knowledge and its management has already been linked to organizational performance and strategy at a theoretical level. This linkage takes a variety of forms and is found in the literatures of evolutionary economics (Nelson and Winter, 1982), the economics of innovation and information (Arrow, 1974), and technology management and the diffusion of innovations (Teece, 1986). Knowledge has long been considered one of the key organizational resources in a resource-based view of the firm (Penrose, 1959). Knowledge has often been described as the only real sustainable competitive advantage (Winter, 1987). At least one writer on knowledge and strategy has expressed bemusement about the broader world's "discovery" of knowledge management, wondering what took it so long. (Teece, 1998a). Still, these arguments were long on theory and short on specific principles or even examples of what types of knowledge led to what forms of strategic advantage.

If knowledge and strategy have only been weakly linked, connections between knowledge management and organizational performance are even more difficult to establish. This has not, of course, kept them from being hypothesized and discussed. There is a burgeoning literature, in fact, that attempts to associate the management of knowledge and "intellectual capital" (one suspects that the term is often used in this context purely for its financial connotations) to organizations' balance sheets and stock prices. The usual formulation of this linkage is to bemoan the fact that knowledge and intellect cannot be found on the formal balance sheet, when they clearly play a major role in an organization's financial success. The economist James Tobin's "Q" statistic, which measures the ratio between a stock's market value and its book value, is often invoked. The more aggressive adopters of this logic (e.g., Edvinsson and Malone, 1997) argue that Tobin's Q is almost entirely a reflection of a company's knowledge, and that traditional balance sheets should be replaced by financial measures that are more sensitive to intellectual capital. More reasoned analyzes (e.g., Sveiby, 1997) simply point out that there is utility in trying to measure some aspects of an organization's intellectual capital, but warn against trying to directly convert knowledge to financial measures. Those few firms that have actually begun to measure activities related to knowledge and intellect have thus far revealed no correlations or connections with their actual financial performance. Furthermore, financial managers at the company most often cited with regard to a high knowledge-induced Tobin's Q, Microsoft, profess little interest in a new set of intellectual, capital-based measures.

In short, despite theoretical linkages, knowledge management is not yet tied to strategy and performance in practice. Like its predecessors, total quality management, reengineering, and electronic commerce, knowledge management as currently practiced is characterized by terminology, approaches, methods, and organizational units that are consciously separate from the organizations served by the concept. Such subconcepts as "communities of practice," "knowledge sharing," "organizational learning," and "knowledge representation" have been advanced by academics, consultants, and leading practitioners, but have yet to enter the realm of business strategy. Few firms employ the word "knowledge" in strategy or planning documents. The very small number of firms that have a "knowledge strategy" have not linked them to business strategy in any specific way. Many knowledge management efforts are thriving in terms of resource use, the amount of knowledge captured and distributed, and even the breadth of visibility and acceptance, but many also remain somewhat isolated from the broader strategic context. They may even be taking place without the knowledge of senior executives, and are often not focused on the most strategic objectives or initiatives senior managers would name.

In those companies where knowledge management (KM) has already carved out its own identity, what we need now are closer linkages to such fundamental attributes of the business environment as its business performance, business strategy and resulting business objectives, core competencies, and employee capabilities. I will examine each of these attributes with regard to how knowledge management can be more effectively integrated within them.

2.2 Principles, Concepts, Techniques, Tools, and Methodologies

In this discussion I'll take a broad orientation toward the relationship between strategy and knowledge management. From an exhortational point of view I will first argue that knowledge management needs a closer tie to business strategy, and vice versa. From a descriptive standpoint I will attempt to provide an understanding of the various ways in which the two concepts can be related. Some of these ways include:

- The role of knowledge and knowledge management in strategy formulation
- Knowledge and knowledge management as a means of achieving strategic advantage
- Strategy knowledge as a particular knowledge management domain
- Core competencies, individual capabilities, and knowledge management
- The relationship between knowledge management strategy and business strategy
- The role of knowledge management in measuring and improving business performance

Knowledge management needs a stronger link to the fundamental strategy of companies. In a recent study of more than 30 knowledge management projects (Davenport, DeLong, and Beers, 1997) most of the initiatives were growing and prospering, but my co-researchers and I concluded that less than 10% were likely to have an important impact on the overall business. The projects improved the efficiency or effectiveness of individual departments or business processes, but strategists and stockholders would have little about which to get excited. The three firms that did have highly strategic knowledge applications were a consulting firm, a research organization, and a firm that provided knowledge services — all, in other words, in the knowledge business. There are few examples of companies where the product isn't knowledge and knowledge management really matters to long-term success.

Further evidence of the relatively weak relationship between knowledge and strategy comes from an examination of the people who are directly involved in business strategy and knowledge management. For the most part these are not overlapping responsibilities in organizations. A quick assessment of the knowledge management personnel in 25 companies that were early adopters of that concept revealed that in only three of the companies (IBM, Monsanto, and Xerox) might the chief proponents of knowledge management be considered "strategists." Only these three might be directly involved in both strategy formulation and articulation of the company's approaches to knowledge management. Much more common are knowledge managers whose backgrounds involve information technology, quality management, the corporate library, and even the financial function. It may be difficult for individuals without any background in strategy to develop a tight linkage between knowledge management and a company's strategy.

Knowledge Management and Strategy Formulation

How can knowledge be linked to strategy formulation? As Fahey (1996) has pointed out both strategy and knowledge are dynamic, multifaceted concepts. A company's strategy may involve its current strategic position, or where it wants its strategy to take it in the future. Future strategy may involve working within an existing strategy context ("playing an existing game") or creating a new game. Strategy-oriented knowledge may involve many different domains, including customers, competitors, industry entrances and exits, suppliers, technologies, regulators and regulations, and product/service substitutes. A company can look at an existing strategy to determine what knowledge will make it successful, or look at its existing knowledge and specify what strategy will best take advantage of it. All strategic knowledge is mediated through the mind of a human strategist. Perhaps all of this complexity is one reason why few firms have established close connections between knowledge and strategy.

Given the various possibilities with respect to knowledge and strategy formulation, companies wishing to further develop the relationship can debate a series of questions. Arguably these questions must be answered before detailed work linking knowledge and strategy can begin. Some of the questions include the following:

- Are we relatively satisfied with our existing strategy at this time, or do we need to develop a new one?
- If we are comfortable with our current strategy, what do we need to do to execute it more effectively?
- What knowledge do we need to carry out these strategy execution activities?
- If we need a new strategy, what about our business environment do we need to know to develop a new strategy?
- If we need a new strategy, is there something we have substantial knowledge of already that could propel the company in a positive new direction?
- For each new strategy alternative, what knowledge would be necessary to advance it?

There is no evidence that such questions have actually been discussed in the context of a particular business. If such discussions did take place, they were probably in knowledge-intensive industries, e.g., professional services, pharmaceutical research, or high technology manufacturing. I would argue, however, that considering knowledge in the context of strategy formulation could be of benefit to all types of firms. Even the most prosaic manufacturing firm can benefit from better knowledge of its customers and markets, or from knowledge-based innovations in operational processes.

Knowledge Management and Strategic Advantage

It is also possible to view the strategy–knowledge relationship in terms of how knowledge and its effective management can confer strategic or competitive advantage on a firm. Indeed, I have seen definitions of knowledge management (not in print but in presentations) in which the use of knowledge for competitive advantage was part of the definition; i.e., knowledge management doesn't exist unless competitive advantage is pursued or even attained. Although no definition is more correct than another (only more or less consistent with established usage), my own view is that competitive advantage should be decoupled from knowledge management for purposes of definition. We can then try to establish as an empirical question the extent of the relationship between knowledge management and competitive advantage.

This relationship involves both strategy formulation and strategy execution. Some of the means by which knowledge management could actually lead to strategic advantage include:

- As suggested in the discussion above of strategy formulation, knowledge management can enable an innovative strategy that would not otherwise be possible. For example, a systems integration firm could reuse both methods and software and thus achieve high productivity relative to competitors.
- Knowledge management could make possible better execution of an important but common strategy throughout an industry. Pharmaceutical firms, for example, all compete on the speed and effectiveness of new drug development processes. Companies such as Hoffman-LaRoche, Johnson & Johnson, and American Home Products are all using knowledge management to try to develop better drugs more quickly. One empirical study of drug development activities in the pharmaceutical industry (Bierly and Chakrabarti, 1996) found that firms with more aggressive and innovative knowledge-creation approaches were found to be more profitable over time than those with more prosaic knowledge strategies.
- Companies can also gain advantage by adding knowledge to the products and services they offer for sale. The knowledge might be bundled with an existing product or service — e.g., offering a case-based reasoning capability on a customer service Web site for a computer company — or could be a standalone market offering. Both Ernst & Young ("Ernie") and Arthur Andersen ("KnowledgeSpace") sell electronic access to their knowledge separately from traditional professional services. It is not yet clear, however, whether customers will find these knowledge-based offerings sufficiently popular to lead to competitive advantage for the service firms.

- Companies could also achieve competitive advantage by using knowledge and knowledge management to perform nonstrategic processes exceptionally well. Knowledge management initiatives based on "best practices," for example, are typically broadly focused and relate to all types of business activities. If a firm can use supplier knowledge to improve its procurement processes, share financial knowledge across financial processes, and even circulate knowledge effectively about human resources or information systems processes, it might conceivably gain advantage over its competitors. To succeed with this approach, however, the knowledge management capability would have to be both broad and highly effective.

In order to actually achieve competitive advantage, a firm must have a good idea of what aspects of the business can lead to advantage, and be clear on what type of advantage it is seeking (revenue or profit levels, growth, market share, etc.). Perhaps the most critical requirement — thus far found only in professional services companies and information technology vendors such as IBM and Xerox — is that those who set strategy are conversant with knowledge management and the strategic opportunities it provides. Since knowledge management in most companies has been a subject discussed largely by middle managers and consultants, the opportunity to convert it into strategic advantage has been lacking.

Strategy Knowledge as a Knowledge Domain

What is strategy itself if not knowledge? Lyles and Schwenk (1992) have illustrated how an organization's strategy is, among other things, a set of knowledge structures that are widely held within an organization. It is possible to articulate this knowledge, show how it changes over time, and classify it into core and peripheral elements. The knowledge would generally be held by senior managers of a firm, although certainly it is possible to distribute knowledge of strategy more widely. In Lyles and Schwenk's analysis of a consulting firm's strategy knowledge, the particular types of knowledge include mission and purpose knowledge, competitor knowledge, client knowledge, industry knowledge, and the relationships between these knowledge domains.

Another example of research illustrating the role of knowledge in strategy compares two firms in the chemical and insurance industries (Earl, 1997). In these companies knowledge is viewed as arising from data and information. The research was one of the earliest to create a model of knowledge management, but has a strong IT orientation. This particular line of research might be viewed as a continuation of the "information technology for competitive advantage" literature in the 1980s, but with a new knowledge orientation.

If strategy is knowledge, or is at least partly a matter of knowledge, several questions immediately arise from the perspective of knowledge management. Some of these questions are listed below:

- How widely is the strategy knowledge disseminated, both internally and externally to a firm? Is a strategy only effective if it is broadly known internally?
- How, and to what degree, should the strategy knowledge be documented? Certainly most firms document their mission and purpose, but not necessarily all the background knowledge that goes into them. Other aspects of a strategy may not be documented at all.
- Does more strategic knowledge mean a better strategy? Are the best strategies those that have the most knowledge backing them up? Lyles and Schwenk suggest that strategic knowledge structures become more complex over time, but is this positive or negative?
- Is it necessary to "forget" old knowledge in order to adopt a new strategy?
- Is the flow of strategic knowledge from the top managers of a company down, or can new strategies arise when knowledge flows from bottom to top?

While these are interesting questions, few if any organizations have adopted a knowledge-centered perspective on strategy. Perhaps as strategists become more aware of knowledge management and its relationship to strategy, we will see more efforts to view firm strategy as a knowledge-oriented problem.

Core Competencies, Capabilities, and Knowledge Management

Knowledge has also been closely associated, albeit frequently in vague terms, with more people-oriented strategy concepts. The notions of strategic capabilities (Stalk, Evans, and Shulman, 1992) and core competencies (Prahalad and Hamel, 1990) have much to do with what an organization and its people know. These terms are used in a somewhat fuzzy manner; the relationship between capabilities and competencies is not clear, for example, in any strategy writings with which I am familiar. And knowledge is not the only factor in capabilities and competencies. That Wal-Mart has the capability or competency to do cross-docking well (a logistical activity frequently discussed in this literature) results not only from a knowledge of how to do the activity, but also the will to do it effectively, the skills to perform it well, and the accompanying equipment. Since no strategist has broken such capabilities down into their components, we cannot know exactly to what degree knowledge is involved or important to strategy.

It may be useful in the context of this discussion to at least define these concepts clearly with respect to knowledge and knowledge management. I define core competencies as the things an organization knows how to do well; knowledge of how to do the thing is a key aspect of a competency, but other factors are also necessary. I will refer to capabilities as being the things an individual knows how to do well; again, knowledge is an important component of a capability, but not all that matters. Individual capabilities can aggregate to organizational competencies given sufficient amounts of resources and management attention. These concepts are critical to the relationship between knowledge and strategy because they bring people into the discussion. One critical aspect of knowledge is that it originates and is applied in the minds of knowers; therefore discussions of knowledge management and strategy must eventually involve the topic of what those knowers actually know.

From a strategy perspective, the key issue around competencies and capabilities is deciding which ones are important to have. I have always felt that the competency and capability discussions in the strategy literature were a bit too retrospective; they tended to look back at firms that had exhibited good performance and explain it in terms of what the organization or its people knew how to do. Honda, a company frequently mentioned in this literature, has done well because of its competency in small engine design; Canon succeeded because of its knowledge of opto-electronics. But how did these and other successful firms decide which competencies and capabilities needed to be amassed or built? Human beings, and by extension the organizations they comprise, can become knowledgeable about a wide variety of topics. If I'm a manager at Honda, do I hire more experts in small engine design, or do I conclude that our engines are good enough and that I need instead to emphasize knowledge of body design, market research, or car recycling? This is the challenge of the competency and capability school of strategy.

The only feasible answers, I believe, involve the use of both strategic vision and human resource operations excellence. Strategic vision, industry knowledge, and market intuition — along with extrapolation of existing customer and market trends — are the only ways to determine what competencies and capabilities will be necessary to succeed in the future.

But knowing what knowledge is necessary is only the beginning. An organization must also have awareness of what capabilities each person within the organization has, and the ability to connect human resource investments to desired skills and knowledge. Most large professional services firms, where individual capabilities are the key to organizational success, have capability databases of growing sophistication. Microsoft has a system to keep track of what system development capabilities (called "knowledge competencies") are resident within the organization (Davenport and Prusak, 1998). When Bill Gates decided that more of Microsoft's people needed Internet development capabilities, the system allowed the company to know the extent of the problem and to monitor progress. Microsoft was also able to tie internal and external educational programs to the specific capabilities needed by the company.

Discussions about competencies and capabilities need not replace broader issues of strategy and competitive advantage, but they are a necessary correction to those who view knowledge as an abstraction or a set of documents to be captured in a repository. The concepts of competence and capability remind us that the most important knowledge is in the minds of people.

Knowledge Strategy and Business Strategy

A final strategic concept to be explored in this discussion is the relationship between a company's business strategy and its knowledge management strategy. Thus far few companies have had an explicit or even implicit knowledge management strategy; they have proceeded opportunistically, and middle-level managers have attempted to manage whatever knowledge was under their own control. To have a well-defined knowledge strategy requires a level of sophistication in knowledge management that only a few firms (again, primarily in the professional services industry) have achieved.

But a knowledge management strategy is ultimately important. It forces an organization's managers to make choices about key aspects of their knowledge environment. Some such choices are the following:

- Is tacit knowledge or explicit knowledge more important to manage?
- What knowledge domain is most important to the firm? Is it customer knowledge, competitor knowledge, product knowledge, supplier knowledge, or something else?
- Where does the knowledge environment most need improvement? Is the firm good, for example, at creating knowledge, but weak at applying and using it?
- Should the firm make or buy its knowledge in specific areas of the business?
- Which aspects of our knowledge environment should we be measuring?
- How will the firm make money on its knowledge?

Clearly no firm can manage all forms of knowledge equally well. Just as organizations made choices about what types of data and information they needed to manage first (and not necessarily wise or strategic ones; the first IT application in business was payroll), they must make choices about knowledge domains, knowledge management process steps, initiatives, and so forth. And if knowledge is to be considered an important organizational resource, the same levels of planning should go into knowledge management as for other key resources, such as labor and capital.

How do firms make the choices necessary in knowledge management? The answer, of course, lies with business strategy. A company or business unit should select its knowledge activities primarily on the basis of how they support or enable aspects of its overall business strategy. The aspects of strategy to which knowledge management can be tied are generally not financial goals, but rather high-level operational strategies. One categorization scheme for such strategies is described by Treacy and Wiersema (1996); they argue that a business can emphasize either operational excellence, product innovation, or customer intimacy, but not do well at more than one such "value discipline." While I'm not convinced that these categories are mutually exclusive, each orientation has implications in terms of knowledge management. Customer intimacy-based strategies, of course, would rely heavily on the accumulation and management of customer knowledge. Product innovation strategies would have at their core both customer knowledge and product knowledge, and would have a strong focus on product-oriented knowledge creation. Operational excellence strategies would be supported by process and best practice knowledge. These different strategic emphases might also dictate the knowledge "subtypes" within a domain; with regard to competitor knowledge, for example, an operational excellence firm would presumably be most interested in knowledge of the production processes and practices of its competitors.

I am not necessarily suggesting that a firm organize an effort to create an explicit knowledge management strategy. Like many functional or resource strategies, they will have little impact unless they are used. What is more important, as Mintzberg (1993) has often pointed out, is that managers engage in a substantive dialogue about strategy. I am simply suggesting that the dialogue begin to include the subject of knowledge and its management, and the key choices that the organization needs to make around the issues.

KM and Business Performance Improvement

Perhaps the ultimate test of any business concept is whether it leads to measurable improvements in business performance. Knowledge management's mark in this respect, as discussed above, has been

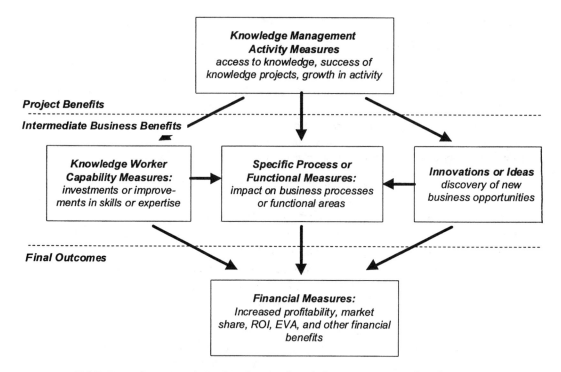

FIGURE 1   Alternative relationships between knowledge management and performance.

relatively light. Despite considerable discussion of the ties between knowledge and indicators of performance (balance sheets, stock price, market value, etc.), few if any companies have thus far been able to establish a causal link between their knowledge management activities and their business performance, regardless of how it is measured.

In fact, establishing this linkage will always be difficult. Many factors help to determine the performance of a business or other type of organization, and attempts to trace causality to any single factor such as knowledge management are fraught with peril. If revenues rise, for example, anything from favorable economic conditions to new products and services to luck can be cited as a rationale. Advocates of specific projects or programs will undoubtedly take credit for improved performance, but these claims must be credible.

One way to establish credibility in relating knowledge management to improved financial performance, or to embark upon a less ambitious approach to relating these two topics, is to employ "intermediate" measures. The measures will ideally reflect an aspect of the organization's performance that, if improved, will lead to better financial performance. A model describing the relationships between various alternative measures is shown in Figure 1. In the model, the most basic set of measures involve knowledge management activities. These might include the number of users of a knowledge management system, the number of "hits" to a knowledge repository, or the satisfaction levels of employees with a knowledge management initiative. Improvements in these measures are not a sign that better organizational performance is being achieved, but only that levels of knowledge activity are improving.

Companies should then try to link knowledge management activity measures with intermediate measures of business performance. These might include process measures, indicators of knowledge worker capabilities, or ideas and decisions. As Figure 1 illustrates, knowledge worker capabilities and ideas and decision improvements may also lead to process improvements. Specific process measures will vary by the particular process to which knowledge management is applied. Table 1 lists some exemplary measures for several different business functions. If both knowledge management and process measures (ideally more than one measure for a given project) are rising at the same time, we can credibly argue that knowledge management helped to cause the improvement in process performance.

TABLE 1   Types of Process Measures

Sales

"Face time" with customers
Speed of ramp-up for new sales hires
Customer evaluations of sales force expertise

Service

Cost per service call
Time per service call
Percentage of service calls requiring field dispatch
Customer satisfaction with the service function

Production

Downtime per production incident
Speed of transfer of production technologies

Research and Development

Time to market
Percentage of revenues from new products

Measures assessing improvements in knowledge worker capabilities involve the human dimension of knowledge management. Sample measures might include retention of knowledge workers, knowledge worker satisfaction, investment in knowledge worker learning, aggregate levels of knowledge worker productivity, and even the tested knowledge levels of employees. Of course, in order to measure these entities, a clear definition of who is a knowledge worker within an organization must be established. As with the process measures, if both knowledge management activity and several knowledge worker capabilities are increasing simultaneously, then we might expect that one trend is causally related to the other.

Some firms are also attempting to measure improvements in the ideas and decisions of their personnel. While the quality or innovation level of ideas is difficult to measure, it is possible to measure the outcomes of ideas. Buckman Laboratories, for example, assesses its knowledge management by measuring the percentage of revenues derived from new products (as noted in Table 1, this can also be a measure of the new product development process). It is also possible — but difficult—to measure the effect of knowledge on decision-making. General Motors, for example, is attempting to implement a program of decision audits, particularly with regard to new car development decisions. While I know of no company that is doing so, it would also be feasible to determine whether the results of decision audits are improving over time.

Improvements in intermediate measures may be sufficient for some organizations to show value from knowledge management. Others may feel the need to link these intermediate measures to financial measures. If indicators of knowledge management activity are going up along with those of processes, knowledge worker capability, or ideas and decisions, and financial performance is also improving, we may be able to build a "chain of credibility" tying knowledge management to better financial performance. But even in this situation it may be necessary to eliminate other obvious explanatory factors for improved financial performance before concluding that knowledge is the cause. Also, both before-the-fact, base-case measures, and after-the-fact measures must be taken if any improvement is to be demonstrated.

It should be apparent that developing credible relationships between knowledge management and performance requires a significant amount of time and effort. Given the difficulties of measurement, a company's knowledge managers should be certain that the measurement activities are necessary and sufficient to achieve credibility with senior executives before embarking upon a measurement program. It may be the case that nonrigorous measures, e.g., anecdotes about how knowledge management led to a big sale, may be sufficient to justify expenditures on knowledge management and to keep top managers satisfied.

2.3   Suggestions for Further Research

The points made in this discussion have been based more on logic and supposition than on empirical research, simply because there have been few organizations that have begun to link knowledge management to business strategy and performance. When managers begin to think in such terms and to make formal and informal linkages between knowledge and strategy, there will be something to research. Since there is already a group within the strategy research community that is interested in knowledge, we will likely see increasing amounts of research at the intersection.

The most basic research question involving knowledge and strategy has been articulated in a summary of knowledge management research questions (Teece, 1998b). He advocates that researchers should "Assemble evidence to test the proposition that firm-level competitive advantage in open economies flows fundamentally from difficult-to-replicate knowledge assets." In short, does knowledge yield competitive advantage? Teece suggests that this research question is methodologically challenging, since it has the measurement of knowledge or intellectual capital within a firm as a prerequisite.

There are some specific research foci that would be particularly helpful to practitioners. They would involve empirical research on the following types of issues:

- To what degree are actual knowledge management projects consistent with the strategies of firms and business units?
- How are firms justifying their knowledge management activities with regard to competitive advantage issues?
- How has knowledge management changed the competitive dynamics of specific industries?
- What specific measures of knowledge management and its relationship to business performance have companies actually adopted, and what kinds of results have they achieved?
- What other examples are there of strategy as knowledge, and how is that knowledge actually managed in organizations?
- What are the means in practice by which knowledge management affects strategy formulation?

This type of research would need to be field-based and would probably have to involve highly participative or ethnographic methods. Since the relationship between strategy and knowledge management will vary considerably across firms, it will be more useful to have substantial detail on one firm than less information about many firms. In short, it is unlikely that statistically rigorous academic research would be of much value in this context. Another implication is that work at the strategy-knowledge intersection should be undertaken not only by strategy academics and economists, but also by researchers with social and behavioral science backgrounds.

What is most important for the field of knowledge management is that actual empirical research be undertaken in a variety of fields. The field for too long has been the province of speculators, casual futurologists, and journalists. We are already convinced that knowledge and knowledge management are important; now we need to understand just what kind of difference they make.

References

Arrow, K. (1974) *The Limits of Organization*. New York: Norton.

Bierly, P. and A. Chakrabarti. (1996) Generic knowledge strategies in the U.S. pharmaceutical industry, *Strategic Management Journal* 17, pp. 123-135.

Davenport, T.H., D., DeLong, and M., Beers. (1997) Successful knowledge management projects, *Sloan Management Review*, Winter.

Davenport, T.H. and L. Prusak. (1998) *Working Knowledge: How Organizations Manage What They Know*. Boston: Harvard Business School Press.

Earl, M.J. (1997) Knowledge as strategy: reflections on Skandia International and Shorko Films, in L. Prusak, ed., *Knowledge in Organizations*. Boston: Butterworth Heinemann.

Edvinsson, L. and M. Malone. (1997) *Intellectual Capital.* New York: Doubleday.

Fahey, L. (1996) Putting knowledge into strategy. Presentation to Ernst and Young "Managing the Knowledge of the Organization" conference, Phoenix, Arizona.

Lyles, M.A. and C.R. Schwenk. (1992) Top management, strategy, and organizational knowledge structures. *Journal of Management Studies* 29, pp. 155-174.

Mintzberg, H. (1993) *The Rise and Fall of Strategic Planning.* New York: Free Press.

Nelson, R. and S. Winter. (1982) *An Evolutionary Theory of Economic Change.* Cambridge, MA: Belknap Press.

Penrose, E. (1959). *The Theory of the Growth of the Firm.* New York: Wiley.

Prahalad, C.K. and G. Hamel. (1990) The core competence of the corporation. *Harvard Business Review* 68:3 pp. 79-91.

Stalk, G., P., Evans, and L.E., Shulman. (1992) Competing on Capabilities: The New Rules of Corporate Strategy, *Harvard Business Review* (March-April).

Sveiby, K.E. (1997) *The New Organizational Wealth: Managing and Measuring Knowledge-Based Assets.* San Francisco: Berrett Koehler.

Teece, D.J. (1986) Profiting from technological innovation, *Research Policy* 15, pp. 286-305.

Teece, D.J. (1998a) Capturing value from knowledge assets: the new economy, markets for know-how, and intangible assets, *California Management Review* 40:3 (Spring), pp. 55-79.

Teece, D.J. (1998b) Research directions for knowledge management, *California Management Review* 40:3 (Spring), pp. 289-292.

Treacy, M. and F.D. Wiersema. (1996) *The Discipline of Market Leaders.* Reading, MA: Addison-Wesley.

Winter, S. (1987) Knowledge and competence as strategic assets, in D. Teece, ed., *The Competitive Challenge: Strategies for Industrial Innovation and Renewal.* Cambridge, MA: Ballinger.

3

Introducing Knowledge Management into the Enterprise[1]

Karl M. Wiig
Knowledge Research Institute, Inc.

3.1   Overview

Knowledge management (KM) is pursued actively by many organizations to improve or sustain their viability and success. Most enterprises introduce KM practices specifically tailored to their needs, environments, and perspectives. Hence, there are many varieties of KM. Some emphasize the use of technology to capture, handle, and locate knowledge — although as a first step many of them concentrate on

---

[1]This chapter is adapted from the author's forthcoming book: *Establish, Govern, and Renew the Enterprise Knowledge Practices* (Wiig 1999).

improved management of information rather than management of knowledge.[2] Some focus on knowledge sharing among individuals or on building elaborate educational and knowledge distribution capabilities. Others foster environments in which their employees can be innovative and creative. Still others pursue building and exploiting intellectual capital to enhance the economic value of their enterprise. Additionally, a few advanced enterprises are able to pursue all of these thrusts. Some have gone even further to create a "knowledge-vigilant" environment to make the enterprise act intelligently in the interest of long-term success and viability.

The successful enterprise's objectives for pursuing KM are quite clear: It wishes to manage knowledge effectively to make people — and the whole enterprise — act intelligently to sustain its long-term viability by developing, building, and deploying highly competitive knowledge assets (KAs) — in people and in other manifestations. It expects that intelligent behavior will lead to proper and effortless handling of routine and simple tasks and that nonroutine, complex, and unexpected tasks will be handled timely, competently, and in the best interest of all concerned with suitable balances between long-term and short-term objectives.

Your Challenge, Should You Choose to Accept It

You are asked to start a KM practice within your organization. Questions immediately arise: "What is KM anyway? Is KM just another fad? What are the opportunities? What are the resource requirements? What are the risks? Where do I start? How do I proceed?" And particularly "What do I do right now?" Many of these questions are addressed by other chapters in this book. By describing common KM building blocks, this chapter focuses on options for how to undertake the KM introduction to ensure long-term business success.

3.2 A Broad Perspective of Knowledge Management

Before we can outline approaches to introduce KM within the enterprise, we must identify which particular "brand of KM" we focus on. Our definition of KM is broad and embraces KM approaches and activities throughout the organization. In our view, the reasons to undertake KM are practical, basic, and far from faddish. It is to enable and motivate people — and the enterprise itself — and provide opportunity and permission to act intelligently and thereby secure success and viability through superior performance.

In spite of the many approaches, one aspect of truly successful KM escapes many. KM is not only a technology, is not only a set of explicit and rigidly systematic activities, is not only a patent method to increase the economic value of the enterprise. Instead, KM is an effort to make the enterprise "knowledge-vigilant." Systematic and explicit KM is designed to create an enterprise-wide, adaptive, contextual, comprehensive, and people-centered environment that promotes continual personal focus on knowledge-related matters. This should occur as part of the daily mix of activities and responsibilities to make everyone — alone or in groups — act intelligently for their own and the enterprise's benefit and sustained viability.

Some aspects of enterprise-wide intelligent-acting behavior are indicated in Figure 1. The underlying premise is that "**Knowledge is a fundamental factor behind all of the enterprise's activities.**" The model outlines elements that fall under the auspices of KM, such as learning and innovating and the effective creation and application of knowledge assets (KAs) in internal operations. It also points to the need for permission, motivations, opportunities, and capabilities for individuals to act intelligently.

---

[2]We distinguish between "knowledge" and "information" and recognize that they are fundamentally different. **Information** consists of facts and other data organized to characterize a particular situation, condition, challenge, or opportunity. **Knowledge** consists of truths and beliefs, perspectives and concepts, judgments and expectations, methodologies and know-how and is possessed by humans, agents, or other active entities and is used to receive information and to recognize and identify; analyze, interpret, and evaluate; synthesize and decide; plan, implement, monitor, and adapt – i.e., to act more or less intelligently. In other words, knowledge is used to determine what a specific situation means and how to handle it.

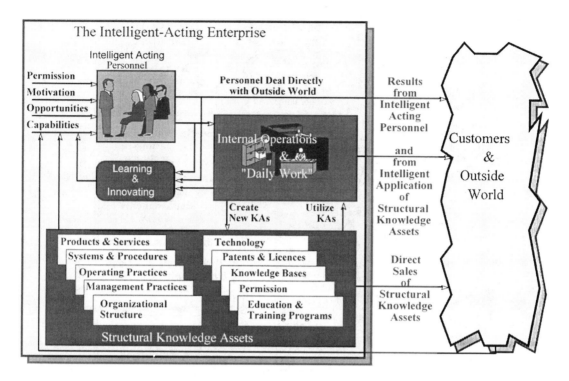

FIGURE 1   Role of individuals, knowledge assets, learning and innovation, and internal operations for enterprise-wide intelligent behavior.

KM is the systematic and explicit management of knowledge-related activities, practices, programs, and policies within the enterprise. Consequently, the enterprise's viability depends directly on:

- The *competitive quality* of its knowledge assets
- The *successful application* of these assets in all its business activities — i.e., realization of the knowledge asset's value

From a slightly different perspective: "**The goal of knowledge management is to build and exploit intellectual capital effectively and gainfully.**"[3] This goal is valid for the entire enterprise, for all of the enterprise's activities, and needs additional explanations since there is considerable complexity behind it.

One important aspect for effective KM is the requirement to deal with the complexity of how people use their minds — that is: think — to conduct work. It concerns what they need to know and understand and how they must possess and have access to the knowledge to act intelligently. The same considerations must also be dealt with on the organizational level.

Two aspects of effective, broad-based KM cannot be overemphasized. The first is that the KM initiatives and activities normally do not lead to more work. Instead, as a result of improved knowledge and knowledge use, often far down in the organization, operations exhibit less rework, less hand-offs, quicker analysis, decision, and execution, particularly of nonroutine tasks, and other desirable traits. The second aspect is that KM activities and initiatives, instead of being new and added functions, must to the largest extent possible be based on and be part of ongoing efforts — often without making these more difficult, time consuming, or effort demanding.[4]

---

[3]Fernando Simões (1998).
[4]Lucier and Torsilieri (1997).

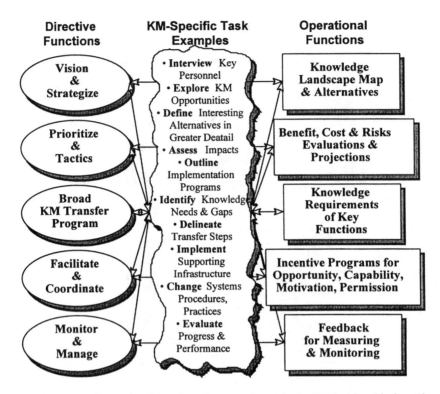

FIGURE 2 The directive and operational functions that are reflected in the KM building blocks with examples of KM-specific tasks.

Building Intellectual Capital

Personal and enterprise intellectual capital building takes place in hundreds of different ways. Examples of building professional intellectual capital range from educating stockbrokers in theories that explain principles behind market mechanisms — to building enterprise knowledge assets by embedding outstanding engineering knowledge into new designs for automotive transmissions. Intellectual capital building in crafts people clearly can range from training employees in better methods such as for assembling electronic boards to educating them in principles for operating and maintaining their equipment. As implied in Figure 2, intellectual capital may be built through innovation. It is also built by importing external knowledge, by standardizing practices, by securing knowledge through patents, and so on. Intellectual capital building requires investment of resources, money, time, attention, etc.

Sharing knowledge, networking to become acquainted with what others know, etc., are examples of investment activities to build intellectual capital for future use. The areas of knowledge required to work effectively only partially consist of the primary professional and craft knowledge that normally is the focus. It appears that "enterprise navigation knowledge" — the understanding of who to contact and how to treat them in all kinds of special situations — constitutes the major part of the typical employee's valuable knowledge. It must be remembered, however, that this personal knowledge is "rented." It is not owned by the enterprise but is only possessed by the individual.[5]

Building intellectual capital *effectively* requires that only the best available knowledge is incorporated in the resulting intellectual capital. It also requires that the most effective approaches and methods are selected and utilized to support innovation, capture new knowledge, transforming it to be archived or

[5]Edvinsson and Malone (1997).

used, and the like. Furthermore, it refers to building intellectual capital to support the enterprise strategy and direction in the most appropriate ways. Numerous methods are available. For example, there are many ways of teaching, many ways of learning, and many ways of capturing knowledge and embedding it in technology, ranging from working alone with pencil and paper to collaborating with others while using the most modern computer-aided design (CAD) tools.

Building intellectual capital *gainfully* refers to investing in, among others, expanding, verifying, and validating intellectual capital in cost-effective ways where the comprehensive, life cycle returns exceed investments. It also refers to allocating the highest priority for investing in intellectual capital items with the greatest expected utility. In the new economy, gainful knowledge building also refers to building personal knowledge assets in an environment which is conducive to knowledge transfer from personal tacit knowledge to structural knowledge assets owned by the enterprise. Personal security becomes a key to success.

Exploiting Intellectual Capital

Whenever work is performed, and whenever goods, services, licenses, and the like are sold or exchanged for value, the value of intellectual capital is realized in some manner: intellectual capital (IC) is exploited. We are particularly interested in how intellectual capital is applied to perform regular work — and that includes both routine tasks and unexpected challenges. Most work is perhaps "routine" and is often based on pre-established principles. Nevertheless, this type of work is based on application of the most appropriate knowledge. Increasingly, work within most enterprises is becoming "project work" that demands individualized approaches and improvisation to deliver the quality desired. To meet such requirements, intellectual capital must be exploited appropriately to allow the desired intelligent behavior. In addition, many ICs can be sold, licensed, used to leverage joint ventures, etc., to the enterprise's advantage.

Exploiting intellectual capital *effectively* refers to using the most appropriate methods for transforming the requisite intellectual capital to the best form to realize its value and to place it at the Point-of-Action (PoA). It also refers to the manner in which the intellectual capital is applied to perform work or to achieve the desired business purpose. For example, if a specific knowledge asset will be used by a customer service representative (CSR) to explore and discuss how best to help a customer, that knowledge must be available at the PoA by being possessed as automatic knowledge in the mind of the CSR. This tacit knowledge must be made available to the CSR through both training and education. Other knowledge, such as how to proceed with a highly unusual implementation of complicated assistance, may be made available to the CSR through a computer-based help system or an expert network or by receiving help over the telephone or in person from a master practitioner.

Exploiting intellectual capital *gainfully* requires that the most appropriate intellectual capital (i.e., "best" knowledge) is used wherever work is performed and that it is used in the best interest of the enterprise and individuals. That is particularly important in high value-added situations, which ideally should always be the case, and can only be achieved in a knowledge-vigilant environment. It also requires that intellectual capital that are sold or licensed be leveraged to the highest extent possible without producing negative impact.

Knowledge Management Building Blocks

We introduce and discuss a set of common building blocks for creating a customized KM introduction program. Our objective is to provide enough detail for practitioners to obtain an integrated overview of the functions and subordinate tasks involved and the methods and levels of expertise required to perform the functions. It is not within our scope to provide enough insight and understanding to be prepared to deliver the detailed work involved.

Together the building blocks form a coordinated and comprehensive KM practice. However, the intent is not for the beginning KM practice to pursue all building blocks, but to select those that form a suitable subset for the particular situation that the enterprise faces.

The major KM building blocks that should be considered for introduction of a new KM practice include the ones listed below. They are presented in approximate order of implementation and discussed in detail later:

1. **Obtain management buy-in**: Pursue management commitment since it has proven essential for success of KM efforts. This stems from the central position that knowledge occupies in the enterprise.

2. **Survey and map the knowledge landscape**: Identify the nature, strengths, and weaknesses of the enterprise knowledge assets and situation in view of enterprise direction and operations and market pressures and opportunities.

3. **Plan the knowledge strategy**: Determine how KM will support the enterprise or business unit strategy and pencil out KM thrusts and expected priorities.

4. **Create and define knowledge-related alternatives and potential initiatives**: Identify opportunities for improvements such as opportunities for revenue enhancement, creation of new products and services, relief of knowledge-bottlenecks and other knowledge-related actions with the support of department and enterprise-level priority setting and outline their expected impacts and benefits.

5. **Portray benefit expectations for knowledge management initiatives**: Delineate expectations to prioritize, guide implementation, and monitor the effectiveness of KM efforts.

6. **Set knowledge management priorities**: Determine priorities for activities based on KM strategy, expectations for net benefits, needs, and availability of capabilities.

7. **Determine key knowledge requirements**: Identify knowledge required to deliver quality work in key or complex positions.

8. **Acquire key knowledge**: Capture knowledge from departing personnel and knowledge required for key critical knowledge functions (CKFs).

9. **Create integrated knowledge transfer programs**: Create comprehensive knowledge transfer programs — for example, by coordinating training programs, creating expert networks, or communicating expert knowledge such as concept hierarchies and mental strategies to practitioners (content as subject knowledge and methodologies as metaknowledge).

10. **Transform, distribute, and apply knowledge assets**: Organize and transfer expert knowledge to practitioners. Reconfigure, deploy, and exploit knowledge through effective use of "best" knowledge in all daily work.

11. **Establish and update KM infrastructure**: Build and maintain generic capabilities, some of which are specific to KM while most are shared with other activities and functions.

12. **Manage knowledge assets**: Create, renew, build, and organize knowledge assets to address priority knowledge opportunities.

13. **Construct incentive programs**: Motivate employees to act intelligently, i.e., be innovative, share knowledge, expend effort to capture knowledge (e.g., lessons learned), ask for assistance when meeting unfamiliar or difficult situations, etc.

14. **Coordinate KM activities and functions enterprise-wide**: Identify KM-related activities and assist them to coordinate, cooperate, and collaborate to build valuable capabilities and practices.

15. **Facilitate knowledge-focused management**: Provide high-level activities to change the enterprise "customer service paradigm," culture, work environment, management philosophy and practices, operating practices, decision rights, work flows and "opportunities to act intelligently," and personal motivators.

16. **Monitor knowledge management**: Provide feedback on progress and performance of KM program and activities.

An overview of the directive and operational functions represented in these building blocks is presented in Figure 2 along with a few examples of representative tasks. The directive, or management-related functions are shown to the left. The operational, or detailed and hands-on functions, are shown to the right and generally correspond to the directive functions, and they share many of the specific tasks.

3.3 Eight Important Agenda Items for the KM Practice

Based on experience in many organizations, we can outline several suggestions for "things-to-do" that must be pursued up-front for successful introduction of the enterprise KM practice. These suggestions are listed below in sequence of importance:

- Develop a **broad vision** of the KM practice and obtain **top management buy-in**:

  KM champions must have a deep and flexible mental outline of how KM will be conducted and organized to support the enterprise's direction, goals, and objectives. This vision provides the foundation and guide for creating KM capabilities and infrastructure supports and for setting priorities.

- Pursue **targeted KM focus** determined from knowledge landscape mapping insights and other opportunities and based on KM priorities that align with enterprise objectives:

  Undertake "bite-sized" and sharply targeted KM initiatives with clear benefit expectations that cumulatively build to implement the KM vision.

- Allow team members to **focus full time on KM** and build **KM professional team**:

  Designate highly competent KM team members to work dedicatedly on KM and avoid the mistake of diverting their efforts by giving them additional responsibilities. This is difficult.

  The KM practitioners and team members must have good understanding of "knowledge" (in contrast to "information"); its role in conducting knowledge-intensive work in diverse situations (i.e., "how people work"), methods for eliciting, acquiring, transferring, and organizing knowledge, etc. These are often new professional areas for the enterprise.

- Install **KM impact and benefit evaluation methods**:

  Impacts and benefits from KM are often indirect and happen gradually over time. Dynamic event chain and other impact assessment approaches must be selected and serve as foundations for cost-effectiveness or EVA analyzes.

- Implement **incentives** to manage knowledge on personal and enterprise levels, collaborate broadly, and act intelligently — innovate, capture, build, share, and use knowledge:

  The enterprise must express its support of KM clearly. Employees on all levels must understand personal benefits resulting from active KM. **Disincentives** must be removed.

- Teach **metaknowledge** to everyone:

  Knowledge workers at all levels exhibit sharp increases in their work focus, effectiveness, ability to develop and take advantage of improved knowledge when taught with procedures that allow them to internalize metaknowledge (mind maps, concept hierarchies, associative maps, etc.).

- Ascertain that implemented KM activities provide **opportunities, capabilities, motivations, and permissions** for individuals and the enterprise to act intelligently:

  To realize the full value of personal knowledge and other knowledge asset manifestations, the use of these assets for delivery of products and services must be highly effective. Effective intelligent-acting behavior can only be achieved when the conditions of opportunity, capability, motivation, and permission are satisfied.

- Create supporting **infrastructure**:

  Build upon all appropriate existing capabilities and gradually add new ones as required to facilitate effective KM, particularly in the chosen target areas.

These recommendations stress the need to work with top management and other levels of management to build understanding and obtain their agreements. They point to the necessity to connect KM opportunities to enterprise direction and strategies and to chart knowledge landscape overview. Of particular importance is the need to work with employees at every level to generate the understandings that KM is in their interest. Practically, it is important to identify "profitable" start-up alternatives, start with a

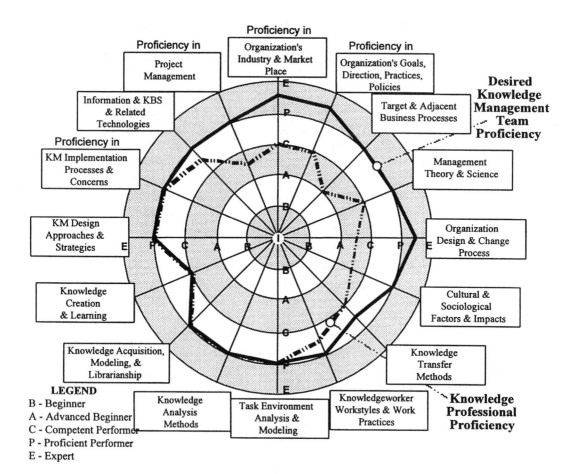

FIGURE 3 Example of KM-related proficiencies for the combined KM team.

manageable "chunk," build expertise in specialist areas and in the general population, and embed the road to broad KM practice within people's daily work.

The initial efforts must be staffed with sufficiently competent individuals and these must be allowed to dedicate themselves to the KM introduction effort.

Knowledge Management Teams Need Broad Expertise

To ascertain proficient introduction of KM in the enterprise, it may be necessary to consider the spectrum of expertise areas that ideally should be available on a multidisciplinary KM team. Sixteen of these expertise areas are indicated in Figure 3. It must be noted that most of these need to be available only as proficiencies in peripheral members of the team. The knowledge specialists or "knowledge management professionals" (KMPs) should be expected to take the lead in only six areas.[6]

A common problem with many KM introduction efforts is that they are undertaken by teams that have not had the opportunity to prepare themselves with adequate understanding of KM theory and practical approaches. Practical knowledge management work requires focus and expertise at several levels: One level is to provide insights and details to set priorities and strategic direction. A second is to understand broad operational requirements to determine needs for infrastructure. A third is to work with knowledge-intensive functions to determine needs and opportunities. A fourth is to deal with

---

[6]Wiig (1995, p.17).

knowledge itself — elicit, organize, encode, or build into KBS applications, among others. The associated work needs to be considered from different perspectives. They are the strategic, the tactical, and the operational. In particular, the KM management and operational team must include perspectives and expertise to deal with central KM issues such as:

- Providing insights to set priorities and strategic direction.
- Understanding broad, enterprise-wide requirements to determine needs for incentives, infrastructure, and other supports.
- Working with knowledge-intensive functions (i.e., how people — and organizations — obtain, create, hold, share, and apply knowledge) to determine needs and opportunities.
- Dealing with knowledge itself — elicit, organize, encode, deploy personal and structural knowledge for direct use or to build it into intellectual capital such as products, services, technology, or KBS applications.

Successful Introduction Requires Dedicated Effort

Another problem often encountered is that the KM champions and their teams are not allowed to pursue the introduction effort full time — or at least with sufficient time budget to meet the requirements for the work to be completed at sufficient depth. It is not unusual to find that the champions are asked to take on the responsibilities for introducing KM without having been released from other obligations. Such partial support most often lead to aborted efforts.

3.4 Knowledge Management Building Block Examples

Below we introduce examples of KM building blocks — potential KM activities — for introducing and sustaining the enterprise's KM practice. It is not the intent that building blocks be implemented rigorously, but that they reflect building of systematic, i.e., integrated and reasonably interconnected, activities that support each other in a functional manner. Again, care must be taken to allow improvisation to reflect the real context of people, work environments, management philosophies, and the needs of the enterprise, its customers, its suppliers, and other stakeholders.

A KM introduction effort may focus on a limited or a wide selection of building blocks and those will frequently be pursued in parallel, iteratively, or periodically. Furthermore, other versions of the building blocks discussed will be created to suit the situation. Approaches to introducing and sustaining KM will vary by constraints and necessity. Capabilities and management perspectives vary, needs are different, and as a result detailed emphases and implementation solutions need to be adapted to the situation at hand.

The general relationships between the various KM building blocks are illustrated in Figure 4. Only major and clearly definable relationships have been labeled. Some broader relationships, such as those associated with the management functions have not been identified, nor have all their points of influence been shown. It is implied in Figure 4 that a KM introduction program will start with the first building block, "Obtain Management Buy-In," and that other activities proceed from there. That may not always be the case if sufficient insight already exists to allow the introduction effort to be targeted and focused on highly value-added efforts or when different perspectives for what the KM practice should achieve are pursued. Similarly, one or more of the building blocks may be omitted when there are sufficient insights to proceed to higher levels.

1. Obtain Management Buy-In

Obtaining management buy-in is part of the process to create a shared vision for how the KM will be undertaken and what it will do for the enterprise. KM champions already have beginning visions to help the process. These can only be realized when the management team can "wrap their arms around" what KM is, what it can do for the enterprise, and how it might be accomplished and realized within their organization.

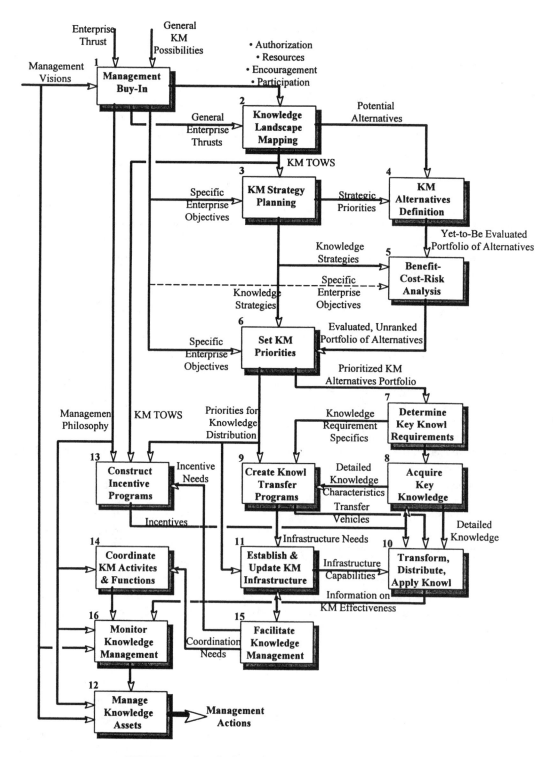

FIGURE 4   Selected relationships between KM building blocks.

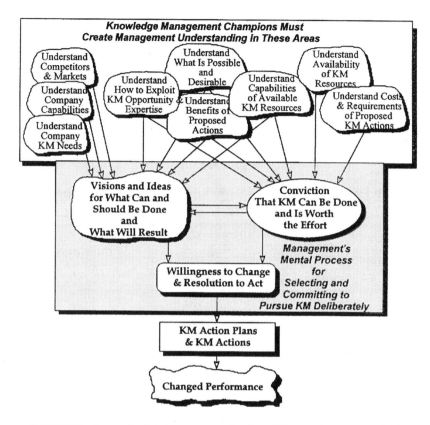

FIGURE 5 Factors for helping managers develop willingness to change and act.

Purpose

For KM to be successful, the importance of having management's commitment and continued support cannot be overemphasized. However, obtaining commitment and support is not simple and must be related to present and future enterprise objectives, needs, and capabilities in realistic terms, often with concrete examples. This author has repeatedly experienced the difficulty of communicating convincingly about KM in the abstract or in general terms. Specific, relevant, and credible (i.e., realistic) examples are required and must be created. Most managers have learned from experience to be conservative when presented with exciting new opportunities. Many new and supposedly promising opportunities have proven difficult to implement and have not produced the expected results. The KM opportunities must be well founded to be acceptable and to work.

General Characteristics

For anyone, and managers are no exception, to "buy into" activities or thrusts that require attention and resources, they need to develop the conviction that it can be done and will be worth the effort. As indicated in Figure 5, to reach that state, they need to develop visions and ideas for what can be done, what will result, and what effective KM may produce as results.

Example of Approach

KM champions must engage themselves actively in providing assistance to managers by helping them understand the promises and requirements of KM. As a minimum, managers need to believe and understand nine different factors as indicated in Figure 5. These are associated with competitive and market pressures and opportunities that explicit KM may affect, enterprise needs and capabilities — in general and as regards KM, and benefits, costs, and risks of potential KM thrusts and activities. Only when managers have obtained these insights, can they reasonably be asked to act.

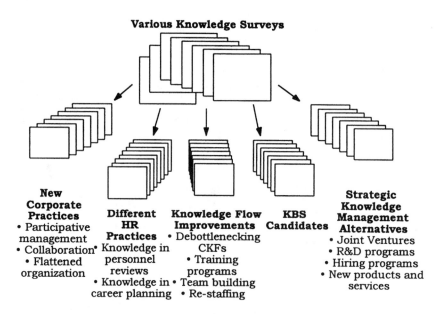

FIGURE 6   Knowledge surveys have many uses.

Specific tools are of great help in delineating and communicating the specifics and potentials associated with concrete KM opportunities. In particular, dynamic benefit evaluation methods, often supported by computer-based graphic simulation tools, are helpful.[7]

2. Survey and Map the Knowledge Landscape

The knowledge landscape map (KLM) is similar to other results from analysis of the enterprise's capabilities, assets, opportunities, and vulnerabilities. However, in nature it is quite different since it addresses aspects of the enterprise that often are hidden from conventional observation and have therefore not previously been charted. The knowledge landscape map effort could be very large if conducted in-depth with a large team. Typically, however, it needs to be performed swiftly and economically — in a few weeks or at most two months with dedicated and competent teams with from three to five members. In the interest of the enterprise, it is often most effective to create a swift knowledge landscape map that leaves holes and later go back to flesh out areas of specific importance and interest.

Purpose

Knowledge landscape maps serve many purposes. As indicated in Figure 6, they and similar surveys have many additional uses beyond providing an overview of the knowledge landscape. The major purpose here is to provide the necessary understanding of the enterprise's knowledge state. Management and KM champions need this overview to lay the groundwork for strategy setting, prioritizing KM projects and activities, and identifying specific KM needs and opportunities. At later stages in the KM practice, a knowledge landscape map is also used to audit and provide feedback for KM governance monitoring.

It can be expected that the knowledge landscape map will identify a considerable list of potential KM actions, many of which will be of a critical knowledge function (CKF) nature.

General Characteristics

A knowledge landscape map typically provides brief descriptions of major knowledge-related assets, programs, activities, and practices, among others. It will also describe their state relative to the competition and existing markets and to general developments inside and outside the enterprise. A taxonomy

---

[7]Examples are tools such as Extend (by Imagine That, Inc.) and PowerSim (by PowerSim, Inc.).

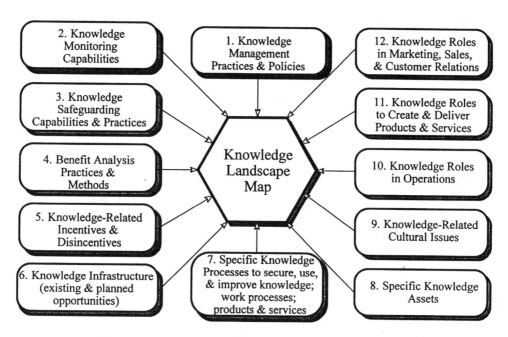

FIGURE 7 Twelve major knowledge landscape map components.

that may be pursued for the knowledge landscape map can be built around the twelve components and the associated examples of knowledge-related characteristics indicated in Figure 7 and Component Group 1 through 12.

KLM Component Group 1. Examples of KM Practices and Policies

- Specific policies relating to KM
- KM organizational structure with roles, responsibilities, and reporting relationships
- Specific KM practices such as:
    - −practices for obtaining missing expertise in unfamiliar situations
    - −knowledge sharing
    - −knowledge capturing and inventorying
    - −knowledge safeguarding
- Priority setting procedures and policies.

KLM Component Group 2. Examples of Knowledge Monitoring Capabilities

- Monitoring of KM effects and performance as predicted from detailed dynamic model
- Overall knowledge asset monitoring system — its structure, components, what it measures, who is responsible, how frequently it is executed, etc.
- Knowledge investment follow-up and continued justification (for example, relating to the investments in ongoing educational programs or existing KBS applications that need resources to be updated and serviced on a continual basis)
- Practices for estimation of performance of knowledge-related activities

KLM Component Group 3. Examples of Areas for Knowledge Safeguarding

- Knowledge sharing with suppliers, customers, outside parties (universities, competitors, etc.)
- Employment contracts
- Patents
- Document controls and office practices

KLM Component Group 4. Examples of Benefit Analysis Practices

- "Bottom-Line" measures to be used (NPV, changes in $/share after taxes, viability measures such as time-weighted $ after taxes for each year over next 10 years, etc.)
- Economic Value Added (E.V.A.) methodologies
- Static and dynamic value-chain quantitative models
- Qualitative analysis framework and impact analysis methods
- General benefit projection models or approaches

KLM Component Group 5. Examples of Incentives and Disincentives

- Individual vs. team performance evaluation for salary increases and bonuses
- Personnel evaluation items such as giving points for self-education, knowledge sharing, seeking assistance for important decisions to ascertain that best knowledge is used, etc.
- Time (or expense) budgets for education
- Lack of budgets for education and knowledge upgrading
- Recognition programs for capturing Lessons Learned and for innovations

KLM Component Group 6. Examples of Knowledge Infrastructure

- Knowledge professional capabilities, management responsibilities
- IT-based infrastructure (networks, groupware, knowledge mining tools, etc.)
- Policy infrastructure (compensation policies, explicit knowledge-related policies)
- Capabilities to coordinate knowledge processes
- Capabilities to develop and maintain knowledge assets (including core competencies)

KLM Component Group 7. Examples of Specific Knowledge Processes

- Processes (for securing, using, improving the knowledge itself, improving the work process, improving the products and services)
- Degree of integration and coordination between knowledge processes such as Lessons Learned systems and practices, learning labs, R&D programs, education and training program design functions, teaching centers and activities, KBS development groups, etc.
- Identifying quality of existing knowledge flows and opportunities to create important new ones

KLM Component Group 8. Examples of Specific Knowledge Assets

- Specific knowledge assets (incl. core competencies):
  - what they are
  - how they are used
  - why they are valuable (in operations, product and service creation and delivery, and their market and customer relations)
  - where they can be found
  - form in which they occur (in peoples' heads, education and training materials, KBS applications, formal knowledge bases, unstructured databases, patents, documents, etc.)
- The state of knowledge assets:
  - competitive and strategic positions, maintainability
  - potential market value
  - degree to which their values are realized
  - unrealized opportunities
  - analysis of threats, opportunities, weaknesses, strengths (TOWS)

KLM Component Group 9. Examples of Knowledge-Related Cultural Issues

- Not Invented Here (NIH) syndrome
- "We strive to be inventive and innovative," "We share knowledge readily" — people gain recognition from superiors, peers, and subordinates
- "Here, we gain status by being the most knowledgeable and therefore keep our expert knowledge to ourselves — we do not share it readily."

KLM Component Group 10. Examples of Knowledge Roles in Operations

- Levels of expertise made available to different areas of operations to maximize quality and minimize costs
- Processes and practices for developing and building operations knowledge
- Processes and practices for applying knowledge (such as to ascertain that only the best knowledge is used for important tasks and decisions)

KLM Component Group 11. Examples of Knowledge to Create and Deliver Products and Services

- Inclusion of explicit and implicit knowledge assets in products and services (advice, technology, etc.)
- Use of technological and knowledge-content building blocks to create new products and services
- The levels of expertise — and the areas of expertise — used to develop new offerings
- Knowledge flows associated with product and services creation and delivery

KLM Component Group 12. Examples of Knowledge Roles in Relation to Customers

- Knowledge in marketing, sales, and customer relations
- Degree to which knowledge about customer situations and customer experiences is obtained and applied to improve products and services and their delivery — to ascertain that customers succeed in their business using "our" products and services
- Use of knowledge for revenue enhancement

Example of Knowledge Landscape Mapping Approach

Mapping the knowledge landscape may be approached in different ways depending on the amount of time and effort available. As indicated above, mapping is best performed by a small, full-time, and highly expert team of from three to five members. The duration should be kept short, from a few weeks to at most two months, perhaps to be repeated later for additional information after more is known.

The knowledge landscape is exceedingly complex and information gathering to construct an overview map can never be exhaustive, even in small organizations, and judicious sampling therefore becomes necessary. Typical approaches include:

- Broad Questionnaires (100+ responses) with from 4 to 8 pages of open-ended and general questions.[8]
- Detailed Questionnaires (50+ responses) with from 10 to 20 pages of specific questions for detailed information.
- Group Sessions consisting of day-long interactive workshops with from 20 to 30 managers.
- Brief Interviews (20+ individuals) consisting of 30 minute interviews of selected people in the organization.

---

[8]Additional discussions of topics, representative questions, group session conduction, and questionnaires are provided in Wiig (1995).

- In-Depth Interviews (30+ individuals) with one to two hour interviews of several managers and knowledge workers.
- Formalized reporting that is part of annual department and division reports.

Normally, several of these approaches are combined to, for example, cast a wide net to obtain balanced inputs. This can be done through questionnaires, followed up with personal interviews, and, at times, also with group sessions.

The knowledge landscape map is typically presented as a stand-alone high-level summary report (different from a presentation supported by narrative). Results include brief descriptions of the knowledge characteristics nature and conditions in view of enterprise's direction and operations and market pressures — often in terms of significant threats, opportunities, weaknesses, and strengths ("TOWS"). A major outcome of the exercise is to generate a preliminary list of opportunities and associated KM-related actions. We also expect that a number of potentially important KM initiatives will be defined as a result of the knowledge landscape mapping process.

3. Plan the Knowledge Management Strategy

The KM strategy must be an integral part of the overall enterprise (or business unit) strategy. It must support the thrust of the organization and match its management philosophy, capabilities (particularly, knowledge professional expertise and availability and infrastructure), and the knowledge-related TOWS.

Purpose

KM affects all aspects of an organization. This includes personnel, education, proficiencies, work quality, effectiveness of the enterprise, culture, customer relations — and very importantly — profitability and enterprise viability. For successful management of all these knowledge-related matters, a sensible strategy is required. Its purpose is to organize, coordinate, and monitor the efforts to ascertain that the desirable results are obtained.

The purpose of a KM strategy is to delineate agreed-on objectives by reflecting the KM vision in a coherent framework. An explicit strategy is required for introducing the KM practice to provide a general and flexible guide for implementers when they create and select concrete KM solutions adjusted to actual conditions and requirements. It is an important communication vehicle for spreading understanding and participation throughout the enterprise. Effective KM depends on coordination of knowledge-related activities, cooperation with many parties, and collaboration with individuals and functions, who can contribute to create better practices.

General Characteristics

Strategies can be presented in many different formats. At a minimum, the KM strategy must express the goals for the KM practice and identify how these goals support the enterprise thrust. The strategy should also express the vision for KM and how the strategy will achieve implementation of that vision. The strategy must be sufficiently clear (but not too explicit or detailed) to serve as foundation for developing tactical steps to introduce and sustain the KM practice. The strategy may include expectations for completion of major milestones to coincide with events inside and outside the enterprise.

The KM strategy will often include definitions of a "customer service paradigm." It outlines in broad terms the service levels and products that external (and internal) customers can expect from the enterprise, particularly "next year" and further into the future.[9] The customer service paradigm becomes a basis for the level of expertise employees need to have to deliver the desired performance.

Typically, the KM strategy will also deal with focus areas, infrastructure capabilities, program coordination mechanisms, needed knowledge management and engineering expertise. It generally includes potential principles for knowledge transfer paths.

---

[9]For further discussion of customer service paradigms and its use in determining KM strategy, see Wiig (1999).

One Approach to Determine the KM Strategy

We can outline an example of a program for determining the KM strategy by conducting the following steps:

- Map the knowledge landscape.
- Understand the enterprise customer service paradigm and general enterprise thrusts and visions and create a KM-specific customer service paradigm.
- Determine how the KM strategy will support the enterprise strategy and the corresponding strategy elements.
- Identify knowledge-related business issues — in the form of opportunities and threats.
- Build on capabilities — for example, R&D traditions, a strong sharing culture, extensive IT capabilities, and CEO support.
- Understand the general value of the most important KM actions and steps determined as part of the knowledge landscape map.
- Outline priority knowledge-related thrusts, KM practices, and focus areas to address TOWS issues.
- Document the strategy to include all desired elements.
- Develop general KM plans — for example, general governance functions, infrastructure requirements, program plans, budgets, coordination requirements with other enterprise activities and the like.

4. Create and Define Knowledge-Related Alternatives and Potential Initiatives

The knowledge landscape mapping activity will identify several KM action candidates, many of which are easily identified to be of greater promise than others. In addition, it is often desirable to pursue other activities to conceptualize and identify salient KM action alternatives. Several are direct outgrowths of the KM vision, some may result from specific studies, others may be compiled by accessing people's ideas throughout the target organization, and still others may be from benchmarking of "best practices," although copying what others do may not be the best alternative.

Many KM introduction situations proceed to pursue the first opportunities that present themselves without looking further to explore if higher value candidates should take priority. That may result in missing important opportunities and may even lead to failure of the KM introduction.

Purpose

The primary purpose of creating and further defining KM alternatives is to examine the enterprise to locate the important opportunities.

Second, when a strategic undertaking — such as KM — is pursued by an enterprise, an optimal approach with a "best set" of activities must be chosen for implementation. A collection — large or small — of realistic KM action candidates must be defined or created to provide a portfolio from which the most valuable and promising alternatives can be selected.

General Characteristics

The KM action candidates that are sought should identify opportunities for improvement in areas such as revenue enhancement, creation of new products and services, relief of knowledge bottlenecks and other knowledge-related actions. Initially, the potential KM candidates, or alternatives, will be defined in relatively high-level terms. An example would be to delineate a critical knowledge function (CKF) in a predefined "one-pager" format. Alternatives, defined as CKFs, can typically be described by the five characteristics indicated in Table 1.

Later, the candidates are normally described in greater detail, with assessments of benefits, costs, and risks, requirements for supporting infrastructure, etc. Figure 8 provides an overview of functional classes of generic KM activities that may be included to provide a broad KM practice. Each of these generic

TABLE 1   The Five Critical Knowledge Function Characteristics

1. Type of knowledge involved in performing a task — *Chemical reactor operating expertise*
2. Business use of that knowledge — *Produce specialty chemicals for the commercial market*
3. Constraint that prevents knowledge from being utilized fully, the vulnerability of the situation, or the unrealized opportunity that is not taken advantage of — *There are too few proficient operators, as a result, many reactors are not run well. (constraint) — Our design knowledge is superb compared with competition (opportunity)*
4. Opportunities and alternatives for improving the CKF — *KBS to capture expert knowledge on reactor operation and make it available*
5. Expected incremental value of improving the knowledge-related situation — release knowledge constraint, take advantage of opportunities to use knowledge differently, etc. — *Decrease operating costs through reduced downtime, operating problems, utility consumption, and increase production capacity and market share*

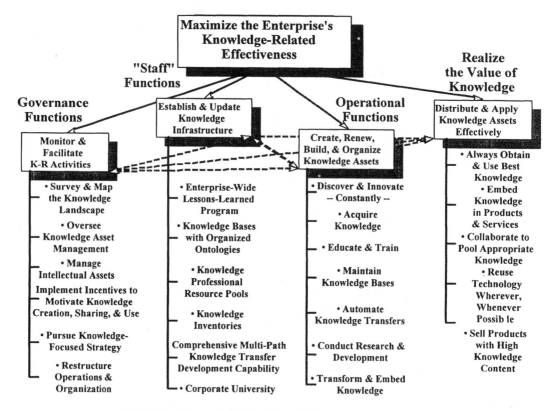

FIGURE 8   Classes of high-level knowledge management activities.

activities includes a large number of specific and detailed KM tasks. Hence, "acquire knowledge" can be performed many different ways, ranging from brief exit interviews about how a departing employee performs his or her work to "shadowing" a departing person for weeks or months to transfer expertise and understandings.

The introduction program must not be too broad. It must be limited with a sharp focus to be manageable and to provide reasonably expedient results. Consequently, it is important to select judiciously among promising candidates to identify only those that are absolutely necessary to initiate the KM practice. When the first steps have been successfully completed, and better insight into KM techniques and methods are obtained, additional alternatives can be pursued.

Several examples of possible KM-related activities are shown in Table 2. It delineates more than 50 commonly observed activities. However, it is imperative not to copy what others do, even when identified as best practices or "best-of-breed," but to engineer KM candidates to best support the particular situations and conditions which they are to support.

TABLE 2 Selected Examples of KM-Related Activities

1. Acquire knowledge
2. Always use best knowledge
3. Automate knowledge transfers
4. Build computer-based education systems
5. Build corporate university
6. Build knowledge bases
7. Build portfolio of knowledge-related actions
8. Collaborate to pool appropriate knowledge
9. Compile knowledge into knowledge bases
10. Comprehensive multipath knowledge transfer programs
11. Conduct research & development
12. Create and organize knowledge repositories
13. Create expert networks — structure, scopes, budgets, access mechanisms
14. Create KBS & educational program development function
15. Create the knowledge strategy
16. Create lessons learned programs
17. Build staffs of technical specialists
18. Determine knowledge requirements for specific tasks
19. Determine knowledge-related benefits
20. Develop and deploy KBS applications
21. Develop educators (trainers)
22. Develop information technology (IT) infrastructure
23. Develop products with valuable knowledge contents
24. Discover & innovate — constantly
25. Create programs for effective knowledge capture
26. Embed knowledge in services
27. Embed knowledge in systems and procedures
28. Embed knowledge in technology
29. Build a program for enterprise-wide formal education and training
30. Establish KM professional consulting team
31. Make available the expertise of field experts & innovators
32. Implement incentives to motivate knowledge creation, sharing, & use
33. Establish knowledge acquisition program
34. Discover knowledge in databases (KDD)
35. Build knowledge inventories
36. Provide incentives to motivate employees to share knowledge
37. Maintain knowledge bases
38. Make knowledge available to customer service representatives
39. Make knowledge available to field service
40. Manage intellectual assets
41. Place high expertise in conceptual sales situations
42. Promote personal innovation
43. Provide best knowledge to workers at all levels
44. Provide companion KBS application products
45. Provide knowledge-based customer services
46. Provide learnings & outcome feedback
47. Pursue knowledge-focused strategy
48. Restructure operations & organization
49. Sell knowledge embedded in technology
50. Sell knowledge products
51. Sell or license patents and technology
52. Sell products with high knowledge content
53. Sell separate KBS application products
54. Set knowledge activity priorities
55. Share knowledge throughout enterprise
56. Survey & map the knowledge landscape
57. Transform knowledge
58. Use external sources for valuable knowledge
59. Utilize technical specialists
60. Validate & verify knowledge

Examples of Approaches

We can generate initial definitions of KM action candidates by collecting information through critical knowledge function analysis (CKFA) about the situations, needs, or capabilities that may require attention and describe them as critical knowledge functions. Table 1 above presented the five critical knowledge function characteristics. Basic attributes of critical knowledge function analysis are outlined below:

- Identifies and characterizes areas of knowledge-related criticality, i.e., bottlenecks, vulnerable situations, opportunities, etc., that warrant management attention.
- Is used to locate knowledge-sensitive areas.
- May result in findings such as needs for knowledge-transfer, KBS applications, staff expansion, knowledge capture, work flow restructuring, etc.
- May rely upon task environment analysis (TEA), knowledge surveys.
- Analyzes functions based on interviews, group sessions, manager introspection, survey results.
- Applies common knowledge overview methodologies such as knowledge flow diagrams, knowledge proficiency profiles (an example is indicated above in Figure 3), etc.

One approach to obtaining a broad and practical portfolio of KM alternatives is to conduct a group session with 10 to 30 individuals selected from area managers, professionals, and other key personnel. These group sessions can be conducted in a short-day format that may start with two hours of presentations to reaffirm management support, reintroduce KM concepts, discuss the nature and give examples of KM opportunities and benefits, preferred benefit assessment methods, and characteristics and formats for describing KM opportunities. After the introductory session, the group is split up into small teams (five or less) to work independently for 60 to 90 minutes.[10] In the separate team sessions, the teams generate alternatives, document them with assessments of their perceived merits — preferably in an automated environment using prestructured formats. Typically, after lunch, the group is reassembled for each team to present their top 5 to 10 ideas in a noncritical ("safe") session. Other teams are encouraged to provide complementary elaborations and expansions that are captured concurrently. It is important that these proceedings exclude criticism and judgments. We have experienced that many "doubtful" alternatives presented in such sessions, by breaking with conventionally accepted thinking, were later found to have great business value.

5. Portray Benefit Expectations for Knowledge Management Initiatives

Describing and estimating the effects, benefits, costs, and risks from introducing KM activities in an organization are methodologically similar to impact assessments in multivariate, multistage dynamic processes such as physical, social, or economic environments. Impact analyzes are complicated and seek to isolate the marginal effects of introducing particular changes, often when other changes are made concurrently, thus obscuring the effects of interest. From another perspective, the portrayal of effects and benefits is comparable to the outlining and definitions required in any type of project planning.

KM benefits are often considered to be intangible, as the results of complex, dynamic event chains are not clearly understood. A few enterprises with good understanding of KM benefits estimate results by representing the evolution of effects with causal value-added process models that include sophisticated quantitative dynamic models based on detailed cost accounting. However, most organizations rely on beliefs or qualitative analysis to prioritize or justify KM activities.

Few, if any, organizations have been able to develop simple quantitative indicators for KM performance measures. Instead, many choose to delineate the expected effects and potential benefits and costs as qualitative, causal-event-chain representations as indicated in Figure 9. Unfortunately, such

---

[10]Our experience indicates that when given a clear charge to identify KM opportunities and describe their nature and expected purpose (benefits), small teams can operate effectively without facilitators and that they avoid inefficient tangential discussions but quickly generate good alternatives, often in a "Round-Robin" manner.

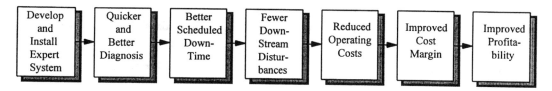

FIGURE 9 Use of a simple *linear event chain* to represent effects and benefits for introduction of an expert system to assist operators of a chemical plant.

linear representations are generally too simplistic, and more complicated models that illustrate parallel effects and iterative processes and impacts may be required.

To identify and assess expected effects and benefits from undertaking KM efforts and align them with different parts of the business process, we find value in using the framework of three value disciplines.[11] These are: (1) pursuit of operational excellence; (2) pursuit of product leadership; and (3) pursuit of customer intimacy. Examples of typical KM benefits for these value disciplines include:

Operational Excellence Benefits

- Less costly customer products and services producing higher net profit — resulting from increased benchmarking and greater sharing of best practices between different groups and inside and outside the organization.

- More timely product deliveries, reduced inventories, less rework, and greater customer satisfaction — by increasing craftspeople's and foremen's knowledge of their own and adjacent processes.

- Greater product consistency leading to reduced operating costs — from increased knowledge by all employees of the effects of product variations on customer requirements, sales, and enterprise profitability.

Product Leadership Benefits

- Higher quality products leading to higher value to customers and better market acceptance, which in turn provides greater profitability and enterprise viability — resulting from better transfer of knowledge from outside sources and new educational programs that provide wider horizons and general understanding in designers and marketing people.

- More innovative and advanced products that open up new market niches with increased sales to increase net income per share and share value — made possible by fostering personal innovation, increased sharing of knowledge between marketing, manufacturing, and product development, and a new research agenda.

Customer Intimacy Benefits

- Increased orders and proposal acceptance from having more knowledgeable sales and marketing people — by transferring exceptional performers' mental models and perspectives to practitioners.

- Higher customer satisfaction leading to greater customer loyalty, less sales and marketing cost per dollar sold, greater market penetration from better service to customers with individual requirements — by pooling knowledge among team members and having instant access to expert networks.

- Greater market penetration and profit margins with individualized product specifications and customer service — achieved by obtaining and acting on in-depth knowledge of product use in customer environments and effects on customer profitability and success.

---

[11]Treacy & Wiersma (1993).

Purpose

There are several reasons for establishing the effects and benefits of potential KM actions. Often, we need to outline these in some detail. We perceive the five major reasons to be:

- To support decision making and priority setting, obtain estimates of the magnitude and timeframe of potential benefits, costs, and risks.
- To enable or facilitate the desired outcomes from KM efforts, delineate the various effects that are sought or expected with identification of ancillary activities that must be considered.
- To delineate the nature of expected and desired events and agree with stakeholders about the descriptions of the event chains and the desirability or associated risks, and provide a graphical (visual) framework to support the collaborative process.
- To promote understanding of desired effects to support effective implementation over the lifetime of the process by describing the events and associated characteristics.
- To follow and monitor the KM-influenced event process to manage it appropriately, and provide sufficient understanding of the expected events by outlining expectations over time in sufficient detail.

General Characteristics

There are two separate objectives against which potential benefits are measured. The first — and often dominant — objective deals with satisfying expectations for short-term improvements or other relatively explicit benefits from operations, product, and services, or customer and market perspectives. In these cases, goals often focus on cost reductions or revenue enhancements. The second objective deals with longer-term perspectives and include sustaining enterprise viability. It also includes concerns for how well a particular KM action will assist in building the broader KM practice and its capabilities.

The objectives are normally "multiattribute" or multivariable. They invariably include financial estimates. They may also include considerations for other enterprise concerns ranging from employee satisfaction to potential risks involved with— in the extreme — how the potential actions will affect the social, economic, or physical environments.

Many of the KM feedback measures listed below have been used to provide insight into the performance of KM programs, both during implementation and later, during operation. Reportedly, none have been found to be of unique value and additional insights are needed. Among examples of KM feedback and measures are:

- Intangible benefits not measured rigorously, only qualitatively
- Surveys to obtain feedback regarding:
  - Customer responses and successes resulting from KM
  - Uses and benefits from KM systems
  - Employee satisfaction resulting from KM activities
- Performance criteria for KBS in support of customer service, including quality of services delivered, customer satisfaction, cost and efficiency, and growth of market penetration
- Quantification including numbers such as users, participants, messages passed, documents created, and the like
- Sophisticated dynamic value process model to estimate returned business value from KM
- Formal quantitative methods to estimate intellectual capital

Examples of Cost/Benefit Approaches

For each — or for an associated group of — KM action alternative(s) the preliminary nature of the sought benefits and characteristics will have been established as part of the initial definition. Several approaches may be appropriate to elucidate the benefits, costs, and risks stemming from introducing a KM activity. One is the conventional cost/benefit analysis that tallies costs and benefits over time in monetary measures to project expectations by net present value, return on investment, or in similar terms.

Other approaches have considerable merit for establishing expected net benefits. Among these are economic value added (E.V.A.) and mathematical programming (linear [LP], nonlinear, quadratic, and mixed integer [MIP] programming) methods that determine the incremental values of introducing new activities in complex, interconnected processes.

Another approach is to provide a qualitative (or occasionally, quantitative) model of the expected event chain as illustrated in Figure 10 for the introduction of a KM incentives program. The value of these models is that they allow characterization of a broad range of effects, mostly speculative, that are the real reasons for undertaking the effort. Conventional methods are frequently not able to accommodate introduction of these effects, and have particularly not been able to support explications of relationships between them. Granted, these models cannot be made precise. They can, however, be highly valuable for supporting processes for reaching agreements on which mechanisms are involved and what the particular expectations are for benefits.

When using this approach, it is normal to explicate expectations for each "link" (or function block such as the "Greater ability to understand customer problems and needs"). The explication takes the form of a short (one page) description of the nature of effects and their desired or undesired impacts, inputs, outputs, and requirements for support.

Risks — or probabilities — of obtaining desired cost effectiveness can be established using formal analysis such as Monte Carlo simulation. However, since the risks often are associated with perceived hypothetical futures, it may be appropriate to assess and characterize risk expectations using subjective and qualitative analysis.

6. Set Knowledge Management Priorities

KM priorities must be established to best fulfill the objectives of the enterprise. Principal objectives typically emphasize enterprise profitability and operating effectiveness. Other objectives may address the support of general development of enterprise-wide KM practice or other capability-building strategic goals. It is the function of priority setting to identify the most important steps to be implemented, either because they have direct end value for the enterprise, or because they are necessary stepping-stones to achieve high priority goals.

Purpose

The main purpose of prioritizing KM alternatives is to provide the basis for selecting which KM alternatives should be pursued first and which should be delayed. Priority setting helps establishing ranking and timing for implementing individual alternatives. It also determines priority hierarchies for activities based on KM strategy, expectations for net benefits, needs for supporting infrastructure, availability of capabilities, and so on. Given the nature of individual options, an implicit, but important, objective of priority setting is to steer the direction of the overall KM practice, while at the same time supporting other enterprise thrusts.

General Characteristics

The results of priority setting define the sequence of implementation, nonfunctional relationships between KM building blocks, and balanced allocation of resources between competing requirements. The priority setting function should include consideration of the enterprise's general priorities for KM and specific management judgments. Given the complex mechanisms and long duration from initial KM actions to final results, it is normally impossible to provide sufficient precision by quantitative analysis to include all factors involved. Hence, management visions and judgment always need to be introduced to temper the interpretation of formal analyzes in this area.

Example of Approach

Ideally, the selected alternatives should maximize the combined utility — the multiattribute utility — defined by the various objectives that the enterprise pursues. Considering the nature of benefits, costs, and risks (BCR) — what is known and how well it is known — priorities can be determined in many ways.

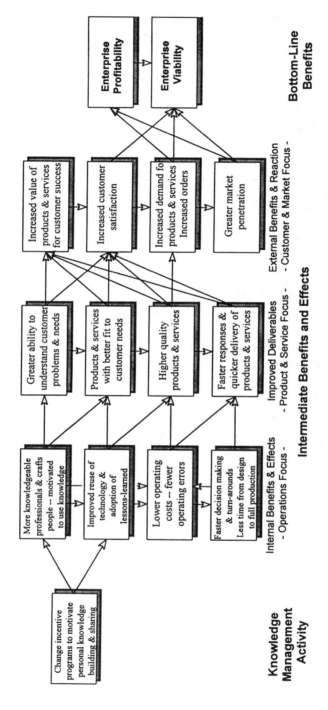

Enterprise Profitability

Enterprise Viability

Bottom-Line Benefits

Increased value of products & services for customer success

Increased customer satisfaction

Increased demand for products & services
Increased orders

Greater market penetration

External Benefits & Reaction
- Customer & Market Focus -

Greater ability to understand customer problems & needs

Products & services with better fit to customer needs

Higher quality products & services

Faster responses & quicker delivery of products & services

Improved Deliverables
- Product & Service Focus -

Intermediate Benefits and Effects

More knowledgeable professionals & crafts people -- motivated to use knowledge

Improved reuse of technology & adoption of lessons-learned

Lower operating costs -- fewer operating errors

Faster decision making & turn-arounds
Less time from design to full production

Internal Benefits & Effects
- Operations Focus -

Change incentive programs to motivate personal knowledge building & sharing

Knowledge Management Activity

FIGURE 10   Example of a simplified effect-benefit representation of expectations resulting from introducing a KM incentive program.

Ranking can be performed by establishing the risk-adjusted utility of each alternative after formal analysis has provided reliable estimates for all relevant factors. It can also be determined by using management judgment alone after all readily available facts and analyzes are made available. Major approaches to setting priorities include: (1) quantitative risk-utility portfolio analysis; (2) qualitative risk-utility portfolio analysis;[12] and (3) Pareto analysis. Relevant approaches to priority setting include considerations of the following aspects:

- The main objective of KM priority setting is, given suitable alternatives, to identify which specific alternatives to pursue and for what purpose.
- Pursue KM directions that support short-term and long-term goals of the enterprise.
- Find KM directions that take maximum advantage of existing initiatives, capabilities, and infrastructures.
- Implement effective changes to redirect the enterprise's culture and employee motivations.
- Make employees use best knowledge.
- Allocate resources to the highest value KM-related activities.

7. Determine Knowledge Requirements

The focus of this building block is operational "knowledge." It seeks to identify **which** areas and types of knowledge are required to deliver quality work by people — or automated agents — in particular work functions.

Delivering the quality of work needed for desired performance, requires possession of, or access to, knowledge beyond certain levels. Determining which area and type of knowledge is required to deliver routine work competently, may in most situations be relatively simple. However, since most professional work — and many types of crafts work — involves significant contributions beyond the routine, establishing the ideal knowledge requirements for key positions may be difficult and may require sophisticated knowledge acquisition methods.

Depending on the situation, a person may interact with different functions and parties outside the particular "job" that is the focus of investigation. These interactions require additional knowledge that often is not considered to be part of the work function. The full complement of needed knowledge should reflect both continued business as usual and new requirements associated with new ways of conducting business. The body of knowledge needed for a particular function may also be used in different ways. Some knowledge may be required for immediate — i.e., automatic or tacit — use, while other knowledge may be accessed for more reflective and explicit use. These requirements have implications for how the different areas of knowledge are made available or possessed by the knowledge workers.

Purpose

Understanding the knowledge requirements for a particular function is a prerequisite for ascertaining that the requisite knowledge is made available at the point-of-action (PoA) at the time and in the form needed. We need to know the knowledge requirements at different levels of depth to hire new people, create new positions or new services, educate and train people, or evaluate individuals.

General Characteristics

Knowledge requirements are determined at several levels of detail and abstraction. At times only the level of a general discipline is required (e.g., M.S. mechanical engineer or Ph.D. economist). More often, expertise in specific skill or knowledge areas is required (e.g., two years experience in risk evaluation of small commercial loans).

---

[12]See Wiig (1995) for an example of a qualitative risk-adjusted utility portfolio analysis.

For important critical knowledge functions detailed understanding of knowledge requirements is needed. As an example, an operator of a power plant water treatment facility needs expertise in, and understanding of, among others, water chemistry, impurity-caused steam tube corrosion, water-borne particle erosion of turbine blades, environmental effects and regulations, the treatment plant operating principles, and behavior characteristics and symptoms. This may amount to hundreds of knowledge elements in all.

Examples of Approaches for Determining Knowledge Requirements

An approach to determining knowledge requirements may start by establishing the work to be delivered, and how that work should be performed. A "customer service paradigm" for a particular job may be used to express various expectations. Among them, expectations for delivered quality and timeliness, routine work and versatility, ability to deal with unanticipated challenges, range of authority, models for conducting work (e.g., teaming, collaboration, accessing expert networks, use of technology), and other performance evaluation criteria. Ideally, the customer service paradigm should also represent business requirements for the next three to five years.

In situations where broad, general insight is required, brief discussions with persons responsible for the work function may suffice. Further discussions with their superiors and counterparts in adjacent functions are often helpful. An approach, frequently employed by human resource specialists to identify the general capabilities needed for different positions, is conventional competency requirements analysis. This method provides well structured overviews of the competency areas required and typically generates organized data that fall within accepted categorization systems adopted by the enterprise. The results of competency requirements analyses are often used directly to determine educational and training needs as well. However, it is often found that a greater, more in-depth understanding of the knowledge requirements is needed for systematic KM. This may include determining needed migration of expert insights (in contrast to skills training or general education) or requirements for how knowledge must be held for use under different conditions.

In situations where specific knowledge requirements need to be understood, that information can be obtained through facilitated small-group discussions. The groups should include experts, function supervisors, and, if possible, representatives from up-stream, down-stream, and adjacent functions. The discussions should normally be limited to a few hours.

In situations where detailed understanding of special expertise is required, in-depth investigative methods must be used. Such methods may involve several (at times as many as five) one-on-one hour-long "ethnographic interviews"[13] of current experts and of others with special insights. It is very important to audiotape — and if possible, videotape — the interviews and transcribe them for analysis of what has been communicated. This is often resisted by the KM professionals, because of the extra work and because they think all important insights can be captured by note taking. Unfortunately, note taking and listening to recordings have proven to be insufficient for capturing deeper meanings behind the examples and stories that often are provided by the interviewees.

In all situations understanding of the knowledge requirements must be considered from several perspectives. It is most natural to consider the requirements from the perspective of the target job or function itself. However, that may be too narrow. The knowledge requirements must also be considered from a perspective of the broader work to be performed, such as the function or general tasks to be performed (e.g., marketing or special project work), the department, or the product line. Additionally, the perspective of future developments and needs must be considered. Extensive research has been conducted to identify the amount of knowledge really needed to perform quality work. An example of one such in-depth study is provided by Lange (1986).

---

[13]See Spradley (1979).

8. Acquire Key Knowledge[14]

The focus of this building block is to obtain and characterize the detailed knowledge required to deliver the requisite work. Its function is to "manipulate the knowledge itself" by obtaining information that describes the actual knowledge in sufficient detail for communication to people or other repositories.[15] The knowledge can be represented as detailed "rules" for incorporation in knowledge-based system (KBS) applications. It can be represented as "reference cases." It can be implicit in natural language statements. It can be embedded in drawings, procedures documents, and numerous other representatives.

Many types of knowledge acquisition have been practiced by individuals and enterprises. Obtaining knowledge through training, education, apprenticeships, or licensing, are well-known activities. When introducing systematic KM within an enterprise, such activities will continue, but often need to be augmented with methods to acquire personal knowledge by using "knowledge engineering" methods. It is that area we emphasize here.

Purpose

"Actual knowledge" needs to be acquired from many sources to initiate the transfer to where it, in the end, can be put to use or otherwise leveraged. Several circumstances require that knowledge be acquired for good business reasons. When individuals leave or are promoted, their valuable knowledge may be elicited for continued use by others. When valuable lessons are learned, these may be captured to be incorporated into the repertoire of practitioners. When outstanding practitioners are recognized to have valuable insights, these may be elicited to be shared with others. When sources of valuable knowledge are found outside the enterprise, that knowledge may be obtained for inclusion in the enterprise's knowledge assets. All these situations represent business opportunities for managing knowledge effectively by acquiring it and making it explicit within the enterprise's realm. In effect, acquiring knowledge is often a significant step in building knowledge assets.

General Characteristics of Acquisition and Acquired Knowledge

Acquisition of personal or explicit knowledge involves processes in which a knowledge source — an expert, database, outside agency or information provider — is queried to elicit detailed information that describes the knowledge of interest. This information is captured and frequently reconstructed, i.e., transformed and represented in different forms, to make it more useful.

When eliciting personal knowledge, we find it valuable to identify the conceptual level at which the individual possesses it. This perspective is important when considering how knowledge must be possessed and made available for use in routine and nonroutine tasks, since these tasks are approached using decisions-in-the-small when cloistered or decisions-in-the-large when collaborating with others. The five conceptual knowledge levels are:

1. Goal-Setting or Idealistic Knowledge — *Vision and Paradigm Knowledge* or "Knowledge WHY"
2. Systematic Knowledge — *System, Schema, and Reference Methodology Knowledge* or "Knowledge THAT"
3. Pragmatic Knowledge — *Decision Making and Factual Knowledge* or "Knowledge HOW"
4. Tacit Automatic Knowledge — *Routine Working Knowledge* or "KNOWHOW"
5. Tacit Subliminal Knowledge — *Emerging or Not-Yet-Understood* or "Glimpsed" Knowledge

---

[14]The field of knowledge acquisition is well developed, with numerous approaches to elicit, organize, and archive or store knowledge. For an excellent discussion of detailed knowledge acquisition, see McGraw & Harbison-Briggs (1989).

[15]We can talk metaphorically about "communicating knowledge" whereas in reality we only communicate information that describes it. Similarly, we cannot share knowledge directly, only share information about it. Knowledge can be communicated to end users in different manners depending on its nature, expected use, and the capabilities of the user. However, in all cases, knowledge — which, as indicated above, is fundamentally different from information — must be communicated as information in some form or other to describe it.

Example of Approach

Knowledge acquisition is the basis for transferring knowledge from experts in knowledge-intensive positions to others who need to be equipped to perform at similar levels of excellence. Experts have internalized their expertise to such an extent that they often have lost their capability to explain what they know. To obtain some of their detailed knowledge (estimates indicate that it may only be possible to make explicit 10% of a person's deeper knowledge), their expertise must be elicited using clever knowledge acquisition methods. Among the methods are ethnographic interviews, observations at work, and simulations of the work environment. In any event, their knowledge may be situational and can then only be accessed in real work situations. Often, it is not possible or desirable to obtain the deeper knowledge. It may be adequate to obtain simpler insights for transfer of expert perspectives to other practitioners. Some examples of knowledge elicitation and acquisition are:[16]

- **Comprehensive "Knowledge Engineering"**

  Knowledge acquisition by comprehensive "knowledge engineering" elicitation sessions where experts' knowledge is acquired in considerable detail. Multiple elicitation sessions over weeks or longer periods are conducted to obtain records of what the expert knows about a particular topic, and how she or he applies that knowledge. The sessions may be unstructured interviews conducted in the expert's work environment with all work artifacts at hand, observations of the expert at work, and follow-up interviews. All are audiotaped and/or videotaped for transcription and in-depth analysis.[17] Of particular value in the sessions is to obtain explicit statements or explained actions about how to interpret and handle specific conditions and situations and their expected variations. The statements or explanations should contain adequate detail to make it clear to others how the conditions or situations should be dealt with. This type of knowledge acquisition is useful for obtaining detailed knowledge to be encoded in knowledge-based systems (KBS) such as expert systems.

- **Eliciting Advanced Concepts and Metaknowledge**

  Knowledge acquisition by eliciting advanced concepts and metaknowledge in the form of "chunking" hierarchies, associations, and thinking strategies from the expert. The added understanding of "metaknowledge" provided through these methods will be in addition to the basic knowledge of how to deal with routine and quasiroutine situations that average practitioners already possess. Normally, the descriptors of the underlying knowledge within a particular topic area, such as evaluating commercial credit applications, can be obtained in a few one-to-two hour interviews, "show-and-tell" work sessions, or simulation sessions — all of which are taped and transcribed.

    This type of knowledge transfer is effective for making practitioners discover "**What to think about**" (the concepts or chunking hierarchy that represent the "things" the expert thinks about), the associations regarding similar, related, or affected areas, and "**How to think about it and deal with it**" (the reasoning strategies). The objective is not to make practitioners think precisely like the experts. That is impossible. Rather, it is to allow practitioners to develop similar and sound approaches and frameworks that allow them to be more effective in delivering their work. It is regularly found that practitioners become more versatile and deal better with nonroutine and unexpected challenges when they receive and internalize metaknowledge of strategies to deal with nonroutine challenges.

- **"Lessons Learned" Programs**[18]

  Knowledge acquisition through "lessons learned" programs is an example of a targeted and systematic approach to obtaining special kinds of knowledge stemming from unique learning situations. Whenever a memorable learning situation has occurred, a lessons learned program

---

[16]Several knowledge-acquisition methods are discussed by Wiig (1995).
[17]McGraw & Harbison-Briggs (1989).
[18]See Wiig (1995) for further information on lessons learned program methods and work sheets.

facilitates capture of the learning, often in a predefined format on paper or directly into a computer. The format may be unstructured, or it may be structured to facilitate transfer into an automated system such as a case-based reasoning (CBR) system.

An important requirement for effective capture is to motivate the individuals involved. When errors have been made, they must be provided with a "safe environment." They must be provided with "slack time" or work relief to capture the incident. They must understand the positive advantages for themselves to engage in the effort.

- **Import External Knowledge**

Knowledge acquisition through "environmental scans" or other methods can identify external knowledge sources from which appropriate knowledge can be obtained directly, bought, licensed, or otherwise imported. Collaboration with suppliers and customers are often valuable sources.

Importing external knowledge is in many instances achieved through "Gate Keepers" within the community of practice (CoP). They act as scouts who find potentially valuable knowledge and transfer it to appropriate repositories or individuals. A very important GK function is to interpret what the new knowledge may mean for the CoP in both the short and long term.

9. Create Integrated Knowledge Transfer Programs

Enterprises have always incorporated and maintained substantial knowledge transfer programs. However, it is frequently found that the knowledge transfer options follow past or traditional paths and are uncoordinated. When this is the case, training programs may be created and conducted by a department that has little communication with KBS development groups or entities that pursue knowledge discovery in databases (KDD) with modern methods. Similarly, organizational structures and practices for supporting knowledge pathways such as "expert networks" and CoP-acquired computer-based training (CBT) capabilities may be missing, unsupported by budgets and incentives, or not be considered, when knowledge requirements and educational programs are determined. For some enterprises, important knowledge pathways, such as "outcome feedback," may also be missing completely.

For many enterprises, the opportunities for new or better coordination of integrated knowledge transfer mechanisms are discovered as part of knowledge landscape mapping. This is particularly the case, when new technologies and different emphases on knowledge-effective work environments are introduced in the enterprise. Other areas of needs for improved knowledge transfers come from knowledge requirements analysis or from quality studies and knowledge-aware business process redesigns (BPR).

Purpose

An integrated and systematic knowledge transfer program allows for the creation of coordinated and comprehensive approaches to support gathering, restructuring, inventorying, and distributing knowledge — i.e., understandings and other intellectual assets — to individuals throughout the enterprise. Expected benefits from creating and illustrating an integrated knowledge transfer program can:

- Determine opportunities and framework for coordination of different KM activities and functions, often in different parts of the enterprise.
- Locate and propose missing knowledge transfer pathways.
- Construct balanced knowledge support programs for targeted work areas and positions.
- Develop broadly shared infrastructure capabilities.
- Portray knowledge transfer plans, implementation programs, and deployment expectations.
- Help gain an overview of the overall knowledge flows to monitor effectiveness and plan further KM efforts.

The integrated knowledge transfer program provides an overview with summarizations of the individual knowledge-related functions and pathways within the program. It purpose is not to plan the program in detail. That belongs to other KM building blocks (No. 10).

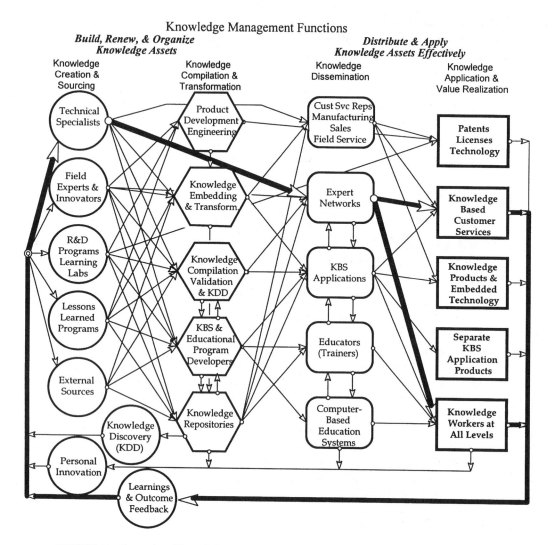

FIGURE 11   Examples of knowledge functions and pathways in an integrated transfer program.

General Characteristics

Knowledge transfer programs are described by their functions and knowledge pathways. The functions are described by what they do, or sometimes by who they are (e.g., "KBS applications" and "knowledge compilation, validation, and KDD"). The pathways are described by which knowledge is transferred (e.g., "dialoguing with clients to provide education in use of our products" or "proposal writing") and the mechanisms by which it is communicated (e.g., "interactive multimedia computer-based program" or "14-day classroom training course").

Knowledge transfer programs are normally divided into stages in the path from knowledge creation and sourcing, through compilation and transformation, dissemination, to application and value realization as shown in Figure 11. This example, a simplified real situation, shows 20 KM-related functions (e.g., "Knowledge Embedding and Transformation") and over 80 knowledge pathways (e.g., the role of "Field Experts and Innovators" in providing knowledge and guidance to "Knowledge Embedding and Transformation").

Example of Approach

The integrated knowledge transfer program rests on information about knowledge requirements in areas of the enterprise and will typically take several years to implement to its full potential.

Figure 11 emphasizes (by heavy lines for pathways) an example of the knowledge transfer for "expert networks." It starts with "technical specialists" who work to support "knowledge-based customer services" and "knowledge workers at all levels." In addition, Figure 11 emphasizes the feedback learnings that "technical specialists" receive from "learnings and outcome feedback." The technical specialists and others obtain systematic communications that relate to how well knowledge performs in customer situations. Performance can be measured in the form of shortcomings, new opportunities, and new ideas, all of which relate to "making the customer becoming successful."

In real situations, consolidated diagrams of twice the complexity of Figure 11 show the major functions and pathways. Many subdiagrams, often separate ones for different business functions and job classes, will show individual transfer programs. These are supported by brief descriptions of the related functions and pathways. Many of the knowledge transfer functions, particularly those for compilation and transformation, will have general applicability for several departments and some experimental ones may be in different stages of development. They will be included in a comprehensive transfer program map that distinguishes between existing, in-progress, and planned activities.

10. Transform, Distribute, and Apply Knowledge Assets

After knowledge has been acquired or otherwise "sourced," it normally cannot be used in its raw form and must be changed — transformed — to become a valuable knowledge asset or to facilitate application. Some transformation functions are institutionalized and require formal methods. However, the vast majority of such KM operational functions are performed as part of individual employees' regular workday, often tacitly as seen from both personal and enterprise perspectives. This is the case, not only in knowledge-vigilant enterprises, but anywhere knowledge work is conducted.[19]

Figure 8 above indicated that KM activities fall into four classes. Governance functions, particularly incentive programs, provide motivation for individuals to accept responsibilities for personal KM participation. Staff functions and infrastructure provide needed capabilities. Operational functions are the focus of this building block and include all knowledge manipulation activities, ranging from discovery and capture of new understandings to making knowledge available to the people at CoPs. The last class deals with realizing the value of knowledge through use or leveraging the knowledge assets in other ways.

As Figure 11 above indicated, operational KM functions flow knowledge through a natural process by first building, renewing, and organizing knowledge assets. Next, they distribute and apply the knowledge assets. The four functional KM areas directly associated with manipulating knowledge are: (1) knowledge creation and sourcing (building block No. 8 above), (2) knowledge compilation and transformation, (3) knowledge dissemination, and (4) knowledge application and value realization.

Purpose

Systematic and explicit KM intends to create an enterprise-wide, adaptive and people-centric environment with personal focus on knowledge-related matters as part of the daily activities and responsibilities. Most knowledge manipulation activities will not be explicitly specified. For this reason, this building block only focuses on specifications of explicit activities that will be implemented as concerted efforts.

General Characteristics and Examples of Approaches

The activities that are made explicit in this building block include extensive knowledge pooling and collaboration; broad knowledge sharing; sale and licensing of knowledge assets; and incorporation of competitive knowledge in products and services. Examples of characteristics for different functions include:

---

[19]Transformation and use of knowledge in different circumstances are discussed by Wiig (1993).

Examples of Knowledge Compilation, Verification, Validation, and Organization

- As knowledge is acquired, it can be compiled manually or automatically. When the compilation and subsequent functions are performed automatically, the process must include considerable embedded knowledge to create valuable and reliable results.

- Collecting and editing of knowledge into a repository involves selecting the appropriate knowledge and verifying and validating that it is complete and correct. The selection may be performed with expected uses in mind by understanding what the topic area requires — or will require in the future. Subject matter experts (SMEs) from the business area need to be part of the validation.

- Collected knowledge destined for repositories (such as knowledge bases, databases), need to be organized according to an established ontology (categorization of "what is") and must match the organization of the repository for ease of use.

    These tasks can best be performed by individuals with understanding of the target topic areas and knowledge organizational principles (e.g., Library Sciences, Epistemology).

Examples of Knowledge Transformation

- Knowledge can be transformed in many ways. Traditionally, it has been collected, compiled, verified, validated, and organized by an "end user." This professional may transform it by embedding it into intermediate products, new customer products, or educational programs. The end user professional may also transform the knowledge by creating personal or shared work tools for use as bases for decision making. This is commonly done by insurance underwriters or securities portfolio managers.

- The transformation of obtained knowledge may be performed at the same time as the compilation and organization but it involves a different treatment. Knowledge is transformed specifically to conform with the format of the target repository.

- Specific transformation is required when knowledge is destined for a knowledge-based system (KBS). In that case, it must be reconstructed (using epistemological rational reconstruction principles) in a representation that supports the particular reasoning method employed by the KBS. Such reconstructions include "crisp" or fuzzy rules to support rule-based reasoning, editing into case formats to support case-based reasoning (CBR), or the like.

- Transformation of another form is required when knowledge is destined for educational or training programs. In that case knowledge must be reorganized to fit the teaching formats and delivery vehicles (e.g., one-to-many personal knowledge transfer or interactive multimedia computer-based delivery).

- If the source knowledge exists in numeric forms (or in advanced cases, as natural language), a selective transformation can take place by using automated knowledge discovery in databases (KDD). This method still requires considerable manual interaction.

- When the obtained knowledge will be used for education in general principles, anchored in concrete examples, the source information must be analyzed to identify the underlying meanings, using hermeneutic methods.[20] This is particularly the case when knowledge is prepared for teaching schemas and scripts, with concrete examples to let recipients internalize and build "systematic," "pragmatic," and "automatic" knowledge and acquire the understanding of how to apply it in practical situations. To facilitate knowledge construction and acquisition, the education may also use conceptual maps.[21]

---

[20]See Shapiro (1987) or equivalent for a discussion of these techniques.
[21]See Wiig & Wilson (1997) for practical approaches to conceptual maps.

Examples of Knowledge Dissemination through People

- For a knowledge-vigilant environment to function, individuals must be motivated to share knowledge, work together in collaborate settings, and rely upon each other to secure delivery of quality work.
- In special situations, when the need is to transfer deep and comprehensive bodies of knowledge, apprenticing and "shadowing" are powerful practices.
- Collaborative teamwork is at times the most powerful knowledge dissemination mechanism. In situations where the body of knowledge to be applied at the PoA is large, it may be impossible for single individuals to be experts in all required areas. In these situations, requisite knowledge can be applied to secure excellent work products along with knowledge transfer from experts to other practitioners through knowledge pooling by allowing individuals with narrower expertise to collaborate.
- Informal or formal person-to-person knowledge sharing has become important and is widely promoted and supported through motivation and recognition. When done in general (prior to the emergence of a need for the knowledge) it is not very efficient overall. When knowledge sharing takes place at the time of need, it may allow both for learning and for application of important and highly targeted knowledge to secure excellent work outcome. In any event, time budgets, such as slack-time must be allocated.
- Expert networks are used to make the knowledge assets possessed by individuals available for distribution to people at the point-of-action (PoA). Expert networks are useful in situations where the person at the PoA has time to consult with an outside party without degrading the quality of delivered work. Expert networks require that mechanisms are put in place to allow access to experts and to give experts time budgets and motivations to provide the requested services.
- Instructor-led educational and training programs have been the standard approach to provide knowledge and understanding. They are expensive and require comprehensive preparation but can in many situations not be replaced. Often knowledge building budgets emphasize this type of knowledge transfer as do measurements of building personal knowledge.

Examples of Knowledge Dissemination by Automated Means

- It is becoming increasingly important to provide knowledge-based systems (KBS) for knowledge worker support, particularly in customer service applications, but other one-of-a-kind applications are also abundant. The technologies come in a wide range and include case-based reasoning (CBR), fuzzy or true-false logic expert systems, neural nets, and natural language processing.
- Fully automatic KBS applications that include embedded expertise to perform business functions ranging from supervisory control of chemical and other industrial processes to inventory and logistics management are also options. Such systems also perform technical functions for support of processes such as CAD/CAM (computer-aided design/computer-aided manufacturing), IC (integrated circuit) design, statistical analysis, and computer software operations (e.g., compiler optimization).
- Computer-based education and training provide highly effective vehicles for communicating embedded knowledge of many types in interactive sessions to students, crafts people, professionals, and others who seek to increase their understanding.

Examples of Application of Knowledge Assets

- Motivated individuals will derive personal benefits from ascertaining that only the best knowledge is applied to deliver quality work. This particularly requires fostering feelings of pride and accomplishment, peer recognition, reduced friction with supervisors and customers, job and monetary security, and so on.

To facilitate such intelligent-acting behavior in motivated employees, mechanisms must be created to give them access to all needed structural knowledge assets and resources. Mechanisms must also be in place to educate and train them to possess routine skills and general personal knowledge to handle time-critical and automatic tasks, and for giving the opportunities and permission to deal with the work challenges that are presented.[22]

Examples of Knowledge Value Realization

The value of knowledge assets is realized when the assets are used to create products or deliver services, or when they are sold or traded for value. Examples are:

- Provide increased knowledge content in customer services by educating customers on how to use, maintain, modify, and exploit the enterprise's products and services for their own success.
- Sale of non-vital patents, licensing, technologies, etc.

It is prudent here to give a note of warning: This author and others have over the last years attempted to identify "best practices" for knowledge manipulation and other KM activities. Whereas patterns and principles of successful practices can be established,[23] it appears to be impossible and undesirable to copy and "mimic" most of the best practices from other organizations with different conditions, cultures, resources, and strategic thrusts. Each implementation must reflect the actual conditions of the local situation.

11. Establish and Update Knowledge Management Infrastructure

Many types of infrastructure capabilities may be required to support effective KM efforts. They include information technology (IT) capabilities such as collaborative environments based on various groupware systems. Examples of IT-related infrastructure functions are presented in the listing below:

- Simple E-mail/intranet/Internet/WWW
- Groupware to support widespread collaboration
- Knowledge inventory system
- Personal and departmental "Yellow Pages" on intranets
- Corporate knowledge map
- Distance learning systems
- Global knowledge sharing system
- Corporate memory databases
- Knowledge navigation tools
- Knowledge creation and transformation tools for knowledge discovery in databases (KDD)
- Communication evaluation and summarization tools
- Intelligent agents
- KBS/AI/expert systems
- Office management systems
- Specialized applications — lessons learned, etc.

---

[22]The great majority of decisions, automatic actions, and setting premises are automatic, tacit, and based on associations and other tacit knowledge. This creates significant consequences for how knowledge workers need to possess or have access to knowledge. Hence, we need to distinguish between Decisions-in-the-Small (DiS) and Decisions-in-the-Large (DiL) as follows:

   DiS — Automatic, nonconscious, tacit; result from instantly applied judgments; performed within a person's mind; Based on premises formed from associations.

   DiL — Deliberate, conscious, explicit; result from active problem solving; may be team-based; in part based on examinable premises.

[23]See APQC/EFQM/KMN (1997).

Other types of infrastructure capabilities may also be created to support KM. Among them are professional support groups and other shared resources.

Purpose

Infrastructure is required to build and maintain generic capabilities, some of which are specific to KM whereas most are shared with other activities and functions.

Enterprise-wide KM activities must be closely integrated with almost every other activity within the enterprise. This demands that a number of resources and capabilities be shared to serve both KM and the other functions.

To make KM as effective as possible, supportive capabilities that may be quite sophisticated must be made available for support. Among them are: (1) bridging activities between KM and other functions (e.g., coordination of education and training programs between business functions, human resources, "training" department, and the KM entity); (2) making it possible to execute selected KM-related activities (e.g., different location/different time communication between individuals); (3) creating KM-supportive environments (e.g., systems and procedures that facilitate KM practices such as episode management or knowledge sharing); and (4) increase the effectiveness and efficiency of KM functions (e.g., making available competent KM specialists).

Infrastructure Capability Examples

One area of considerable importance is the infrastructure that makes it possible for individuals and functional entities to assess where they stand relative to knowledge needs, what the opportunities are for building knowledge, and for performing given this knowledge. This infrastructure element may be absent, informal, or fractured. Without such a capability, it is difficult for people and entities to prepare themselves proactively and KM becomes ineffective. Other approaches to KM infrastructure capabilities include:

- Implemented information technology capabilities (as indicated in the listing above).
- Shared vocabulary (dictionary of terms) in the form of a formalized and structured enterprise-wide ontology.
- Specialized procedures such as lessons learned programs with budgets, standards, formats, and IT back-up.
- Corporate university system with broad curricula and general capabilities to transform knowledge, repackage it, and transfer it to recipients.
- Knowledge inventory systems that identify location and availability of expertise. Such systems can be conventional database and query systems, or lately they have effectively been implemented using personal intranet homepages.
- Organized knowledge bases (KB) and KB upkeep functions including libraries and associated staff.

12. Manage Knowledge as Assets

Knowledge assets (also called intellectual assets) must be divided into two separate categories. The first comprises knowledge assets that are owned by the enterprise (e.g., technology, implicit and explicit structural knowledge, etc. as indicated in Figure 1). The second comprises knowledge assets that are "rented" by the enterprise (i.e., knowledge/expertise/understandings that are possessed by employed individuals who may leave with their personal knowledge assets at their own volition). Knowledge assets are investments for the future and may be considered to have no value in themselves. They only realize their value when sold directly or used to create goods and services that are traded for value. In this sense, knowledge assets are similar to physical assets whose value also represents future opportunities.

From this perspective, the knowledge asset value (i.e., its opportunity value for support of enterprise success — profitability, viability, etc.) is a function of how well it can support the enterprise in its internal and external operations. For commercial enterprises, the knowledge asset value largely depends upon its competitive quality and must be considered from operational, tactical, and strategic perspectives to determine their value.[24]

Purpose

Consistent management of knowledge assets involves considerations for the need to build (i.e., create new, renew, update, purge) and utilize or exploit knowledge assets, to provide requisite supports, and to monitor, steer, and motivate execution of the desired activities. Consistent knowledge asset management must take place at every level of the enterprise. Ideally, each individual and functional group (department) and the enterprise as a whole should manage their knowledge assets explicitly and systematically to promote sustained intelligent behaviors based on competitive and high-quality knowledge assets.

Examples of Knowledge Asset Management Activities

Comprehensive programs for knowledge creation (e.g., extensive R&D); financial and time budgeting for knowledge gathering, organizing, codification, and archiving; and broad educational and training programs.

13. Construct Incentive Programs

This function should begin soon after management buy-in has been obtained and immediately after knowledge landscape mapping has indicated needs for incentives and potentials for removal of undesirable disincentives. The incentive program function is perpetual.

Purpose

This building block may be the most important to implement after obtaining management buy-in. Its purpose is to motivate employees to act intelligently, i.e., to be innovative, share knowledge, expend effort to capture knowledge (e.g., lessons learned), and ask for assistance when meeting unfamiliar or difficult situations. In general it seeks to make individuals take personal responsibility for knowledge vigilance.

General Characteristics

Our behavior is influenced by many factors. This means that the knowledge-vigilant enterprise must consider which factors influence desired behaviors positively or negatively. Among the factors to account for are:

- Situational factors — associated with insufficient time, manpower, knowledge. Often observable and openly discussed.
- Cognitive factors — ideological commitments, perplexing complexity of the challenge, familiarity with a particular decision-making style excludes more appropriate ones.
- Affiliative factors — often hidden and not recognized. "Rig" acceptance by steering decision; obtain approval by those who will implement; preserve group harmony; continue propagation of desirable cultural features; avoid punishment; maintain status, power, compensation, or social support.
- Egocentric or self-serving factors — hidden and inappropriate and covering a wide range such as greed or desire for fame or emotions like anger and elation. "What's in it for me?"
- Cultural factors — pertains to "how things normally are done here" and "what is considered acceptable behavior."
- Acceptance factors — behavior to facilitate acceptance by others. "We like to be liked."
- Psychological cost factors — perform tasks to minimize exerted intellectual effort. Hence, for example, procrastinating when we do not quite know how to perform a task.
- Well-being factors — decisions are often made to make one feel better. "I like to help my customers!"
- Job security factors — influenced by how implications of our decisions may affect our job.
- Self-preserving factors — associated with an individual's basic values.

---

[24]Excellent discussions of management of knowledge and intellectual assets are provided by Stewart (1997) and Edvinsson & Malone (1997).

Knowledge vigilance can be impeded by cultural factors such as:

- Concentration on financial goals only with drive toward short-term (quarterly) results and fulfilled requests.
- Management perception that organization possesses knowledge and people are expendable.
- Competition between business units.
- Top management funding priorities favor more tangible structured efforts such as standard processes and systems.
- "Not invented here" syndrome and "It's not my job."
- Knowledge hoarding for personal/professional gain.
- No emphasis on sharing; not measured/no set goals.
- Innovation is highly valued and this detracts from using other's knowledge because their innovations will not be seen as "original."
- Lack of broad understanding of enterprise direction and strategies.
- KM is perceived to be more work.
- Nature of work prevents people from spending time together.
- Fast-paced, dynamic business environment translates into the same at company.
- Learning/sharing is within established teams/organizations, not across enterprise.
- Hierarchy, organizational structure, disincentive reward systems, functional and cultural barriers.

Among positive cultural factors that support knowledge vigilance are:

- Aggressive company goals that are conducive for KM.
- Proactive management reacting to changing environment and industry.
- Strong teaming culture that supports exchange of ideas.
- Cross-functional execution of business initiatives.
- Effective champions who promote change, help teams withstand outside distractions.
- Very strong leadership support from chairman and CEO.
- Openness and honesty, sincere service attitude toward membership.
- High trust culture for shared learning.
- End-value ("bottom-line") focused.
- Strong desire to learn and create. Willingness to change if people drive changes to their work. Pecking order is based on education and experience — book and experimental learning/knowledge — very entrepreneurial with interest in new technologies.
- Broad belief in the value of continuous learning.
- Recognition that excellent customer service requires more knowledge than humanly possible — therefore, sharing knowledge is critically important.
- Culture is based on finding the best answers from everywhere. Knowledge is key to sales, service, quality, etc.

Example of Approach

Start by identifying cultural impediments and other disincentives. Once identified, find ways to remove or negate these, but stay in line with the enterprise management philosophy and direction. Some desired changes may be easy and quick to bring about. They should receive high priority to call attention to effective KM. Others will be at odds with current management practices and will be harder to change. It should be noted that if management is unwilling to change them that is an indication that effective KM may not be possible.

14. Coordinate Enterprise-Wide Activities and Functions

In any enterprise, there are considerable overlaps in responsibilities and functions, and for good reasons. No individual person or function can possibly understand or pursue all requirements or needed perspectives of a complex environment. These issues are particularly relevant for the management of areas such as human understanding (personal tacit knowledge) and the relations between the state of structural knowledge assets and the potentials for enterprise success or failure. As a result, coordination of these activities and result-oriented collaboration between knowledgeable representatives from the different areas become prerequisites for effective KM.

Purpose

Coordination is required to ascertain that direction, resource allocations, and schedules, among others, within and between enterprise KM efforts support the overall mission to the largest extent possible and with a minimum of conflict, friction, and loss of energy. One aspect of the coordination is to ascertain that different efforts work together toward appropriate goals and how they, and their functional units, will benefit from cooperating and collaborating. A particular challenge is associated with ascertaining that coordination efforts avoid excessive focus on process to the detriment of result orientation.

Example of Approach

Several tasks are required within this building block. The first is to identify KM-related activities and their team members. This task is often difficult, since it is common that different entities are unaware of similar efforts next door. The next task is to provide means for the entities to coordinate, cooperate, and collaborate to build the desired KM capabilities and practices. Other tasks focus on helping the different efforts to determine how their work will support the overall KM vision and therefore the enterprise's bottom line benefits.

Examples of functions that are related to KM and require coordination between KM and operational entities are overlapping KM and human resources (HR) efforts, KM and IT efforts, and KM and planning efforts. These and many others represent needs for concerted coordination.

15. Facilitate Knowledge-Focused Management

Organizational structure, systems, and procedures may be changed to facilitate managing and operating the enterprise, when an explicit knowledge focus is adopted. Of particular importance are the facilitation of conditions and creation of policies and practices that make it possible for individuals to act intelligently by using the knowledge they possess and the knowledge assets they have access to. In particular, individuals need to be placed in situations where they have opportunities to exercise their capabilities to make large and small decisions to deliver quality work. They must also be provided with permission to take on independent responsibility to, for example, improvise to deal with nonroutine situations in the best long-term interest of the enterprise.

Purpose

The new focus inherent in KM — "how people work with their minds, not only with their hands" — leads to many changes within the enterprise. To realize the desired KM benefits requires that KM efforts and activities be facilitated on the personal level as well as on all group levels.

General Characteristics

Facilitating KM requires high-level activities to change an enterprise's "customer service paradigm," culture, work environment, work flows and "opportunities to act intelligently." Operating practices, decision rights, personal motivators, and communication of management philosophy and practices may also need to be modified.

Examples

Some functions that should be put in place early include: (1) establishing KM as "business strategy"; (2) providing knowledge-fostering opportunities (innovation, importing knowledge, acquisition, organizing, sharing, and particularly, use of best knowledge); (3) performing regular knowledge landscape mapping; (4) providing knowledge-related incentives to motivate employees.

16. Monitor Knowledge Management

Purpose

To ascertain appropriate and effective KM on the highest level, the enterprise management must monitor the process of introducing KM, and later establish the effectiveness of the KM practice itself. More importantly, individual units, even individuals, need to monitor how well they are able to pursue KM practices, manage knowledge assets, and identify opportunities for improvements by introducing corrective changes.

General Characteristics and Approaches

Assessing the effectiveness and impacts of KM is not simple. In most enterprises, introduction of KM is part of a broader enterprise rejuvenation program where many initiatives are pursued at the same time. Their effects tend to blend, and it is difficult — in most cases even theoretically impossible — to obtain a macro measure of the performance of one of these initiatives.

A tool of value for obtaining feedback is the detailed description of expectations as discussed for Building Block No. 5 above.

There are many approaches for providing feedback on progress and performance of KM efforts and activities. These include: (1) conventional cost-benefit analyses; (2) economic value added (E.V.A.) assessments; (3) quantitative informational measures; (4) "enterprise monitoring system" with knowledge focus; (5) regular (e.g., annual) knowledge landscape mappings; (6) special studies; and (7) qualitative or anecdotal feedback in partially understood systems (within the theoretical limitations of the systems science concepts of "observability," "identifiability," and "controllability").[25]

3.5 A Knowledge Management Introduction Program

Introduction of KM into the enterprise must always be shaped by the enterprise's thrust, capabilities, condition, and environment. In general, it takes several years before a full-fledged KM practice has been spread across even a medium-sized enterprise. Several aspects of KM introduction take time. The implementation of the complete range of incentives, with dismantling of disincentives and introduction, are slow processes, which may take several years before agreements are established and procedures are put in place. The creation of internal KM professional capabilities often involves finding and hiring additional people and allowing existing employees to pursue advanced degrees in the new fields.

In spite of the long planning horizon required for full KM practice implementation, a beginning KM program can be conceived and implemented within months, once the vision has been created, a target area has been selected, and the decision has been made to allocate resources. Benefits will follow and, if the target area has rapid dynamics, should be observable soon thereafter.

Given descriptions of the KM building blocks and a perceived introduction horizon of less than one year, we can outline a potential introduction program for a general enterprise KM practice. It should be noted that this is an arbitrary illustration, not a specific recommendation. The example of the beginning KM introduction program is illustrated in Figure 12 and includes the six major steps:

- Build management understanding and commitment to pursue KM.
- Map perspectives of the knowledge landscape.

---

[25]See Austin (1996).

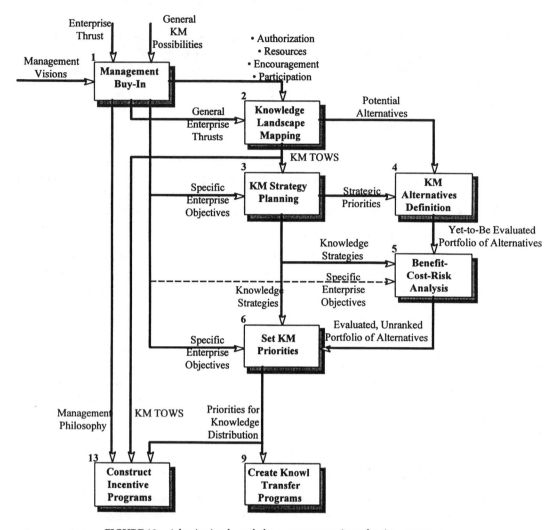

FIGURE 12   A beginning knowledge management introduction program.

- Plan the enterprise KM priorities, focus, and strategy.
- Identify desired KM benefits.
- Adjust KM priorities.
- Create KM-related incentive programs.

3.6   Concluding Remarks

We presume that effective KM is of vital importance for any enterprise with a quest to ensure viability and survival. Accepting this premise, then, requires considerations for how the KM practice can best be introduced to result in maximum value. There are dilemmas in these considerations. On one side, it is desirable to introduce KM quickly to gain early bottom-line results by targeting the most promising opportunities and making available the most capable proficiency to ensure the best possible approaches and yields from the efforts. On the other side, it is desired that KM introduction should introduce minimum disturbances and distractions from daily work and other efforts, consume minimum resources, be able to utilize people who are not urgently needed elsewhere, and expose the enterprise to minimum risks. Most senior managers would also prefer that approaches to KM introduction would be sufficiently well understood to allow delegation of its implementation and management, thereby reducing the need for their own involvement and attention.

Given these competing objectives, it is not surprising that many KM efforts fall prey to the four common problems that Lucier and Torsilieri (1997) point to, all of which are correctable: (1) "No specific business objective," (2) "Incomplete program architecture," (3) Insufficient focus upon strategic objectives," and (4) "Top management sponsorship without active involvement."

References and Related Reading

American Psychological Society (1998). *APS Observer Special Issue: HCI Report 6 — Basic Research in Psychological Science: A Human Capital Initiative Report.* February 1998, Washington, D.C.: APS.

APQC/EFQM/KMN (1997). *Knowledge Management and the Learning Organization: Best-Practice Report.* Brussels: EFQM.

Austin, Robert D. (1996). *Measuring and Managing Performance in Organizations.* New York: Dorset House.

Edvinsson, Leif & Malone, Michael S. (1997). *Intellectual Capital: Realizing your company's true value by finding its hidden brainpower.* New York: Harper Business.

Fayyad, Usama; Piatetsky-Shapiro, Gregory; & Smyth, Padhraic (1996). "The KDD Process for Extracting Useful Knowledge from Volumes of Data." *Communications of the ACM, 39* (11): 27-34.

Lange, Robert D. (1986). *Competencies Required for the Design and Implementation of Manufacturing Systems for Advanced Composite Structures.* Denton, TX: University of North Texas, Ph. D. Dissertation.

Lucier, Charles E. & Torsilieri, Janet D. (1997). "Why Knowledge Programs Fail: A CEO's Guide to Managing Learning." *Strategy & Business,* Fourth Quarter, (9): 14-28.

Shapiro, Stuart C. (1987). *Encyclopedia of Artificial Intelligence.* New York: Wiley.

Singley, Mark K., & Anderson, John R. (1989). *The Transfer of Cognitive Skill.* Cambridge, MA: Harvard University Press.

Spek, Rob van der & Spijkervet, André L. (1995). *Knowledge Management: Dealing Intelligently with Knowledge.* Utrecht, the Netherlands: Kenniscentrum CIBIT.

Spradley, James P. (1979). *The Ethnographic Interview.* New York: Holt, Rinehart & Winston.

Stewart, T. A. (1997). *Intellectual Capital: The new wealth of organizations.* New York: Currency Doubleday.

Suchman, Lucy (1995). "Making Work Visible." *Communications of the ACM, 38* (9): 56-65.

Treacy, Michael & Wiersma, Fred (1993). "Customer Intimacy and Other Value Disciplines," *Harvard Business Review.* January/February, 84-93.

Wiig, Elisabeth H., & Wilson, Carolyn C. (1997). *Visual Tools for Developing Language and Communication: Content, Use, Interaction.* Chicago, IL: Applied Symbolix.

Wiig, Karl M. (1993). *Knowledge Management Foundations: Thinking about Thinking — How People and Organizations Create, Represent, and Use Knowledge.* Arlington, TX: Schema Press.

Wiig, Karl M. (1995). *Knowledge Management Methods: Practical Approaches to Managing Knowledge,* Arlington, TX: Schema Press.

Wiig, Karl M. (1999). *Establish, Govern, and Renew the Enterprise's Knowledge Practices.* Arlington, TX: Schema Press.

Wiig, Karl M.; Duffy, Neil; Viedge, Conrad; Cook, Jonathan; Radley, Robert; Jooste, Adriaan; Jackson, Carl; Harley, George; and Brooking, Annie. (1997). *Leveraging Knowledge for Business Performance.* Johannesburg, RSA: Wits Business School.

Section II
Knowledge Management: People and Measures

4

People Who Make Knowledge Management Work: CKO, CKT, or KT?

Angela Abell
TFPL Ltd.

Nigel Oxbrow
TFPL Ltd.

4.1 Context and Background

The considerable interest in knowledge management (KM) as a key initiative within organizations is well reflected in the large number of surveys undertaken in the past twelve months by academics, consultants, and practitioners of different approaches and the implementation of KM in organizations. Facts and figures are emerging from all quarters, but few provide generic models for staff structures, skills, and competencies required to plan, design, and deliver a successful KM organization. Not surprising, given that the essence of KM is that it reflects the ethos and value of each organization. Its objective is to leverage and utilize the uniqueness of the organization — to capitalize on the mix of people, processes, services, and products that define its identity and place in its competitive market. The approach has to reflect that uniqueness in order to maintain it and the changes that KM undoubtedly brings to the organization must build on its own strengths, not that of other KM organizations. But change does not happen without intervention by people, and the question for all organizations following this route is who should these people be. The evidence from the early adopters of the concept is that like all radical approaches mistakes and failures have occurred along the way, and that these are generally a result of concentrating too much on the technical and process aspects of KM rather than on culture and people. As many of those implementing KM have observed, KM is 80% about people and cultural change rather than technical development. But processes, infrastructure, management style, and organizational values

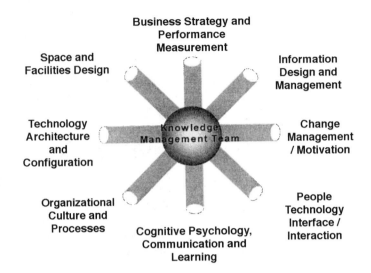

FIGURE 1

are all part of the culture, and the job of the executives who are charged with bringing about this change is to integrate the hard and soft elements of organizational life while focusing on the end result.

So who brings about the change? A KPMG research study in the U.K. early in 1998 indicated that of 100 leading companies almost all thought that KM was "here to stay" and many were taking steps toward developing an approach, but only a quarter were actually at the implementation stage.[1] Bain & Co.'s annual review of management techniques in 1997 surveyed 870 U.S. companies and found that only 27% were "using knowledge management," and of these only a quarter were using it in any substantial way.[2] The KPMG research suggests that many organizations are unsure of how best to get started and half of the sample considered the lack of necessary skills to move from the planning to implementation stage a barrier to effective KM. Certainly the combination of skills required to plan, design, and implement KM program are daunting (Figure 1)

One of the challenges of implementing the KM vision is the size of the undertaking and the potential impact on every area of the organization. The concept of the hybrid manager is comfortable when viewed as individuals who can multitask and undertake a variety of roles handed down to them. The concept is more difficult when the role that is being created is responsible for changing the nature of how everyone works, from board level down. The move from vision to implementation requires leadership, commitment, and involvement from the Board, but in addition the Board needs to identify the skills required in the new role and empower that role to enable the change. In his address to the Council of Human Resources Boards in 1996, Michele Darling, Executive Vice President, CIBC Human Resources,[3] argued that the recognition of the "collective knowledge (accumulated, shared, and evolving) plans and runs the business" and it is a "validated platform for action." Calling the action something new facilitated novel and different approaches. It "unchannels conventional thinking." This concept of providing a new role, with a new title, as a mechanism for emphasizing that KM is a new corporate approach rather than an initiative headed by an established function — someone's pet project — is gaining ground, and a host of roles and titles is emerging.

The emergence of new posts to champion a radical change or key project is not new. In the past there have been directors of total quality management, business process redesign, change management, and of many other titles reflecting the current focus. Even without the KM initiative, roles have emerged recently

---

[1]Knowledge Management: research report. KPMG Management Consulting, 1998.

[2]*Wall Street Journal*, 11 November 1997.

[3]The Knowledge Organization: a journey worth taking. Vital speeches of the day v 62 no 22 Sept 1, 1996 pp693-697.

FIGURE 2   Knowledge management strategy.

that indicate a move in organizational focus away from the management of traditional operations to a wider, holistic perspective. The focus on learning as a driver of change, and the emergence of learning organizations has mirrored, complemented, and in many ways overlapped with knowledge management initiatives. Chief learning officers and vice presidents of learning are becoming recognized posts in organizational structures. Other, less explicit posts, are also impacting on the KM scene: Futurist-in-chief, for example, together with heads of communities of practice and directors of strategic development. What all these posts have in common is their lack of traditional role models and established professional or industry practice. Wiig said in 1996, "We need to develop a cadre of knowledge professionals with a blend of expertise that we have not previously seen."[4] Tom Stewart writing about chief knowledge officers in *Fortune* magazine said in January 1998 that "the job is still so new that few people who hold it know quite what it is."[5] Even learning about the job from people in other relevant posts is not easy, as identifying who is out there is far from simple. In the same article Stewart reports Daniel Holtshouse, Director of Knowledge Initiatives at Xerox, as finding the undertaking of a census of the CKO population far from easy as the titles and responsibilities vary. His estimate at the beginning of the year was that approximately a fifth of the Fortune 500 companies employ someone who, in role if not always in name, is a CKO. The KPMG research in the U.K. found that only 5% of the organizations with a knowledge management initiative had a chief knowledge officer.[6]

There are not, as yet, any knowledge management graduates or MBAs, although courses and programs are in hand, and there are very few experienced chief knowledge officers to be headhunted. There is not even an accepted route to follow, or necessarily an obvious profession or function from which to select the embryonic CKO. The very essence of knowledge management is teamwork, a mix of skills and experience, a new approach to organization development, and a new focus on the management of people.

4.2   Making It Work

A knowledge management strategy impinges on all areas of the organization and requires a corporate approach to make it work. (Figure 2).

Chief Knowledge Officers (CKO)

Role

When Victoria Ward was Chief Knowledge Officer of NatWest markets, she answered the question of what does a CKO do all day by defining her role as:

---

[4]On the Management of Knowledge. Karl M Wiig http://revolution.3-cities.com/boneman/wiig.htm
[5]Is this job really necessary? Thomas A Stewart. *Fortune* 12 January 1998 pp72-73.
[6]Ibid.

- Cartographer — mapping expertise and making connections
- Geologist — drilling into specific areas and applying tools
- Sparkplug — igniting an awareness of the need to change
- Architect — designing the physical and cultural environment

At Coopers and Lybrand the role of and responsibilities of the CKO are defined as:

- "CKO's lead the effort to establish and maintain, and continuously improve knowledge based culture and processes within the organization."
- "Creating innovative ways to leverage our intellectual assets."
- "Ensuring that we grow those assets faster than our competitors."
- "Understanding how those intellectual assets continuously create greater value in terms of the quality and responsiveness of client service."[7]

The individual specifications of the role vary from organization to organization but the generic definition would include:

- The identification of corporate knowledge and the barriers that prevent its collection and utilization. This includes the identification of cultural and organizational factors and what Tom Stewart calls "teasing out the difference between prudence and vain, ambitious and slothful hoarding," (Stewart 1998).
- The creation of an infrastructure that facilitates and encourages individual development, group learning, and corporate sharing.
- The introduction of processes to trap and interpret, package and present, and integrate into the work processes and culture of the organization.

Knowledge management cannot be imposed on an organization from above. It is the one initiative whose success depends on gaining the support of the whole organization throughout the process. It is, however, a change program, and its implementation can only work with the full backing of the most senior management. The changes impact on the way the organization works, on its value system and management style. And these changes need to be embedded in the organization for long-term return. Knowledge management is not a quick fix. For these reasons the first step in the journey is a champion at the top whose first job is to convince his (or her) peers to support a strategy that has substantial investment implications with the potential, but unproven, for even more substantial return on investment. This step is the first in the team building approach that lies at the heart of KM. As Darling expresses it in his address, "The knowledge asset depends for its function on the effective interaction and cooperation of people working together. Its maximum performance is therefore critically dependent on not just what they think and how they think as individuals but how and what they think as a team" (Darling, 1996). This is a board-level exercise and calls for board-level skills — negotiation, persuasion, charisma, and the effective presentation of the case.

Once past this hurdle the most common approach at the initial stage is the creation of a knowledge team representing all the major functions of the organization, typically human resources, organizational development, information systems, and the functions that are core to the operation of that organization. In some cases research and development will play a crucial role, in others the office of best practice. The job of the team is to plan the development and implementation of the KM program. It may elect to identify appropriate KM initiatives and then undertake a coordinating role. There is, however, evidence that a new role, as mentioned previously, will emerge to provide the sustained energy and drive needed to create and sustain the impetus needed. Although KM is not a "quick fix," the approach needs to progress quickly enough for the benefits to become obvious to people who are being required to change

---

[7]Knowledge management: the key to success in the 21st century. Ellen M Knapp, Vice Chairman, Coopers and Lybrand. TFPL European Business Information Conference, Lisbon 1998

Three areas of knowledge focus

FIGURE 3

the way they work. Without the post of a KM champion and designer it is likely that some aspects of the program will be neglected, while others may be hijacked or take unwarranted prominence. Someone needs to watch over a change program.

And this is where the CKO comes in, although as already noted the job title may not be the same. A list of some of the posts that have been created to champion and initiate knowledge management approaches appear as an appendix to this chapter, and they are far from exhaustive.

The role of the CKO is to turn a concept into reality, although the term "knowledge management" causes confusion from the outset. It is evident that you cannot manage knowledge. What you can do is to manage an environment that optimizes knowledge, that encourages information sharing, knowledge creation, and teamworking. An environment that enables creative and supportive interaction between people; that stores, codes, and makes available information in a way that adds value to the individual's work and benefits the organization; and that creates a community of trust and common purpose. To achieve a knowledge environment there needs to be a focus in three areas: preparing the organization, managing the knowledge assets, and leveraging knowledge. (Figure 3).

Preparing the organization is, for most organizations, where the initial focus will lie. It is primarily about changing the culture — changing the way people work and building a trusting and sharing community. This requires reexamination of corporate values, the expressed and the perceived; the creation of "space" in which people meet, work, and interact; a focus on individual and corporate learning; assessment of business processes and organizational structures; and an evaluation of reward and motivation systems.

Managing the knowledge assets requires an understanding of workflows and business processes within the organization. It is as much about "connections" as "collections." Connecting people to people, people to information, and providing a means to develop the tacit knowledge required to effectively utilize information are the vital elements in knowledge management. The effective management of stored intellectual assets — such as best practice, proposals, expertise, and intellectual property — will depend on a good understanding of how knowledge is created and used.

The leveraging of knowledge is, of course, what knowledge management is about. Identifying potential benefits, understanding the dynamics of the organization in order to enable those benefits to be realized, and making it happen are the ambitions of every CKO. Measuring success, which is one of the biggest challenges of KM, and demonstrating and promoting success are essential, not only to top management but to everyone in the organization. Identification of benefits, of "quick wins" — projects or initiatives that will have large impact early on in the process — and the building of appropriate measures to monitor benefits are keys to KM success. What counts in a knowledge-based organization is the ability to create, identify, store, and access knowledge, to interpret, apply, and use information, all for the benefit of the clients of the organization.

These three areas of focus, while different in nature, need to be orchestrated simultaneously. In preparing the organization, expectations will be raised, and it is vital for the momentum of the whole initiative for there to be visible benefits that match the expectations. Focusing on managing knowledge assets and achieving valuable benefits must therefore be part of the KM initiative right from the start.

These tasks call for leadership and teamworking skills; for an understanding of, and enthusiasm for, information and communications technology plus people and social skills; an understanding of the business and profit-generation and the ability to take a long-term view. It is, in fact, a task that needs a person who can face many ways at the same time and still hold the wider picture within view. They need the ability to sell and persuade, to negotiate and win confidence, to lead and support. They need to be able to analyze what needs to be done, harness the skills within the organization to manage the projects that affect the change and provide the backing, and facilitate the involvement and contribution in the development process from all parts and levels of the organization.

The CKO role is that of project and change management and is likely to be a vital role for a period of three to five years. After this time, if the CKO has been successful the elements of KM will be integrated into the business and a chief knowledge team will succeed the CKO. There are some CKOs who feel that the whole processes should be run by a dedicated team without a star leader and the culture of some organizations makes this attractive. However, even in public sector organizations, where projects are more likely to be run through teams, a project manager acts as coordinator and focus, and in most organizations the leadership role is necessary to kick start the process. There are parallels here with total quality management (TQM). When TQM first became an issue, senior executives were appointed to review processes and working practices and to implement quality procedures and standards. Once they had become an integral part of the way the organization worked the senior management posts disappeared.

Skills and Background

Who are the people moving the vision forward? Where do they come from and what skills do they bring with them? In 1997 we undertook research to identify CKOs and their equivalent positions in Europe and North America, to establish their backgrounds, and to find out the skills and competencies that they were building in their teams. Almost without exception all CKOs were from "the business." They identified themselves as "a banker," "an engineer," "a consultant." Their original academic discipline seemed less important than their ability to identify with and understand the business and its environment. In terms of their most recent functions in the organization before taking on the CKO role, they were predominantly from change management and organizational development roles, or from one of three main backgrounds — IT, human resources, or a core business function.

In some organizations the driver has been a focus on the individual and corporate learning — the learning organization and the CKO role reflect this focus. Organizations such as CIBC, Motorola, General Electric, Unipart, and Anglia Water are examples of this approach, with their own "in-house universities" promoting a learning environment.

"Leaders in learning organizations expect their employees to take company time to pursue knowledge that is outside the scope of their immediate work. They encourage lateral, cross-functional transfers that force employees to learn and develop new skills and perspectives with new colleagues. The result is great sharing of information and potential for challenging tradition, that is the learning boundary, by bringing different points of view into an organization"[8]

But their common strength is their ability to take a holistic view of the company, and to understand the mix of hard and soft skills necessary to create, sustain, and utilize the knowledge base.

---

[8]Marker orientation and the learning organisation. Stanley F Slater et al. *Journal of Marketing*, July 1995 v59, (12) p63.

The skills required by a CKO can be summarized as those strategic skills required by any leader of a change management program. This skill set may change as the project progresses from vision to planning.

Skills to Develop the Vision	Skills to Plan the Program
• Business knowledge	• Organizational development
• Political understanding	• Information and IT strategy
• Risk analysis	• Financial planning
• Influencing skills	• Communication
• Leadership	• Innovation
• Creativity	• Risk management
• Presentation skills	• Flexibility and openness on all issues
	• Managing across boundaries and beyond limits
	• Helping individuals to self-manage
	• Having the ability to release the full potential of people

The brief case studies below illustrate that CKOs identify knowledge management as a "people business" and acknowledge that the development of people within a sharing culture is key to the success of the system. The other crucial elements are information technology, an understanding of the company's business and its processes, and information management. IT enables employees to connect with each other, to share expertise and experience, and to connect to the corporate memory. The business understanding enables systems to be designed around the information flows that are crucial to business. The identification, acquisition, organization, and distribution of relevant information is at the heart of effective knowledge management. Its utilization is what gives the company its competitive edge.

4.3　Thumbnail Sketches of CKOs

The Human Resources Perspective

Elizabeth Lank, Program Director, Knowledge Management, ICL

While head of management development in ICL and involved in a number of change management/organizational development initiatives Lank researched and wrote a book on learning organizations (*The Power of Learning — A Guide to Gaining Competitive Advantage*, IPD 1994) that focused on the practical "how's" of enabling an organization to learn and adapt. "It was then that I realized that the way in which you create, capture, and share knowledge was a key aspect of this. I felt this was a significant opportunity for a company like ICL, which was undergoing a transition from being a product company to a service-led company. It was increasingly our knowledge about information technology that was the key asset, rather than just the technology itself.

However, it became clear that if we really wanted to make a difference in our knowledge management capability, it required some focused effort. Therefore, in the first week of Keith Todd's tenure as CEO of ICL in January 1996, we put the business case to him to appoint a full-time person on this issue." The result was the role Program Director, Knowledge Management. Lank has a broad business background (including an MBA from INSEAD), which she believes has allowed her to position knowledge management as a strategic imperative, not just another good idea. Lank's job title recently changed to Director, Mobilising Knowledge to reflect her enabling role in creating the environment rather than trying to "manage knowledge" itself.

"In terms of where we are with Project VIK, the team is working with a number of ICL businesses in helping them embed knowledge management as part of their way of working. In November 1996 we launched a very successful cross-company intranet information service called Cafe VIK, which has given employees across the world much faster access to company information and expertise. We are developing our 'Knowledge Asset Map' — i.e., being much more clear than previously on which knowledge assets warrant investment."

From IT Perspective
Ellen M. Knapp, Coopers & Lybrand

C&L International, through its member firms, operates in 750 offices in 140 countries. Ellen Knapp is a member of the firm's Management Committee and its Board of Partners and serves as Chairman of C&L's International Technology Management Group. She has recently been confirmed as the CKO for the new merged company Price Waterhouse Coopers (PWC). One of the U.S.'s primary authorities on the strategic use of both intellectual capital and technology, Knapp is charged with creating competitive advantage and catalyzing organizational change through the development and application of innovative technology and knowledge-based products and services.

Ms. Knapp has been a Senior Partner in C & L's Management Consulting Services and National Director of Information technology Consulting for the U.S. firm.

She has received worldwide recognition for her work in advanced technologies and has published numerous professional papers.

In designing C&L's knowledge management system she has drawn on her expertise in business processes and current management thinking on organizational development and management.

The Business Knowledge Perspective
Victoria Ward, former Chief Knowledge Officer, NatWest Markets

Victoria Ward was the first Chief Knowledge Officer in the City of London. This new role, created in January 1997, had a mandate of trying to identify and exploit the full breadth and depth of intellectual capital within the firm. Victoria believed her track record of 13 years in the financial futures industry, together with an educational background studying avant garde art and philosophy, made her ideally suited to the CKO role.

"You can't run a futures business without understanding your product and it is therefore crucial in knowledge management to understand information and how information flows." Coming from this background, Victoria's approach was to design knowledge management products that helped people communicate rather than design processes. Victoria has since left the Bank to set up her own consulting business.

Gordon Petrash, Global Director of Intellectual Asset and Capital Management, Dow Chemical Company

Dow is a global organization with a mission to develop the tools necessary to implement a corporate intellectual management vision. This vision focused on maximizing the value of Dow's existing intellectual assets and on developing global, business-driven intellectual asset management plans.

Coming from a background in engineering, manufacturing, R&D, and global business management, with experience of international assignments, Gordon Petrash took Dow's vision to "develop management processes that will help maximize the creation of new value creating intellectual assets" forward by first concentrating on their tangible intellectual assets — their patents. By applying a rigorous process of assessment using global, multifunctional teams he was able to demonstrate both cost saving through the identification of nonexploitable assets and increased value through the identification of licensing opportunities, reuse of knowledge contained in the patents, and reassessment of market opportunities.

Petrash left Dow in mid-1998 and is now a consultant with PWC.

The Information Director's Perspective
Jacqueline Rees, Director of Information, Thomas Miller Partnership

With a research degree in modern languages and an academic background, Jacqueline Rees joined Thomas Miller, a global insurance group, and spent 12 years as a manager of a U.K. syndicate in a core business. She identifies herself as an insurer who understands the business and the business application of IT. She reports to a Partner and Board member with responsibility for knowledge management; she is responsible for the business application of IT, or as she puts it, "responsible for the I in IT." What Miller wanted was someone who primarily understood the business and secondly was competent with technology. Her information team reflects this approach of selecting the appropriate personality with a business focus, and is made up of people who have grown up in the centre.

Miller designed and implemented a knowledge management system led by a Partner. As well as Rees as Director of Information they appointed a Director of Learning responsible for the personal development plans of all employees. While the management of knowledge resources through electronic, written, and oral means was, and remains, a key element, the development of an information-sharing culture and learning organization was considered a key platform.

When the London-based headquarters was reorganized to create an open plan environment, the hub was left for a business intelligence unit and learning centre. This physical entity is seen as encouraging and enabling personal development as well as the management of internal and external resources.

Karl Kalseth, Director of Information Management, Norsk Hydro

With a degree in librarianship and business administration, Karl Kalseth brings experience of information systems, information management, and business process reengineering to a team considering how to leverage information assets. Their corporate staff for information systems (CIS) is headed by a senior vice president reporting to one of the senior vice presidents, and the directors of the CIS staff represent information management, infrastructure, and business processes. Although Kalseth has a professional background in librarianship and business administration, his view is that senior staff concerned with the impact of information and knowledge on the business can come from any area — IT, human resources, business units, or library or information functions. The critical feature is, he says, their ability to understand the information flow and its impact on the business.

Sandra Ward, Director of Information Services (U.K.),
Glaxo Wellcome Research and Development Ltd

Dr. Ward has a scientific, education, and information background. She has spent most of her career in the pharmaceutical industry and is an internationally acknowledged expert in information management in the industrial R&D environment. Current interests include the exploitation of IT for information management and knowledge networking.

Her various roles within the Wellcome, then Glaxo, and now Glaxo Wellcome organizations have been focused on the management of information in all forms. In her present role she directs the management of records, library operations, external information analysis, text systems and photographic and graphics design services for research and development staff in the UK.

One of Dr. Ward's strong beliefs is the relevance of the information professional's skills to knowledge management. "The information profession has continually focused on helping people to use existing knowledge, and effective knowledge management is aimed at helping employees to stand on each other's shoulders instead of on each other's toes." She says the questions that any organization should ask in order to assess their knowledge management rating are — do you:

- Have a sharing culture?
- Know what knowledge you have and where it is?
- Make it easy to find?
- Learn from your mistakes?
- Capture and share best practice?
- Organize your information and knowledge?
- Exploit knowledge effectively?
- Reward knowledge sharing?
- Transfer knowledge easily to new staff?

The role of the CKO, she insists, is to ensure that the answer is yes to all of these questions, and to ensure that knowledge does not walk out the door when staff leave.

4.4   KM Teams

The skills required in the KM teams that move the process forward are those that will be required by any team working on a time-limited project as a functional or professional work group, or as part of any group of people committed to achieving the same end result. The prime requirement is commitment and the selection of the teams will be ideally that of self-selection. Teams of joiners, rather than of the coerced, are more likely to be the effective ones. However, in the real organization there will be a mix of self-selection, careful weeding, and explicit lobbying. Not all self-selectors will provide the best input; it is a chance for the habitually vocal to push their views forward on yet another platform. Similarly, the least vocal may be the people who have the innovative ideas or see the picture in a more reflective way. The art of team building needs to be a core skill of the CKO or the CKO team.

The Design Team

As the KM program progresses to design and implementation stages the inclusive approach — bottom-up design and implementation — sees the introduction of KM teams. At the design stage the teams are concerned with the identification of projects within the overall framework of the KM strategy and these teams need to be cross-functional. The bottom-up approach needs to involve everyone and the design teams need representation from every area of the organization. Booz Allen & Hamilton, for example, had an "infrastructure team focused on capturing Intellectual Capital and developing a system to exchange, use and collaborate." The team was cross-functional, including consulting staff, information professionals, and corporate systems and technology staff who "focused on defining and implementing processes and information technology and the people to support both."[9] A U.K. government agency more recently developing a KM system has a series of cross-functional teams each focusing on specific aspects of the program (information audit, quick wins, communicating the benefit, etc.), and coordinated by a central project team.

The cross-functional nature of these teams ensures that all areas of the organization are involved but the nature of the organization will dictate its emphasis. The representation of the core business units will focus the design of the processes and systems on the achievement of business objectives. Their interaction with other units and support functions facilitates the lateral thinking that enables a corporate rather than parochial approach to be considered. The leadership of the CKO or the CKO team is crucial in facilitating the team building that will underpin this process. The point at which their ability to harness the skills, experience, and knowledge of their colleagues across the organization determines the success or failure of the KM approach. The skills of negotiation, persuasion, and presentation are essential, but perhaps more so are those of identifying and acknowledging other people's strengths and weaknesses, their concerns and ambitions, and the ability to build and support the design teams. This requires confidence and tact, with the ability to listen and reflect while still retaining sight of the overall plan — a breadth of vision plus the ability to deal with a number of smaller horizons. The collective skills required of the design team, in addition to those associated with their business function, will include:

General	Detailed
Communication skills	• Understanding of internal communication processes, verbal and written • The ability to assess communication channels suitable for the organizational culture • An understanding of how communication processes can contribute to a change in culture
Presentation skills	• An ability to present to groups at all levels • The ability to present complex changes in an understandable way
Information and communication technology (ICT) literacy	• The ability to identify IT&C implications for proposed processes and systems • An understanding of the strengths and limitations of available systems
Information management expertise	• An understanding of the range of potential inputs and outputs • A knowledge of how information is used • An awareness of the activities involved in the organization of information — capture, coding, and dissemination • An understanding of information design

[9]Knowledge management: the way forward? Presentation, 1996.

The Implementation Team

While only a few firms worldwide have moved into full implementation programs, it is clear that the mix of skills required to implement, maintain, and continually revise procedures, processes, and approaches calls for a new mix of expertise and experience. The leveraging of knowledge is demonstrated by the way these teams pool and enhance each other's skills and how the synergy of different skills is recognized and made explicit. At the end of this chapter there is a list of some of the job titles that are to be found in organizations building knowledge teams, but these don't necessarily reflect the skills employed to undertake the role. In some cases the role calls for individuals to expand their core skills and learning in order that their core skills make more of an impact on the organization. In other cases it is more an acknowledgment of where the skills contribute to a change in knowledge building and sharing.

There are many models. At Buckman Labs, for example, all staff contribute to technical discussion via threaded discussions on the corporate Web. The collation of the information and knowledge exchanged in the discussion is undertaken by a subject "expert" who undertakes this as part of their normal role. It may be a technical expert or a salesperson, or, as in at least one instance, a member of the information team with a specific interest in the topic. In other instances the role of "knowledge navigator" is taken by a departmental or group secretary, the person who knows where the vital resources of that department are kept and who the experts are. They play a role in contributing information about their group's resources and become the main point of contact for their access. At ICL they have tackled the problem of getting information out of people's heads, of turning tacit knowledge into explicit knowledge by using graduate journalists to interview staff at the end of projects or assignments in order to record lessons learned and to identify best practice.

One aspect of knowledge management that will impact the skills requirement of all team members is the concept of virtual teams or knowledge networks. "Virtual teams" suggests an information technology-based approach, but a virtual team is much more than people connecting electronically to work on the same project. Certainly IT has made the virtual team easier to develop and maintain, but its essence is not the technology. It is the concept that a group of people, who are not necessarily geographically located, or working within the same department or work group, may still be working together toward a common goal. They are not directed by one manager, or have one set of departmental loyalties, but are motivated by an agreed set of aims and objectives to which they all contribute.

Tindale in her 1997 presentation describes two such teams.

"Northeast Utilities Co. Earlier this year the Hartford, Connecticut–based electronic company created a "Central Information Group" to identify, categorize, and share knowledge across the organization. This group includes representatives from IS, accounting, human resources, treasury and others."

"Fidelity Inc. The Boston-based brokerage group has formed a 20-member volunteer team "looking at ways to use information and learning. The K'aizen Learning Network" grew from corporate total quality efforts and includes representatives from every major department. It has already produced bench-marking databases for internal and external use, a draft "information architecture" and a Notes-based communications vehicle about the project called "Fidelity Facts."

The development of "knowledge networks" is a common implementation step with a similar approach. Again the technology is a facilitator of the approach, but the concept of people in all parts of the organization contributing to, maintaining, and sharing knowledge through a network of databases, discussion forums, experts, and knowledge modes is the essence of the network. Associated with this approach is the concept of "Communities of Practice" or "Communities of Competence." In some organizations these become a formal strand of a KM or learning organization approach; in others they are forming naturally as communication channels enable people through the organization to identify others with shared interests, irrespective of role, hierarchy, department, or place in the global group. These communities bring together key knowledge workers who form part of the knowledge network, and in some cases the implementation team.

Such teams and networks have to work by consensus and therefore require human interaction, trust, and understanding. The ability to network and build relationships then becomes one of the core skills of the KM team player.

Whatever the approach, the objective is to build a corporate approach that recognizes that knowledge must be built and owned by the business unit but shared and utilized by the whole organization. With this in mind it is becoming possible to identify what skills senior management and CKOs look for in their KM teams. Some of these are the same as required in other stages of the approach — change and project management, for example. But a raft of other more specific skills comes into play. The TFPL research in 1997 identified the following skills and competency requirements and a set of critical skills that allows every member of the team to make a full contribution.

Implementation Skills	Critical Skills
• Knowledge of the business	• Understanding of the business
• Networking	−how is knowledge used
• Communicators and team players	−where does it come from
• Confidence and a willingness to take risks	−what does it mean
• Skills in IT&C applications and tools	• Understanding of their core skills
• Creativity	−what value they add to the business
• Negotiation skills	−what potential there is to add value
• Information management skills, tools, and techniques	−how to initiate their application
• An understanding of internal and external sources	• Communication
• Lateral thinking	−networking and relationship building attributes
• Team management skills	−communicate the value of knowledge
• Subject knowledge	−the ability to deliver information in a form in which it can
• Training, coaching, and mentoring skills	be used
• Problem solvers rather than problem creators	• Strategic skills
	−an understanding of internal politics
	−the confidence to work within the political arena
	−global awareness
	• Capacity to absorb large chunks of information
	−digest and apply it to achieve solutions

Information Skills

"One good librarian is worth a 100 databases" (Tindall)[10]

The above statement emphasizes the point that knowledge management is about more than technology and databases. It is about connecting people to experts, people to information, and the utilization of that information. It is also about understanding how people learn and making it possible for them to do so, how organizational structures and infrastructures affect the building of knowledge, and how organizational procedures, rewards, and values affect the sharing of knowledge. But at the heart of every organization is its need to manage the information it owns and acquires, to utilize it for competitive advantage, to interpret it in a way that assists decision making, and to build a secure information-based organization to underpin the KM environment. In the earlier stages of KM the emphasis was on understanding the business imperative and the issues that need to be addressed in order to build a KM environment. As the approach develops, the need to understand the information management issues has become more apparent and the skills of the information professional are becoming a vital part of the process. In her presentation to the European Business Information Conference in 1998 Ellen Knapp said these skills were becoming essential and were in short supply. Elizabeth Lank, also in 1998, in a talk to the Confederation of British Industry, said that librarians were going to be in demand.

While the word *librarian* conjures up for many the skill of organizing the acquisition of and access to external resources, it has increasingly become one title among many representing a range of skills that are to do with the management and utilization of information. While this profession exploits and needs

---

[10]Profiting from a groupwide knowledge management program. Christine Tindall, Technical Director of Knowledge Management, Ernst & Young. Presentation 1997.

to understand the applications of information and communications technology their focus is on the I of IT. The content, not the conduit. The content of the wagons, not the railroad.

In common with many of the management consultancies that were the early adopters of KM, Booz Allen found that their head count of information professional staff increased considerably as the approach gained ground. Their global community of information professionals (IPC) operates across national boundaries and includes information specialists and knowledge workers.

- "The IPC was established to make the acquisition, management, structuring, maintenance, and dissemination of knowledge and information within the Firm easier, more comprehensive, and more cost effective."
- In the IPC, information, management, and research is a team effort, involving professionals in a variety of roles:

    Practice information managers
    Researcher specialists — practice-specific and general
    Technical services specialists
    Editors/abstractors
    Information assistants
    Managers and team leaders

Their remit includes providing a conduit to the knowledge system, working with teams to capture and structure intellectual capital, to train users in the in-house online system, and to act as change agents for institutionalizing the knowledge program.

Many organizations are taking a similar approach, leveraging the contribution of the traditional library and information management skills by integrating them firmly into the business and adding additional skills to the team. At the same time the virtual library or distributed information approach is overtaking the concept of the central information department. Information specialists are becoming part of the knowledge management environment rather than the custodians of libraries and corporate records. They are to be found in core business teams, in business development functions such as competitive intelligence, and in the development programs of government departments, health care organizations and the not-for-profit sector.

Information Skills in a KM Environment

So what do these skills bring to the knowledge environment (Figure 4)?

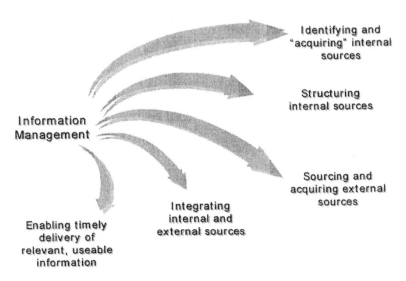

FIGURE 4 Information skills in a KM environment.

Identifying and "Acquiring" Internal Sources

A core platform for the KM environment is the corporate knowledge map that shows where knowledge and information are created, acquired, stored, and utilized. It identifies how information flows between individuals, departments, the corporation and its suppliers, partners, and clients, and the bottlenecks and barriers that distort the flow and inhibit knowledge creation. It includes the implications of infrastructure and organizational structures. Such a map, once drawn, is valuable for planning a KM environment and invaluable provided it is maintained and monitored. The knowledge audit covers all aspects of the KM approach, but the information audit maps the crucial information flows, identifying the silos of information that exist, their value, and whether they are long-term assets or support the operational aspect of the organizations. The information audit identifies the business process requirements and the corporate information assets, determines their value, and assesses the quality.

The audit allows priorities to be set for the management and "acquisition" of those assets and allows "owners" to be persuaded to allow access to "their" resources. The map demonstrates where the pieces of the information puzzle fit together and helps explain the ultimate goal of sharing information and knowledge resources. The audit also identifies access issues; what formats are appropriate where; what level of access is appropriate; and the "differences between prudence and vain, ambitious, or slothful hoarding" (Stewart). The audit also helps identify where the emphasis needs to lie — on collection or connection — in any part of the organization's activities.

Structuring Internal Sources

Making information assets accessible in a meaningful and helpful way, in a way that facilitates their utilization, requires management and effort. The building of vast databases does nothing for the process unless the information they contain

- Is reliable, accurate, and up-to-date — quality assured
- Can be found and accessed as part of a person's everyday activities — the navigation is user friendly — two clicks on the mouse
- Has meaning for the person who accesses it — structure and coding is context friendly
- Is valuable — produces some real benefit

This requires the design of content management tools to structure the information and the establishment of information capture routines, both of which must reflect the way the information "owner" works and the way the information "user" works. It may also require the information or knowledge manager to train others to identify useful data and information and to design coding and indexing systems that information owners can apply to their data.

Sourcing and Acquiring External Sources

Knowledge management has to date tended to focus on internal information flow and in some cases information flow between themselves, their clients, and suppliers. But much of the information being utilized in the knowledge generation comes from the external world and this too needs managing and preferably integrating with internal information sources. External information that is purchased from information vendors requires regular review, especially at a time when there is rapid change, new services, increased competition, and complicated pricing negotiations. The convergence of information and communications technology means that these commercial information services can be supplied direct to users' desktops, but this also requires regular maintenance, management, training, and support. The speed of the competition in the vendor marketplace, the global nature of many organizations, and the global availability of these information sources has meant that organizations are anxious to distinguish between their real needs, what they need to access and the most appropriate sources in order to add value, and what simply contributes to information overload.

The growth of these information sources, the increased sophistication of the staff using them, the demand for reliable sources to be integrated into the organization's internal information delivery platforms, and the capability of the available technology have resulted in more customization of products and of distribution channels, formats, and content. It is no longer unusual for the same information to be available in a variety of delivery platforms, including print, to be supplied intranet ready, to be packaged to meet individual customer requirements, and to come with varying levels of indexing and search tools.

While these developments must be beneficial in providing knowledge workers with access to the store of the world's knowledge, they also bring problems. Selection of sources is not only about cost, but quality, reliability, relevance, and usability, all of which determine value. A host of cost benefit, technical and legal questions face any organization wishing to exploit these resources. The information specialist expertise includes:

- Knowledge and a continual evaluation of external sources and services
- Supply management, including the negotiation of corporate contracts, delivery formats, and copyright, plus the monitoring of performance, costs, and use
- The tracking of new suppliers and services

Integrating Internal and External Sources

While the issues of supply and access management may differ for "internal" and "external" sources, the use of information in a KM environment requires that the two types of information are integrated with each other and the tasks to which they apply. The identification of project-based information requirements, the management of its supply in appropriate formats and in user-friendly applications, is part of the information manager's remit. This may include the provision of tailored, evaluated information sets to specific users based on a mix of internal and external inputs, and contained in spreadsheets, databases, or presentations. Or it may be the availability of the tools to facilitate the mix of information from a variety of internal and external sources, when required; or the provision of general information sets integrated to reflect corporate interests together with ease of access.

The design of tailored research services and reports to meet user-defined product needs, the management of news services, competitor intelligence support, and technical monitoring services all require an understanding of context and significance. It requires that the information that flows through the organization is managed to trigger and alert as well as inform and control.

Enabling Timely Delivery of Relevant, Usable Information

The philosophy of information management has long been "the right information to the right person at the right time."[11] Add to this "in the right format" and you have a summary of the role of the information professional. Research and analysis of market conditions in the potential areas of development; a synthesis of competitor response or developments for a specific focus; the regular supply of information and intelligence to meet a specific user requirement — all these require a mix of information skills. An understanding of how and when information is used, its significance to a process, the most relevant sources and the best means of delivery. These, together with the development of "triggers" consolidated from diverse information inputs and delivered at the most appropriate time, are key to leveraging information to build knowledge.

There are many elements to achieving this result. It may be through the provision of easy access to central information resources, in which case the provision of a help desk to help navigate the resources would be one approach. As the use of information and explicit knowledge sources becomes increasingly fundamental to all job training and coaching in research techniques, sources and internal knowledge management tools will become key. Equally, the provision of facilities to enable peer interaction in face to face exchange of information is part of the information delivery mix. While the traditional library and

---

[11]Ranganthan

information skills play an important part in facilitating desktop delivery of information and in managing mediated information services their utilization across the range of information within the organization increases the value derived from their employment. The utilization of skills of the professional information manager's staff in all aspects of knowledge management will help ensure that information, intellectual assets, and explicit knowledge resources become part of the central infrastructure of the organization, and that the building of such resources are effective and relevant.

The skills of the information professional are:

Strategic	Tactical
• Business knowledge	Information systems design
• Information management	Coding, structuring, cataloguing/indexing
• Information policy	Thesaurus development
• Information strategy	Document and records management
	Search-and-retrieval techniques
	Source expertise
	Research — selection, evaluation, analysis
	Presentation
	Communications
	Training

IM Skills

Information Audit

Information Sourcing	*Organization*
identification	Collection
evaluation	Structuring
supply management and negotiation	retrieval
Information acquisition	filtering
Research & analysis	analyzing
	design
	editing
	Abstracting
	Expert searching of electronic information
	One in five on-line searches gives problems
	Knowledge of printed and people sources
	Analysis of potential information inputs
	Selection of appropriate sources
	Filtering and analysis

The information specialist brings a fundamental skill to the knowledge environment. They are able to create an information map for the organization, build and maintain information resources, and work with colleagues to assist the utilization of that knowledge. These skills have always been available, but their value has only become apparent when integrated with true knowledge teams. Information management alone cannot change culture, win business, or influence the performance of the organization. The organization cannot, however, develop a KM environment without opening up information flows and providing access to appropriate information, nor without managing its store of explicit intellectual assets. Information management is an existing skill that has been harnessed to good effect by some CKOs and will become increasingly important as the KM environment develops (Figure 5).

Sandra Ward of Glaxo Wellcome says that information specialists are explorers, organizers, controllers, advisors. These roles complement the roles identified by Victoria Ward — cartographers, geologists, spark plugs and architects. The two sets of skills remain together for a substantial part of the knowledge management skill set. By extending this set to include learning, organizational development, and human resource skills the organization gains a KM team which can plan, design, and deliver.

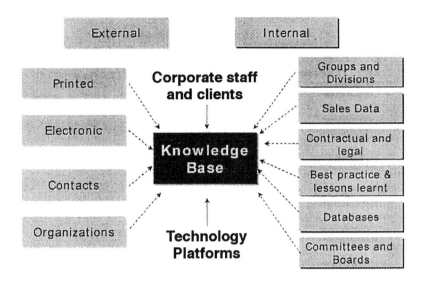

FIGURE 5   An information map in a knowledge environment.

APPENDIX A: Job Titles and Roles

KM Leaders	KM Teams
Change Program Director	Transformation Officer
Director of Innovation	Practice Knowledge Specialist
Futurist-in-Chief	Intellectual Asset Appraiser
Senior Vice President, Strategic Capabilities	Intellectual Asset Manager
Director of Intelligence	Process Competency Knowledge Management Analyst
Director of Competitive Learning	Knowledge Management Program Manager
Vice President of Learning	Principal Investigator, KM Research Program
Director of Organizational Learning	Manager, Knowledge Development Team
Chief Learning Officer	Skills Programs Manager
Leader, Learning and Change	Learning Person
Corporate Director of Intellectual Capital	Knowledge Engineer
Director, Intellectual Asset Development	Leader, Knowledge Resources
Senior Partner, Intellectual Capital	Knowledge Manager
Corporate Director, Intellectual Capital	Practice Information Managers
Director of Human Capital	Research Specialists
Director of Information Strategy and Architecture	Internal KM Consultant
Director of Worldwide Development for Information Resources	Technical Service Specialists
Vice-President, Information Center	Editors/Abstractors
Chief Knowledge Officer	Information Assistants
Director of Knowledge Management	Knowledge Assistant
Director, Knowledge Management Services	Knowledge Specialist
Director, Knowledge Center	Knowledge Officers
Knowledge Management Strategist	Marketing Tutor
Director, Knowledge Network Development	Knowledge Engineer
Director of Knowledge Management and Asset Reuse	Knowledge Editor
Competency Leader in KM	Knowledge Broker/Analyst
Knowledge Management Champion	Knowledge Shepherds
Associate Director, Knowledge Services	Knowledge Gatekeepers
Virtual Teamwork Project and Leader of Knowledge Management Team	Knowledge Navigators
Knowledge Strategies Director	Knowledge Asset Manager
Head of Knowledge Creation	Knowledge Navigators
Program Director, KM	Webmaster
Managing Director of Knowledge and Practice Development	Information Service Provider
Technical Director of Knowledge Management	Knowledge Sponsor
Head of KM Development	Knowledge Owner
Vice-President, Knowledge-based Business	Journalists

5

A Look Toward
Valuating
Human Capital

Jay Liebowitz
*University of Maryland-Baltimore
County*

Kathie Wright
George Washington University

Knowledge management has been gaining worldwide attention in recent years. Many organizations have already created a new position of Chief Knowledge Officer to help their organization better manage, share, create, secure, and distribute their knowledge-based assets. As this field matures, it is critical to develop some measures (and methodologies) for valuating knowledge assets. This chapter will survey some of the leading work in this area, describe factors affecting human capital growth and their relationships, and will then describe the authors' methodology for valuating human capital. Last, future directions necessary to improve the current state of the art in developing valuation methodologies for human capital and knowledge-based assets will be discussed.

5.1   A Survey of Measures and Methodologies for Valuating Knowledge Assets

Several leading organizations and consultants have been developing measures and techniques for valuating knowledge assets. Some of this work is highlighted below [1-8]:

Skandia

Financial Focus: Gross premium income; insurance result
Customer Focus: Satisfied customer index; customer loyalty; market share
Human Focus: Number of employees; average age; empowerment index
Process Focus: Operating expense ratio; premium income/salesperson; net claims ratio
Renewal and Development: Training expense/employee; sales-oriented operations

Skandia combines these different perspectives: financial, customer, human, process, and renewal and development (innovation). Intellectual capital is measured with 21 indicators and efficiency of use of the intellectual capital is measured with 9 indicators.

Dow Chemical

- Projected costs until expiration
- Percentage of annual intellectual asset management (IAM) costs of R&D budget
- Ratio of NPV apportioned to intellectual assets (IA) to net present cost of R&D per period
- Percentage of competitive samples analyzed that initiate business actions by purpose
- Percentage of "Business Using"
- Percentage of "Business Will Use" more than five years since priority filing
- Quantitative Value Classification as a percentage of projected costs (e.g., what percentage of portfolio costs are for defensive cases, potential license cases, key cases)
- Classifications completed

Buckman Laboratories:

- Percentage of company effectively engaged with customer (target: 80%)
- Percentage of revenues invested in knowledge transfer system
- Number of college graduates
- Sales of new products less than five years old as a percent of total sales

Karl-Erik Sveiby (author of The New Organizational Wealth: Managing and Measuring Knowledge Based Assets):

- Developed the Intangible Assets Monitor to focus on external structure, internal structure, and competence of people:

Intangible Assets Monitor (Sveiby):

EXTERNAL STRUCTURE INDICATORS:
    Indicators of growth/renewal:
      Profitability per customer
      Organic growth
      Image enhancing customers
    Indicators of efficiency:
      Satisfied customers index
      Sales per customer
      Win/loss index
    Indicators of stability:
      Proportion of big customers
      Age structure
      Devoted customers ratio
      Frequency of repeat orders

INTERNAL STRUCTURE INDICATORS:
    Indicators of growth/renewal:
      Investment in IT
      Structure-enhancing customers
    Indicators of efficiency:
      Proportion of support staff
      Values/attitudes index

Indicators of stability:
   Age of the organization
   Support staff turnover
   Rookie ratio
   Seniority

COMPETENCE INDICATORS:
   Indicators of growth/renewal:
      Number of years in the profession
      Level of education
      Training and education costs
      Marking
      Competence turnover
      Competence-enhancing customers
   Indicators of efficiency:
      Proportion of professionals
      Leverage effect
      Value added per employee
      Value added per professional
      Profit per employee
      Profit per professional
   Indicators of stability:
      Professionals turnover
      Relative pay
      Seniority

Sveiby's approach is good for assessing the total value of the knowledge assets owned by a company, but is less useful for knowledge management at the "lower level" knowledge assets.

Tobin's q:

This is the ratio between market value stock price times outstanding shares and replacement value of physical assets. This helps quantify the value of knowledge at the "global" level providing an "objective" basis.

The World Bank (Country Knowledge Assessment Study):

Creation of knowledge:
   Public expenditure on education relative to GNP
   Public expenditure on education, absolute
Assimilation of knowledge:
   Gross enrollment rate
   Secondary education
   Tertiary education
   Literacy — newspaper readership
   Adult literacy rate
   Mean years of schooling

Roger Bohn (University of California, San Diego):

Stages of knowledge:

1. Complete ignorance
2. Awareness
3. Measure
4. Control of the mean
5. Process capability

6. Process characterization
7. Know why
8. Complete knowledge

Human Resource Accounting Approaches:

1. Acquisition Costs — value human assets by accumulating all of the associated costs incurred before the time the firm can realize the benefits of the employee.
2. Records replacement value, an estimate of the costs of replacing an employee with someone of equal value
3. Discount the expected future salaries of employees to the present and report it as an asset.

Jeff Wilkins (InReference, Inc.):

Value of a knowledge asset = the sum of the cost-based value and the added value, summed over all relevant processes in which it is a resource.

Ed Mahler (Dupont/E.G. Mahler and Company):

- Net training per employee plus net R&D

Canadian Imperial Bank of Commerce:

- Human capital (the skills individuals need to meet the customer needs)
- Structural capital (information required to understand specific markets)
- Customer capital (the essential data about the bank's customer base)

George Harmon, President of Micord Corporation:

$$Iv = (At-An)-(Lt-Ln)-(Ig+If+Ir+Id+It+Is+Iu)$$

where
Iv = value of the particular information
At = the assets derived from the information at time of arrival
An = the assets if the information did not arrive
Lt = the liabilities derived from the information at time of arrival
Ln = the liabilities if the information did not arrive
Ig = the cost to generate the information
If = the cost to format the information
Ir = the cost to reformat the information
Id = the cost to duplicate the information
It = the cost to transmit or transport the information (distribute)
Is = the cost to store the information
Iu = the cost to use the information, including retrieval

Montague Institute (Compilation):

- Relative value: progress, not a quantitative target, is the ultimate goal
- Balanced scorecard: supplements traditional financial measures with customers, internal business processes, and learning/growth
- Competency models: by observing and classifying the behaviors of "successful" employees and calculating the market value of their output, it's possible to assign a dollar value to the intellectual capital they create and use in their work
- Subsystem performance
- Benchmarking

- Business worth: evaluation focuses on the cost of missing or underutilizing a business opportunity, avoiding or minimizing a threat
- Business process auditing: measures how information enhances value in a given business process
- Knowledge bank: treats capital spending as an expense (instead of an asset) and treats a portion of salaries as an asset
- Brand equity valuation: measures the economic impact of a brand (or other intangible asset) on such things as pricing power, distribution reach, ability to launch new products, etc.
- Calculated intangible value: compares a company's return on assets with a published average return on assets for the industry
- Microlending: substitutes intangible "collateral" (peer group support, training, and the personal qualities of entrepreneurs) for tangible assets
- Colorized reporting: Suggested by the Securities and Exchange Commission to supplement traditional financial statements with additional information (e.g., brand values, customer satisfaction measures, value of a trained workforce, etc.).

Nuala Beck (Canadian economist):

- Knowledge ratio: expresses the number of knowledge workers as a percentage of total employment in an industry, individual company, or organization (measures the "Corporate IQ")
- Return-on-knowledge assets: the number of knowledge workers to profit earned
- Patent-to-stock price ratio: the ratio of the number of patents divided by the price of a company's stock
- Research-to-development ratio: the ratio of research dollars spent to the development dollars spent
- Research and development-to-patent ratio: the ratio of R&D investment to number of new patents issued

NCI Research:

Step 1: Calculate average pretax earnings for the past three years.
Step 2: Go to the balance sheet and get the average year-end tangible assets for the same three years.
Step 3: Divide earnings by assets to get the return on assets.
Step 4: For the same three years, find the industry's average ROA (return on assets).
Step 5: Calculate the "excess return." Multiply the industry-average ROA by the company's average tangible assets. Subtract that from the pretax earnings in Step One.
Step 6: Calculate the three year-average income tax rate and multiply this by the excess return. Subtract the result from the excess return to get an after-tax number — the premium attributable to intangible assets.
Step 7: Calculate the net present value of the premium. You can do this by dividing the premium by an appropriate discount rate, such as the company's cost of capital. This will yield the calculated intangible value (CIV) of the organization's intangible assets. A weak or falling CIV might be a tipoff that one's investments in intangibles aren't paying off or that one spends too much on bricks and mortar. A rising CIV can help show that a business is generating the capacity to produce future wealth.

Thomas A. Stewart:

Defer a portion of salaries, treating it as an investment. To do this, calculate how much of an employee's work is devoted to current-year tasks and how much to seeding the future (training, planning, research, business development, etc.). Thus, all the salary of a clerk is expensed, and most of the pay of a new hire, who accomplished more learning than doing, would also be banked. In the lab, researchers' entire salaries are capitalized.

5.2   Factors Affecting Growth of Human Capital

If we concentrate on those "knowledge assets" associated with only "human capital," we need to first develop a list of factors that affect human capital growth before we can develop a valuation methodology for measuring human capital. The following factors contribute to human capital growth:

- Formal training of employees
- R&D expenditures of the organization
- Morale (benefits, compensation, conferences, travel, vacation time, etc.)
- Formal education (i.e., degrees) of employees
- Mentoring and on-the-job training of employees
- Research skills
- Creativity and ingenuity
- Entrepreneurship and intrapreneurship skills
- Industry competition
- Half-life of information in industry
- Demand and supply of those in the field
- Retention rates of employees (studies indicate that retention is 20% of what we hear, 40% of what we see and hear, and 75% of what we see, hear, and do)
- Formalized knowledge transfer systems (e.g., lessons learned databases or best practices guidelines) institutionalized within the organization
- Informal knowledge transfer systems (e.g., speaking often with top management, secretaries and assistants to top management, attending company events, the "grapevine")
- Interaction with customers and users
- Stimulation and motivation (e.g., challenging assignments, giving responsibility and authority to the employee [employee empowerment], etc.)
- Physical environment and ambiance (e.g., nice office, reasonable resources, etc.)
- Internal environment within the organization (e.g., reasonableness of demands by management placed on the employees, etc.)
- Short term (2-4 years) and long term (5 years or more) prospects, from the employee's perspective, of the organization's viability and growth.

Generally speaking, if these factors increase in a positive way, then human capital should expand (albeit with the possible exception of "industry competition" increasing). The converse is also true in that if these factors (discounting industry competition) take a negative turn, then human capital growth is likely to diminish or be stymied.

We can best group these 19 factors into the following categories:

Training and Education (T&E):
–Formal training of employees
–Formal education of employees
–Mentoring and on-the-job training

Skills (S):
–Research skills
–Entre- and intra- preneurship skills
–Retention rates

Outside Pressures & Environmental Impacts (OP&EI):
–Industry competition

−Half-life of information in industry
−Demand and supply of those in the field

Internal & Organizational Culture (I&OC):
−R&D expenditures of the organization
−Formalized knowledge transfer systems
−Informal knowledge transfer systems
−Interaction with customers and users
−Physical environment and ambiance
−Internal environment within the organization
−Short term and long-term goals

Psychological Impacts (PI):
−Morale
−Creativity and ingenuity
−Stimulation and motivation

Some possible relationships and correlations could be posited between these five groups of factors affecting Human Capital Growth (HCG):

$$HCG = T\&E + S + OP\&EI + I\&OC + PI$$

If T&E goes up, then S should generally increase. If T&E goes down, then S should generally decrease.

If I&OC is in a stressful state, then PI should go down. If I&OC is in a favorable state, then PI should increase.

If OP&EI increases, then it may be necessary to increase T&E and S expenditures. Also, if OP&EI increases, then it may create pressures on I&OC and thus possibly stifle PI. Alternatively, one may argue, that if OP&EI increases, then it may force employees and management to be more creative (to handle the competition) and thereby increase PI.

5.3 Human Capital: A Proposed Valuation Model

Human capital, the notion that a trained and skilled workforce provides a strategic advantage to an organization, is not new. Even before recent advances in technology pushed the economy from the Industrial Age firmly into an information-based economy, business management strategists have understood that the wealth of an organization is not comprehended solely in its working capital and physical assets. But as information-driven processes pervade all business sectors, the development of human capital — in the form of workers who not only understand technology, but who can apply it across contingencies — has rapidly increased as a proportionate component of production in relation to traditional manufacturing and service costs. Unfortunately, managers are lacking an essential tool, the ability to capture and measure the costs of developing human capital, and a methodology for allocating those costs to the work performed.

Traditional accounting models emphasize wealth creation by focusing on working capital and the physical, "tangible" assets typically used in the manufacturing environment. Intangible assets such as organizational costs, trademarks, patents have been incorporated into the balance sheet, but this list does not include human resource valuation, or other forms of intellectual capital, except as part of the catch-all "goodwill." Periodically, attempts have been made to develop a human resource valuation model, capturing costs related to acquiring and developing a skilled workforce, and to allocate these costs to the work and services performed. But without a consistent framework for external validation — without which it is impossible to measure progress within a firm, much less generalize across companies — or a cost-effective means of capturing and manipulating the necessary information inputs, these models have not generated widespread acceptance.

Recent advances in information technology have significantly improved the capacity to capture, store, and manipulate information to the extent that organizations are now able to analyze cost-related data at an increasingly sophisticated and detailed level. Given that there is renewed impetus for evaluating and maintaining information on the value of human capital (as well as other "intellectual capital") assets for decision making purposes, the next necessary step is to develop a framework that supplies an external reference for valuation and measurement.

To be useful, a proposed human capital valuation model, must provide information benefits that outweigh the preparation costs and be easily integrated with traditional accounting information. The purpose of this chapter is to review traditional accounting practice with regard to physical as well as intangible assets to determine if any of these conventions can be extended to human capital. Activity-based costing concepts are introduced as a possible methodology for driving asset recognition with a simple amortization method based on average asset service life as the means for cost allocation.

Robert de Hoog [14], one of the pioneers in knowledge management from the University of Amsterdam, also uses activity-based costing to compute the value of knowledge items. He defines the value of a knowledge item as the revenues generated by a product that can be ascribed to the knowledge item, summed over all products in the process using it as a resource.

The return on a knowledge item is the value of a knowledge item as defined above minus the costs incurred for using the knowledge item as a resource summed over all products using it. Measuring the value of and the return on a knowledge item requires measurement of the two quantities: revenues per product and cost per product.

5.4   Traditional Asset Valuation Concepts

The basic definition of an asset is that it is a cost incurred in the present that will provide an anticipated benefit by either generating a future cash inflow or by avoiding a future cash outflow. As an illustration, consider the fable of the wise and prudent purchasing agent. The agent, who is employed by a large retail gift shop, hears on the evening news that impending legislation will impose significant tariffs on the goods she regularly imports. Because she is a concerned employee, she immediately requests a purchases budget increase, and having convinced management to accept her proposal, stockpiles the affected items.

Ideally, the value of the inventory purchased should reflect the cost of those items that will be sold immediately, as well as the discounted value of the remaining inventory at its eventual replacement cost. But this is not practical for two reasons. First, the uncertainty as to the eventual replacement costs precludes a reliable valuation in the present — perhaps the legislation will not be passed, or an alternative source could be found that will minimize the cost differential. Second, the exercise of deriving the present value (assuming that one has a guarantee of not only the future replacement costs, but also an accurate estimate of the flow of inventory to meet customer demand) is in many cases greater than the informative value of the calculated cost.

Because of these issues, accounting convention uses historical cost as a surrogate. This works fine as long as the replacement cost does not fall below the original purchase price. If it does, the original purchase no longer provides a future benefit, and the excess must be written off (expensed) immediately. Similar conventions operate for long-term physical assets through the use of depreciation and amortization schedules. If technological advances render equipment as obsolete, or reduce its replacement value to less than the remaining book value, a larger proportion of the original cost may be expensed over each year of a shorter remaining service life.

The fact that asset replacement costs often increase over the historical ("book") cost does not affect the original asset valuation. The differential is simply included as an increase in realized profits over what might have otherwise occurred. Thus, the purchases made by the wise and prudent agent will simply allow the gift shop to maintain its profit margins during a period of rising import prices. If the competitors are forced to raise their retail prices because of poor purchase planning, the gift shop may ultimately realize additional profits from increased customer demand. In a related context, if the value of the asset is derived from some other attribute than the timing of the purchase, as for example, a better design that improves usage, this added value will not be realized until the asset is used to generate revenues.

Generalizing from these examples, we can derive three criteria for an asset: (1) it must provide future benefits in the form of added cash inflows or the avoidance of cash outflows; (2) given that there is uncertainty as to the timing and potential realization of future cash flows, there must be a generally accepted surrogate that has recognized monetary value; and (3) it must be owned or controlled by the organization. Previous research [9] has argued that these three criteria can be applied to human capital as an intangible asset. Again, criteria 2 poses the greatest difficulty, since there is yet no consensus on what could serve as an adequate surrogate for human capital costs.

The recordation of assets is driven by the recognition of certain generally accepted events: the creation of manufactured inventory, the purchase of equipment, the generation of capital lease agreements are all manifest as increases in physical assets. Decreases to assets are more problematic; ideally, the decrease, or expensing of an asset should correspond to the time frame in which its service potential is realized. Since in the case of long-term assets, there may be no recognizable event to drive the expensing of the asset, various depreciation and amortization schemes have been developed to apportion the value of the asset over time.

One of the arguments against capitalizing intangible assets is that of the uncertainty of the anticipated benefits and the difficulty in deriving a valuation. Nevertheless, it should be clear from the examples of physical assets described above that historical cost has proven to be an acceptable proxy for asset valuation in other cases where the uncertainty is just as great. Of the few generally recognized intangibles, goodwill is based entirely on the purchase price of an acquired business over its net asset base — with no guarantee that the new management will be able to realize the gains that were assumed as part of the negotiations before purchase.

In terms of its informative value, the inclusion of intangible asset costs within the financial statements may in itself be a sufficient symbolic representation of expected future benefits without an explicit linear relationship. Amir and Lev [10], in their study in the cellular communications industry, identified a strong correlation between the number of FCC geographic area licenses held by the cellular company and their stock returns. In this case, a nonmonetary measure — number of licenses — is regarded by the stock market as a more reliable indicator of future wealth than either the earnings levels or earnings changes over time. Assuming that there is a linear relationship between the monetary cost of the license as an asset and its geographic range, could a financial indicator serve a similar purpose?

Given that this preliminary evidence suggests that nonfinancial indicators contain information for at least one subset of decision-makers (investors), one might explore whether knowledge-intensive industries have similar indicators relating to the development of their employees. The research question, then, is: do the historical costs associated with the development of human capital serve as an adequate proxy for use in a determination of future wealth?

5.5 Intangible Asset Recognition

Thus far, efforts toward creating a taxonomy of intangibles has been on a firm by firm basis; most notably by the Skandia Insurance group, as documented by Edvinsson and Malone [4]. Of these, four unique categories emerge: human capital, customer capital, process capital and innovation capital.

The latter three categories are generally characterized as institutional knowledge assets that have been synthesized as part of the organizational structure. Human capital alone involves control without direct ownership, thereby creating additional uncertainty as to the eventual realization of benefits.

As part of their effort, Skandia developed over 100 metrics to be included in their Intellectual Capital navigational report. A few of these are included below:

Capital asset
Example cost indicators

Human

Replacement and acquisition costs, generalized training and employee development costs, percentage of outsourced personnel resources, development of cross-functional team structures, internal control and ethics

Customer

Creation and development of external relationships, brand loyalties, customer service expectations, market share

Process

IT and communications infrastructure, logistical efficiencies, administrative procedures

Innovation

Renewal and development costs, change in product development and delivery cycle, adoption of industry quality standards

The metrics, however, create a twofold problem. First, they are an amalgam of both quantitative and descriptive measures without a common basis of measurement. Because several of the metrics are descriptive, they are infused with a subjectivity that is difficult to generalize across organizations. Internally, there may not be an impediment to understanding how, for example, the average age of the employee base relates to the training dollars spent per employee, but for external reporting or benchmarking across organizations these measures are problematic.

The second, and more important issue is, when are these measures to be recognized? The current practice is akin to the original balance sheet approach; the value of the various categories are derived and presented as a "snapshot" in time, without further information as to the projected timing of the cash flows. The same issues of external validation and consistency encompass the methodology for asset recognition and expense. What should drive the capitalization of these various categories of intangibles — and more significantly — how should we expense them?

Several disparate sources have proposed utilizing the concepts derived from accounting activity based costing models [11–13] for asset recognition. Activity-based costing has recently become popular as a means of refining the allocation of indirect costs across products and services. Indirect costs, such as depreciation, indirect labor, management, support services, are combined as resource pools, each one relating through a value-added activity to a specified goal. For example, engineering and related costs may be pooled on the basis of engineering changes as the value-added activity driver for product quality. While traditional cost accounting allocates costs across a uniform measure such as direct labor or machine hours, activity-based costing allocates resources to products on a per activity basis. To extend the example above, the engineering quality costs may be allocated per engineering change request. The underlying premise is that by aligning the products with the costs of the activities that create them, there is a more sophisticated and meaningful basis of analysis with which to make decisions.

Activity-based costing generally requires an extended management commitment for implementation. It requires not only a detailed analysis of the resource and activity drivers, but also sophisticated systems support to capture and maintain the detailed transactions. The result, however, extends the set of possible accounting events that drive cost recognition and could potentially be used as a methodology for identifying both activities and related costs that add value in the form of future benefits.

5.6 Activity-Based Valuation: A Proposed Model for Human Capital

From the previous discussion, it is possible to derive the following set of theoretical assumptions for incorporation in a model for the recognition and expensing of human capital:

- That human capital meets the criteria for definition as an organization's intangible assets. Although it is impossible to derive the value of human capital in absolute terms, it is possible to use a nominal valuation mechanism that will be informative to both internal and external users.

- That in order to provide a common basis for measurement within, as well as across organizations, and to integrate the valuation of intellectual capital with other financial measures, the monetary unit is the most appropriate unit of measure.
- That because of uncertainties relating to the eventual realization of these assets, it is difficult to value them in terms of future cash inflows; however, the valuation of these assets based on the historical costs associated with generating them is a suitable surrogate, and is in accordance with accounting convention.
- That increases to the categories of intellectual capital are identifiable by means of the activities that are associated with producing future intangible benefits. These activities have costs associated with them, which can be used for valuation purposes.
- That the accounting conventions of depreciation and amortization (which are already used for accepted intangible assets) can be extended on a conceptual basis to a model for expensing intellectual capital assets.

Using training and development costs as a representative example, the first step is to identify activity drivers with a long-range human capital objective. Under current accounting rules, training costs are expensed during the same financial period in which they are incurred. The activity-based inflow model would differentiate between those costs associated with activities generating a short-term benefit (e.g., specific training to be used immediately) as opposed to long-term objectives (generalized training in which the benefits extend across the employee service life). These latter costs would not be expensed, but rather incorporated (capitalized) as inflows to the balance sheet human asset accounts, with the expectation that they will ultimately extend and enhance the service life of the employees.

With the assumption that the simplest expensing mechanism will be as informative as a more complex methodology, an amortization scheme is proposed, based on average years of service:

Period Expense = (Activity-Based Valuation Costs)/Average Service Life

Activities that decrease the average service life (e.g., downsizing, early retirement, other efforts that ultimately increase turnover rates) will decrease the average service life, which, in turn increases the expenditure of human capital in any given period. Conversely, activities that tend to extend average employee service life (development activities, employee benefits, etc.) will also increase the average service life, resulting in a smaller human capital amortization allocation in each subsequent time period.

Again, although this paper focuses on human capital, this model can be extended across all four categories of intellectual capital. In each case, the inputs to the asset valuation will need to be derived from the activities intended to provide future benefits. Some possible intangible asset valuation drivers are listed below:

Human
−Generalized employee training
−Employee benefit plan eligibility
−Employee relation participation
Customer
−Development of customer databases
−New marketing initiatives
Process
−Development of expert systems
−Logistical support systems
−Reorganization and reengineering efforts
−Business recovery planning
Innovation
−Research and development initiatives
−Specifically identified strategies

Although this is a general framework, within which each organization has considerable latitude as to what should be included as future-related costs, the external validity concerns are met by relating these back to actual incurred expenditures (including accrued obligations). The advantages of the approach lies in its simplicity and the avoidance of a propensity to be over-optimistic in projecting the potentially realizable benefits of intangible assets, by grounding the valuation in historical costs.

The resulting model is normative — in that it promotes longer employee service lives as a desirable objective of the organization. It is also predictive through the observation of changes in the intangible asset accounts. For example, a decrease in the human capital asset account balance could result from either a decrease in the expenditures on activities designed to promote future benefits received from employee services — or, a decrease in the average employee service life, resulting from a downsizing initiative or increase in employee turnover. It is not the valuation in its absolute terms that is informative, but the changes in valuation that occur from period to period that provide the predictive power.

As stated before, this model is a simple one, but it does address some important concerns. First, it avoids the impossible objective of developing an externally valid reference for valuation by using the common accounting convention of historical cost. This enables the valuation of human capital to be integrated into traditional accounting models utilizing currently accepted costing methodologies. By using the average service life of the corresponding asset objective, the normative objectives of improved decision-making capability are achieved. Finally, by examining changes in asset valuation, the fact that the valuations may not be realized in an absolute sense does not diminish the predictive capability of the model.

5.7 Future Work and Research Directions

Measuring intellectual/intangible assets will continue to be an important element for the "knowledge organizations" of tomorrow. Being able to motivate, capture, share, and distribute the knowledge and human capital in the firm is a challenge whereby those who meet this opportunity will be ahead of the competition.

According to the December 1997 study conducted by the Corporate Knowledge Center, only 15% of 202 "big" (annual sales topping $1 billion) U.S. companies accept the concept of a "knowledge organization." Of the companies that have yet to embrace the concept, 78% said they plan to. The researchers recognized just 4 of the 202 companies as knowledge organizations.

According to the Creating and Leveraging Intellectual Capital Conference sponsored by the American Society for Training and Development (ASTD) in June 1997, "very little progress has been made on the formidable task of measuring the business results of these (knowledge sharing) initiatives."

Larry Prusak of the IBM Consulting Group indicates that "you may not be able to measure knowledge, but you can measure outcomes, which are knowledge proxies." According to the ASTD conference previously mentioned, one form of measurement (or evidence) for intellectual capital that many executives said they were using is the story. IBM, for example, set up a success story program — successes in innovation, learning, and sharing as a result of IBM's programs as a measurable benefit of their work in intellectual capital. Business measures are also used at IBM like results of employee surveys via the E-mail system, time saved as a result of knowledge sharing initiatives, win/loss ratios, the numbers of competency groups that exist, and the vitality of the programs. How well a company deals with a crisis is another form of measurement.

References

1. Bontis, N., "There's a Price on Your Head: Managing Intellectual Capital Strategically," *Business Quarterly*, Vol. 60, No. 4, Summer 1996.
2. Brooking, A., *Intellectual Capital*, International Business Thomson Press, London, U.K., 1996.
3. Bukowitz, W. and G. Petrash, Visualizing, measuring, and managing knowledge, *Research-Technology-Management Journal*, Industrial Research Institute, Washington, D.C., July-August 1997.
4. Edvinsson, L., and M. Malone, *Intellectual Capital*, HarperCollins, New York, 1997.
5. Liebowitz, J. and L. Wilcox (eds.), *Knowledge Management and Its Integrative Elements*, CRC Press, Boca Raton, FL, 1997.
6. Liebowitz, J. and T. Beckman, *Knowledge Organizations: What Every Manager Should Know*, St. Lucie/CRC Press, Boca Raton, FL, 1998.
7. Montague Institute, "Measuring Intellectual Capital," Limited Edition — Newsletter on the Web, http://www.montague.com, 1997.
8. Sveiby, K., *The New Organizational Wealth: Managing and Measuring Knowledge-Based Assets*, Berrett-Koehler, San Francisco, CA, 1997.
9. Flamholtz, E., *Human Resource Accounting*, Jossey-Bass, San Francisco, 1985.
10. Amir, E. and Lev, B., Value-relevance of nonfinancial information: The wireless communications industry, *Journal of Accounting and Economics*, 1996. pp 3-29.
11. Wilkins, J. Understanding and valuing knowledge assets: Overview and method, *Expert Systems With Applications: An International Journal*, Vol. 13, 1997.
12. Roos, G. and J. Roos, Measuring your company's intellectual performance, *Long Range Planning*, Vol. 30, No. 3, 1997, pp 374-384.
13. Steffy, B. and S. Maurer, Conceptualizing and measuring the economic effectiveness of human resource activities, *Academy of Management Review*, 1988, pp 271-286.
14. van der Spek, R. and R. de Hoog, Tutorial on methods and techniques for knowledge management, 4th World Congress on Expert Systems, Mexico City, March 16, 1998.

6

Performance Measures for Knowledge Management

Carl R. Moore
Science Applications International Corporation

6.1 Introduction

Most knowledge organizations have the majority of their intellectual capital invested in skilled, knowledgeable staff. These staffs have become known as knowledge workers, and the work they perform is known as knowledge work. In fact, the rise of the knowledge worker and knowledge work has become one of the most popular subjects in print. Every journal, magazine, and newspaper regularly prints articles pointing out the paradigm shift from manufacturing to services as the economic basis of developed nations. With this paradigm shift has come a consequent change from manual labor to knowledge work as the key asset in the economic equation. A key aspect of knowledge management therefore is managing knowledge workers and controlling knowledge work. However, despite the volumes in print about knowledge work as the foundation of a new and better economy, there is a serious lack of understanding about how to measure and manage knowledge work and knowledge workers. The techniques developed to measure and manage manual labor and the assembly lines and processes of the manufacturing economy don't readily import into the service economy and its population of knowledge workers. New approaches to metrics, management, and cost estimation are needed based on an understanding of the dynamic complexities of knowledge work.

Background

In 1950 only about 17% of the American workforce was in information-based occupations. By the 1980s more than 65% of American workers worked primarily with information as programmers, teachers, clerks, secretaries, accountants, stockbrokers, managers, insurance agents, bureaucrats, lawyers, bankers, and technicians (Naisbitt, 1984). Using 1967 data, Marc Porat estimated in his Ph.D. dissertation, published by the Department of Commerce, that the information sector in the United States produced 25.1% of the national GDP and generated 43% of all corporate profits (Porat, 1977). This still growing increase in the quantity and importance of information and information workers in the modern economy has placed a burden on management, however. Phrases such as "re-engineering the corporation" and "process improvement" have become buzz words of consultants successfully selling their services to organizations who recognize they are not managing as well as they would like, but aren't sure why. In many cases this failure to achieve the quality of management desired is the result of trying to apply outmoded methods of management, with origins in traditional manufacturing industries, to modern knowledge-based work without examining or understanding the critical differences between knowledge work and the "labor" of traditional economic and management equations.

Definition of Knowledge Work

The *American Heritage Dictionary* defines knowledge as the state or fact of knowing; or familiarity, awareness, or understanding gained through experience or study. Another cognitive aspect of knowledge is the set of mental models that people develop over time to describe various properties and behaviors within a given domain that describe the real world in the context of that domain. Knowledge is also the ability to detect relationships between objects and events, and reason about those relationships to arrive at judgments, decisions, and conclusions. It is the ability to discern patterns and make sense out of a sea of information.

Knowledge is stored in books, software, organizational processes, products, systems, and documents. However, by far the most common and replete knowledge repository is the brains of humans. In fact, despite the enormous advances in information sciences and the vast holdings of libraries around the world, the knowledge most often called upon in business, government, the military, or for private purposes is that stored in the heads of people in the form of memory, skills, experience, education, imagination, and creativity.

A scan of the business pages in any newspaper or magazine shows that 50% or more of the fastest-growing companies in the United States can be termed knowledge companies in that they sell the knowledge of their employees rather than manufacture a product or deliver a service. These employees are typically highly educated and are creative problem solvers. Their job is to solve complex problems for their clients that the clients are unable or don't have the knowledge resources to solve for themselves. None of these problems are identical, so the work cannot be duplicated and resold on a mass market basis. Each job is essentially custom one-off work. Brian Arthur, an economist at Stanford University and the Santa Fe Institute, describes the shift to knowledge work as a change from selling "congealed resources" – a lot of tangible material resources held together by a little bit of knowledge – to selling "congealed knowledge" – a lot of intellectual content in a physical slipcase (Stewart, 1997).

In knowledge work, then, the primary resource is the knowledge of the employees and their creativity in applying that knowledge, singly and collectively, to custom solutions to client problems. These solutions cannot be packaged and sold over-the-counter on a mass market basis like toothpaste or television sets. Also knowledge workers don't provide a packaged set of services such as Jiffy-Lube or McDonalds. Instead, knowledge work often requires a close long-term relationship with clients in order to exchange detailed information about the client's problem domain and requirements.

Some of the key aspects differentiating knowledge work from other forms of labor include:

- The primary raw material in the knowledge work process is information.
- The primary product of the knowledge work process is information to which value has been added by the knowledge and problem-solving skills of the knowledge worker.
- Knowledge work is mentally rather than physically intensive.
- There is a heavy reliance on the knowledge and creativity of individuals, even in collaborative group settings.
- Knowledge workers are not easily interchangeable or replaceable like assembly line workers due to different levels of personal knowledge, innate problem-solving skills, creativity, personality, and style.

The Need for Knowledge Work Metrics

The need for knowledge work metrics falls into two related categories — project management (measurement and control of knowledge work projects) and quality control.

The first category, project management, requires metrics that will measure the size, effort, and duration of a knowledge work project. These metrics will form the basis both for estimating cost and schedule when developing the project budget, as well as measuring progress during development against the budget baseline. These metrics also form the basis for productivity improvement plans for the knowledge work organization over time.

The second category, quality control, requires metrics that measure the number of errors or defects that occur over the phases of a knowledge work project life cycle. These metrics form the basis of quality assurance plans for the project as well as quality process improvement plans for the knowledge organization over time.

The two categories are related because the number of errors in a knowledge work project is directly proportional to the amount of rework that must be done for the knowledge work to meet its specifications and be accepted by a client. And the more rework that must be done, the less likely will a project be finished on time and within budget. Quality and productivity are as intimately related in knowledge work as in any other industry or manufacturing process.

These needs for knowledge work metrics in project management and quality control point out that the metrics are management tools, not knowledge work tools. Consequently the most useful and effective metrics will be those designed with knowledge managers in mind, not knowledge workers. For example, a knowledge worker may be concerned with the physical size of the finished product since there may be limited media storage onto which the product must fit. Management however is concerned with a size measure that is associated with the amount of effort required to implement the knowledge work. This can be termed the development size as opposed to the physical size. Physical size of a finished knowledge work product has no direct relation to the amount of labor required to develop the product. The development size is directly proportional to the amount of labor expended. Hence the development size is both a predictor of cost as well as a ruler of progress, making it a good management metric, whereas physical size has no value as a metric to knowledge managers. In like manner, to be effective management tools, other measurement, control, and quality metrics must be designed with management in mind, not necessarily knowledge workers.

Knowledge Work Metrics

Unlike most knowledge work fields, the software development industry is a field of knowledge work in which a great deal of research has been performed in the area of performance metrics. While the research was directed at improving the productivity and quality of software development, much of the research results and conclusions can be generalized for knowledge work overall. The following sections of this chapter take advantage of this research in software development to discuss performance metrics for knowledge work.

As learned from the software industry, there are three key areas of performance that require metrics in order for managers to monitor and control the performance level of their knowledge organizations. These three key performance measures are productivity, delivery, and defect density (Symons, 1991). Productivity is the amount of effort required to produce a knowledge work project of a given size; delivery is the time required to develop a knowledge work project; and defect density, or quality, is the number of defects or errors in a knowledge work project of a given size.

Productivity

Productivity is a measure of efficiency. Efficiency in any system is defined as the ratio of output to input, or:

$$Productivity = \frac{Output}{Input} = \frac{Product\ Size}{Work\ Hours}$$

or inverting this relation provides the basic relation between effort and system size:

$$Work\ Hours = \frac{Product\ Size}{Productivity}$$

Delivery

Delivery is a measure of knowledge work project duration. It can be defined as:

$$Delivery = \frac{Product\ Size}{Elapsed\ Weeks}$$

Defect Density

Defect density is a measure of organizational effectiveness in developing the knowledge work project. It can be defined as:

$$Defect\ Density = \frac{Number\ of\ Defects}{Product\ Size}$$

From these definitions the metrics that must be collected in the workplace can be identified. These are:

Product size (S)
Work hours, or effort (E)
Elapsed weeks, or duration time (t_d)
Number of defects (Q)

The focus of this chapter is on the performance metrics of size, effort, and duration. A discussion of defects and quality measures is deferred to other works that specialize in quality management.

6.2   Knowledge Work Measurement and Metrics

As early as 1967 Peter Drucker observed that one of the great management challenges of the coming decades would be making knowledge work productive. But productivity as a traditional business measure has proven difficult to apply to knowledge work for a number of reasons.

Knowledge Work Is Intangible

In traditional labor-based industries, such as manufacturing, mining, farming, or construction, the inputs and outputs to the industrial process are tangible products. The primary inputs to the knowledge worker's job are information products, whether for software programming, systems design, legal advice, financial consulting, project management, teaching, advertising, or specialized medical care. These inputs to the knowledge work process are all intangible. Similarly the outputs of the knowledge work process are often also intangible, consisting of reports, software code, or highly specific advice or direction.

Intangible, information-based inputs and outputs to a business process cannot be directly quantified. For example, how much information is contained in an economic report that serves as an input to a financial analyst's work? Also there is no standard unit of measure for information-based inputs and outputs, since each iteration of an input and/or output differs from previous and subsequent iterations. Another difficulty in knowledge work is that the process of solving complex problems for clients is usually complex. This requires collaborative efforts from different specialists, each of whom are knowledge workers themselves. The result is there are often critical inputs to the overall knowledge work process that are not recognized as such and are therefore not accounted for in attempts to measure the knowledge process. Also there can be feedback loops in the complex knowledge work process, which can have dramatic effects on the process but are also often overlooked because they are not obvious in the complex work process. The result of this complexity, difficulty in measuring inputs and outputs, and unrecognized inputs and effects, is that traditional management measures such as efficiency and productivity fail for knowledge work because the inputs and outputs to the metric equations cannot be accurately and objectively measured.

Measuring and Forecasting Knowledge Work

The relationship between the size of a knowledge work project and its cost is of vital interest to knowledge managers. Investigations into estimation theories and the development of specific measures and tools have examined both the theoretical relationships between project size, duration, and effort, as well as specific metrics and estimation formulas based on those metrics. While the basic relationship between effort and duration has been understood for over 30 years, starting with Peter Norden's research at IBM in 1963, the metrics and estimation techniques based on these relationships have had a controversial history.

The earliest cost and schedule estimation techniques for knowledge work were based on adaptations of techniques used in the manufacturing industry. In the software industry, continuing with software as a knowledge work case study, these methods involved measuring the average productivity rate of the production of software code. The estimated size of the project was then divided by the productivity rate to give an estimate of the staff-months of effort. This effort level was then divided by the budgeted staff level to compute the time to completion for the project.

The downfall of this technique was the false assumption that the work to be done was a simple product of constant manpower multiplied by schedule time, and that these two factors could be manipulated by management at will. In fact, there are still many software and other knowledge work organizations that base their budgets and plans on the idea that staff can be increased until the computed project duration meets the required delivery date for the project. But as pundits have often pointed out, you cannot have a baby in a month by getting nine women pregnant.

Frederick Brooks, in his book *The Mythical Man Month*, pointed out that manpower and time are not interchangeable. Productivity is highly variable and subject to many different influences. A simple industry-wide productivity standard cannot be derived that requires only simple modifications to be applicable to a specific software or other type of knowledge work project.

The Rayleigh Curve

Peter Norden of IBM showed in the early 1960s that R&D projects consist of distinct overlapping phases that have a well-defined Rayleigh curve shape in relating manpower to duration (Norden, 1963). This pattern, illustrated in Figure 1, is predictable and consistent, and hence can be exploited for project planning. Also the long tails on the Rayleigh curve explain why projects tend to slip, since when only two thirds of the duration is complete 90% of the effort has been expended.

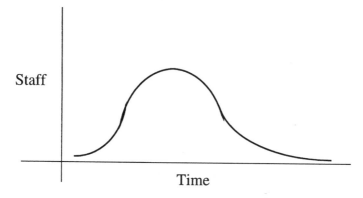

FIGURE 1   Rayleigh curve relationship between time and staff for R&D projects.

Lawrence Putnam later applied Norden's work specifically to software projects and found the same consistent Rayleigh curve pattern relating manpower to duration. Putnam also discovered distinct relationships between the size of a software project, as measured in lines of code, the nature of the development environment, and the complexity of the project in determining a software project's duration and effort. These findings were developed into the Putnam, Norden, Rayleigh (PNR) estimating formulas, which are still popular estimating techniques in the software industry.

Edmund Daly of GTE independently discovered the same type of phases and cyclic behavior between effort, duration, and size in telecommunications projects, supporting the discoveries of Norden and Putnam (Daly, 1977). Similar results were found by later researchers including Boehm (Software Economics and the COCOMO model) and Symons (Mark II Function Point Analysis). This similarity of project structure and behavior over time between the software and telecommunication industries supports the generalization of findings to knowledge work overall

PNR Formulas: Relating Time, Effort, and Size

The Putnam Norden Rayleigh (PNR) equations and estimating method grew out of research conducted by Larry Putnam beginning in the late 1970s on software projects developed at the Rome Air Development Center. Putnam's work was a follow-up to earlier work conducted by Peter Norden on research projects in which the effort and duration relationship were shown to resemble a bell curve, or Rayleigh curve. (Hence the name of the technique.) Putnam discovered that there is a high correlation between the size of a software project as measured in source lines of code (SLOC), the effort required to complete the project and the fourth power of the time required to complete. Putnam's equation for software estimation is (Putnam and Myers, 1992):

$$effort = \left[\frac{size * B^{\frac{1}{3}}}{productivity\_parameter}\right]^3 * \left(\frac{1}{time^4}\right)$$

where:
effort	= total effort expended to complete the project
size	= the finished size of the project in SLOC
B	= a special skills factor that increases with system size
productivity_parameter	= a number which represents the efficiency of the overall development environment for the project
time	= the length of time the project took to complete

For any given system the terms in the brackets are a constant. By examining over 750 systems where the terms in the brackets came out close to equal, meaning systems extremely similar in size, complexity, and efficiency of production, Putnam found six sets of similar systems. The average value of the power ratio for time was for all six sets of projects in the sample set was found to be 3.721, with a standard deviation of 0.215. Consequently the probability of the true power function for time, as it relates to effort on a software project lying between 3.5 and 4.5, was found to be 84%, making the relationship between effort and the fourth power of time an excellent approximation for estimating purposes.

The factor *B* in Putnam's model is a special skills factor. As the size of the project increases this factor also increases slowly to account for the increased need for integration, testing, quality assurance, documentation, and management skills as the size of the software volume grows. Table 1 shows the increase of B with system size.

The process productivity parameter (*PP*) is a single measure of all the factors in a given organization, and for a specific project, that affects software development productivity. It's an indirect measure of productivity, however, since it is derived after the fact from product metrics and as such incorporates all the factors that affect productivity, but does not separate them or specify a measure for each singly. As

TABLE 1 PNR Special Skills Factor Values

Size (SLOC)	B
5–15K	0.16
20K	0.18
30K	0.28
40K	0.34
50K	0.37
> 50K	0.39

such the productivity parameter is a macro measure of the total development environment. Low values are associated with poor work conditions, poor tools, weak leadership, unskilled, untrained people, ineffective methods, or a high degree of complexity. High values have the opposite attributes and a low degree of complexity.

Derivation of *PP* for an organization is made by examining historical data for completed projects. Putnam's software equation is rearranged for this type of backfire analysis as follows:

$$PP = \left(\frac{size}{\frac{effort}{B}} \right)^{\frac{1}{3}} * \left(time\right)^{\frac{4}{3}}$$

The productivity parameter can then be calculated for different categories of projects, and the appropriate parameter then selected for the new project being estimated. Quantitative Software Management Corporation (QSM) computed *PP* for all systems in their database and found the values clustered around specific discrete numbers based loosely on application type. A scale of integer numbers was then assigned as a productivity index (*PI*).

A graph of the productivity parameters from the QSM database against its integer productivity index is illustrated in Figure 2.

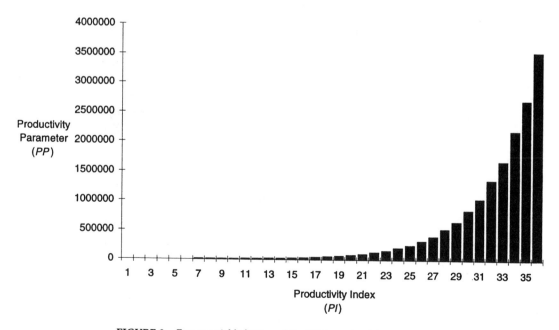

FIGURE 2 Exponential behavior of the PNR productivity parameter.

As shown, the behavior of *PP* is exponential. Consequently each improvement within an organization of one productivity index number will bring with it an exponential increase in productivity. On the other hand, conditions, such as system complexity, which cause a drop of a single productivity index number result in an exponential drop in productivity.

A second key relation in the PNR estimating method is the manpower buildup parameter (*MBP*). *MBP* is a measure of the rate at which staff can be added to a project, and consequently affects the duration in terms of schedule compression, as well as the point within a project where peak staffing requirements occur. *MBP* is defined as:

$$MBP = \frac{effort}{time^3}$$

The value of *MBP* was computed by Putnam for all systems in QSM's database. The computed values ranged from single digits into the hundreds. Putnam assigned six discrete values as indices to represent the entire range of *MBP* in an estimation problem. Each value corresponds to an integer called the Manpower Buildup Index (*MBI*). These values are shown in Table 2.

TABLE 2 PNR manpower buildup parameter

MBI	MBP	Rate of Buildup
1	7.3	Slow
2	14.7	Moderately slow
3	26.9	Moderate
4	55	Rapid
5	89	Very rapid
6	233	Extremely rapid

The *MBP* used to estimate a project provides insight to the development style of an organization. Low values reflect long, slow staff buildups. High values are often reflective of attempts to compress a project schedule by adding staff at a rapid rate. An *MBI* of 1 represents a slow buildup with no attempt at schedule compression. This may be typical in a highly complex project, such as real-time embedded systems, where problems must be solved one at a time sequentially. A value of 1 may also reflect limited resources, either in terms of people or budget. An *MBI* of 6 on the other hand comes close to a rectangular staffing curve with essentially no ramp-up. Full staffing is planned from day one, a technique termed in the industry as the "Mongolian horde" approach to project management.

An *MBI* of 6 is reflective of attempts to compress development time as much as possible. However, this compression comes with a cost. Table 3 illustrates the impact on a 30,000 line COBOL project with fixed *PI* = 11 of changing the *MBI* from 1 to 5.

TABLE 3 Impact of Schedule Compression on Cost

MBI	Development Time (months)	Effort (Staff-months)	Cost	PI
1	16	55	$458,800	11
2	14	80	$666,700	11
3	13	120	$1,000,000	11
4	12	180	$1,500,000	11
5	11	235	$1,958,000	11

As shown, reduction in duration of 31% (from 16 to 11 months) increases the cost by nearly 427%. Using the definition of *MBP* Putnam's software equation can be reorganized for estimation as follows:

$$effort = \left[\frac{size * B^{\frac{1}{3}}}{PP} \right]^3 * \left(\frac{1}{time^4} \right)$$

$$MBP = \frac{effort}{time^3}$$

$$time = \left(\frac{1}{MBP} * \left[\frac{size * B^{\frac{1}{3}}}{PP} \right]^3 \right)^{\frac{1}{7}}$$

For estimation purposes the values of B, PP, MBP, and the size of the project will be fixed, allowing solution of the equation for the estimated duration of the project (*time*). With the duration known the original form of the equation can be solved for the total effort required.

Generalizing to Other Forms of Knowledge Work

While the specific PNR equations were derived for software development, they contain a great deal of invaluable information that can be generalized for all knowledge work.

- Knowledge work projects tend to follow a Rayleigh curve as labor is expended over time. This means knowledge managers can expect staffing to ramp up to a peak between one third and one half of the project duration, and expend around 90% of the planned labor in the first two thirds of the scheduled time. This, of course, leaves knowledge managers with the challenge of actually completing the remaining one third of the schedule with just 10% of the planned labor.

- The effort, or amount of labor required to complete a knowledge work project, is proportional to the inverse of an exponent of the project duration. Consequently labor rises sharply for a given knowledge work project when attempts are made to artificially compress the schedule. Experience in many areas of knowledge management have borne out the high cost of trying to shorten project timelines. A five times increase in project cost to achieve a 30% reduction in schedule is not uncommon. And quite often the costs are incurred, but the schedule compression fails when unexpected technical difficulties are encountered.

- The key variables in estimating knowledge work project labor and schedule are the size of the project and the organization's productivity in the knowledge work being done. However, both size and productivity are extremely difficult to measure in knowledge work due to the intangible nature of the work. Also there is no universal measure that can be used in all forms of knowledge work. Software is frequently measured in terms of lines of code contained in a project. But other types of knowledge work would require size metrics, and associated productivity metrics, which reflect the technology and information being used in the knowledge work. For example, lines of code as a knowledge work project size measure is useless when considering a project to develop a new medical procedure. As will be seen in the following sections, inaccurate results, and hence costly project overruns, are frequently the result of incorrect size and productivity measures. In most cases these incorrect measures can be traced to a continuing lack of understanding of the nature of knowledge work and the dynamics of knowledge organizations.

Software as an Example of Knowledge Work

As was previously discussed, software development is an example of knowledge work that shares the same metric difficulties as other forms of knowledge work, but one in which much research and practical experience has been gained in trying to measure the work process in order to forecast and manage it.

Software development therefore makes a good case study of knowledge work in light of metrics and process measurement.

Software's Troubled History of Measurement and Cost Control

"The history of software development has been characterized by large cost overruns and very significant slippages in pre-established schedules."[1]

Software is an example of knowledge work with a severely troubled history of measurement and cost control. Software development, like other forms of knowledge work, is difficult to measure and hence difficult to manage and estimate. Also, as with other forms of knowledge work, software development is mostly intangible in its processes and progress, and hence is difficult to measure accurately or reliably using traditional work metrics of productivity and schedule.

The software industry is now past its fiftieth anniversary since the development of ENIAC in 1943 and is maturing as an industry. But software development is still risky in terms of accurately estimating cost and schedule for a project. This risk was painfully apparent to the software industry beginning a relatively short time after the introduction of general purpose computers to business and government. Software projects were far more often being delivered late and with significant cost overruns than being delivered on time and within budget. And as the size and complexity of the software projects grew, the schedule slips and cost overruns grew worse. A GAO study showed that more than 50% of the software projects examined showed large cost overruns, and more than 60% had large schedule slips. A similar study performed for the Department of Trade and Industry in the United Kingdom covered 60 companies in all industrial sectors, and 200 software projects; 55% of these projects reported cost overruns, and 66% showed schedule slips. The report's general conclusion was that software development was a high-risk activity.

The cost of software is growing each year. Barry Boehm estimated 1985 expenditures on software as approximately $11 billion for the U.S. Department of Defense, $70 billion for the U.S. overall, and $140 billion worldwide. By 1990 these estimates had increased to $20 billion, $125 billion, and $250 billion, respectively. Boehm estimated the annual rate of growth in software expenditures at about 12% (Boehm, 1987), making 1998 estimates approximately $49 billion, $309 billion, and $618 billion, respectively. These figures are likely conservative.

In 1989 Butler Cox PLC of London analyzed its database of 344 large software projects going back to 1986. The average cost overrun, in staff-months, was 37%, and the average schedule slip was 32% (Woodward, 1989). While Butler Cox's specific findings cannot be claimed to be representative of the entire software industry, they do suggest the potential magnitude of the cost and schedule overrun problem. With approximately $276 billion being spent on software in the U.S. in 1997, based on Boehm's forecasts, a 37% average cost overrun represents a potential $102 billion risk in the software industry. The need for more accurate cost and schedule estimating methods and metrics is obvious. Few organizations can long withstand consistent cost overruns of 37% and schedule slips of 32%. The recent debacles reported in the press at both the FAA and the IRS, with major multibillion dollar software projects being found to be years behind schedule and far over budget, show this problem continuing to the present day.

The primary cause of these schedule and cost overruns is inaccurate and unreliable metrics and cost estimation methods. Of the ten most serious risks in software development listed by Capers Jones in his research, the first is inaccurate metrics (Jones, 1994), and the resulting inaccuracy of software cost and duration estimates. Businesses and other organizations cannot adequately allocate resources and develop plans without accurate, reliable forecasts of costs, schedules, and resources required for software projects.

---

[1]Putnam, Lawrence H., and Myers, Ware, *Measures for Excellence, Reliable Software On Time, Within Budget,* Yourdon Press, Englewood Cliffs, NJ, 1992, page 8.

Current Software Metrics and Estimation Practices

The three most popular estimating methods in practical use today are the Constructive Cost Model (COCOMO), the Putnam Norden Rayleigh (PNR) model, and function point analysis (FPA). These models are all formulaic, derived statistically from past performance measures taken from multiple software projects. While these models have proven useful, they are also frequently woefully inadequate as business forecasting tools because their accuracy is unreliable. And as technology in software progresses, the models are increasingly less reliable, since their foundations are in the older and swiftly obsolescing technology. For example, both the COCOMO and PNR models are based on counting lines of software code. While these two estimation models are still in wide use, the technology on which the project size metric is based has changed considerably. Lines of code as a measure of software project size is rapidly become obsolete and unworkable.

Lines of Code (LOC) as a Knowledge Work Size Measure for Software Development

Line-of-code metrics are perhaps the easiest metric to measure. Lines of code can be counted manually from a source code listing, or automatically using a simple program which reads the source code files and counts the lines of code. Consequently, LOC metrics are cheaply collected and stored for future estimating use. As a result both of the ease of collection, as well as the length of time this earliest of metrics has been in use, most large databases of project sizes are based on LOC metrics.

However, LOC metrics present a number of disadvantages. A problem induced in LOC metrics by high-level languages is that language level has increased more rapidly on the functional side than on the data definition side. Consequently, the ratio of executable source statements to data definition statements in a program has been skewed in high-level languages toward the data definition side. For example, a program written in assembler might have 30% data definition statements; the same program written in COBOL might have 40% data definition statements; whereas the same program written in BASIC might have 50% data definition statements. A modern spreadsheet language might consist of up to 80% data definition statements.

A further problem with LOC metrics is the increasing degree of ambiguity in counting lines of code as the level of the language increases. Assembler is relatively unambiguous in counting the lines of code. COBOL, FORTRAN, PL/1, and other third-generation languages[2] have greater power but are more ambiguous to count in collecting objectively measured metrics for comparison. Some fourth-generation languages are still not measurable, since standards of what to count and what not to count as a LOC metric have not been universally defined. Fifth-generation graphics-based languages are not measurable by LOC methods, since "lines of code" has no meaning in the context of programming with graphic symbols.

An additional major problem with using LOC metrics for sizing software projects is the variation in line-counting methods. There is no standard way to count lines of code, resulting in wide variation from organization to organization and the inability to apply historical data across the industry. Capers Jones has identified line counting variations at both the program and project levels (Jones, 1986). At the program level variations in line counting include:

- Count only executable lines.
- Count executable lines plus data definitions.
- Count executable lines, data definitions, and comments.
- Count executable lines, data definitions, comments, and job control language.
- Count lines as physical lines on an input screen.
- Count lines as terminated by logical delimiters.

---

[2]Machine language was the first-generation language. Assembler was the second-generation language. High-level procedural languages such as FORTRAN and COBOL are considered third-generation languages. Unstructured English-like languages are considered fourth-generation languages. Emerging graphics-based languages are thought of as the fifth generation of programming language.

At the project level line counting variations include:

- Count only new lines.
- Count new lines and changed lines.
- Count new lines, changed lines, and reused lines.
- Count all delivered lines plus temporary scaffold code.
- Count all delivered lines, temporary code, and support code.

At the program level, the variations in counting lines of code result in a range as high as 5 to 1 between the most conservative and most liberal methods of counting. And since most software productivity researchers who use LOC as their underlying sizing metric don't report how they counted lines of code, there may be as much as a 500% uncertainty factor in published software productivity literature based on lines of code (Jones, 1986).

Measuring Productivity and Forecasting Cost

Current approaches to software estimation have several shortcomings that are illustrative of the difficulties in current knowledge work metric systems:

- They focus only on parts of the cost of software development.
- The models are difficult to extend to include other cost drivers or influencing factors.
- The models are difficult to understand.

The PNR method is illustrative of some of the shortcomings in current software metric and estimation techniques, and likely of shortcomings in metric and estimation techniques for knowledge work in general. PNR shows good results based on the projects from a proprietary project database, from which the formulas comprising the method were statistically derived. The two key factors in the method are the size of the software project and the productivity parameter. The productivity parameter is a macro measure of the productivity of the developing organization and is derived from a backfire analysis of previously completed projects; that is, solving the PNR formulas using known project sizes, duration, and effort expended on completed projects in the past to arrive at an average measure of productivity for the organization. However, the predictive capability for the method is based on the assumption that conditions in future projects will be the same or equivalent as those in past projects from which the method was derived and/or upon which the formula's have been calibrated for a particular organization. This is generally not true for a number of reasons, chief among them:

- Software projects, as with most knowledge work, vary in their degree of complexity.
- Staff, and the attendant skill mix, does not remain constant from one project to another.
- Tools and methods frequently change.
- New technology is introduced.
- Software projects vary in their nonfunctional requirements (reliability, performance, etc.).

The software field is a growing and rapidly evolving field. Consequently, it is rare to find a project with the same characteristics that impact productivity as an earlier project. As a result of these differences from one project to another, basing productivity projections on past performance frequently results in inaccurate cost and schedule forecasts. Despite this frequently observed inaccuracy, the backfire technique of deriving productivity is heavily used in many knowledge work fields.

The Relationship between Different Skills and Knowledge Assets in Knowledge Work

Software development, like other knowledge work, is part technology, part process, part organization, and part human factors. A true view of cost impact and forecast requires accounting for all four areas. And all four areas change dynamically, some quickly, some slowly.

The existing software estimation models mentioned focus on the project size and a productivity factor. But the productivity factor is an artifact of the estimation process, not of the software development process. Productivity, as currently used in software cost estimation and scheduling, is an amalgam of many other factors measured indirectly and after the fact. Productivity in software development is the result of the dynamic interaction between many contributing factors. Many of these factors represent skills and knowledge assets that are necessary building blocks in the knowledge work process of software development. Understanding the relationships between these various skills and knowledge assets is critical to successful measurement and management of the software development process. Using this understanding, a better means of estimating software effort, cost, and schedule can be derived using improved more precise and encompassing direct measures of contributing factors rather than the current indirect estimate of macro-productivity used currently.

Factors

All three models, PNR, COCOMO, and FPA, are in agreement that the primary driver for software cost is the size of a program. However, each looks at other driving factors in a different manner. The COCOMO model, developed by Barry Boehm and described in detail in his book *Software Engineering Economics*, lists 15 cost drivers organized in four categories (Boehm, 1981):

Product attributes

- Required software reliability (B1)[3]
- Database size (B2)
- Product complexity (B3)

Computer Attributes

- Execution time constraint (B4)
- Main storage constraint (B5)
- Virtual machine volatility (B6)
- Computer turnaround time (B7)

Personnel Attributes

- Analyst capability (B8)
- Application experience (B9)
- Programmer capability (B10)
- Virtual machine experience (B11)
- Programming language experience (B12)

Project Attributes

- Modern programming practices (B13)
- Use of software tools (B14)
- Required development schedule (B15)

Boehm further identifies 10 cost drivers that were not included in the COCOMO model (Boehm, 1981). These non-included cost drivers are:

1. Type of application (B16)
2. Language level (B17)

---

[3]The designators in parentheses following each identified factor are used for tracking the source factors through the modeling process. It allows a reader to see how and where duplication is eliminated from parallel research through combining source factors into the factors used in the developed network model.

3. Other size measures: complexity, entities, and specifications (B18)
4. Requirements volatility (B19)
5. Personnel continuity (B20)
6. Management quality (B21)
7. Customer interface quality (B22)
8. Amount of documentation (B23)
9. Hardware configuration (B24)
10. Security and privacy restrictions (B25)

While Boehm did not include these factors in the COCOMO model, he agrees with other researchers that theses factors do have an impact on both duration and cost of a software project.

Symons (1991) lists 19 general application characteristics that impact cost and duration in the Mark II FPA method:

1. Data communication (S1)
2. Distributed function (S2)
3. Performance (S3)
4. Heavily used configuration (S4)
5. Transaction rates (S5)
6. On-Line data entry (S6)
7. Design for end user efficiency (S7)
8. On-line update (S8)
9. Complexity processing (S9)
10. Usable in other applications (S10)
11. Installation ease (S11)
12. Operations ease (S12)
13. Multiple sites (S13)
14. Facilitate change (S14)
15. Requirements of other applications (S15)
16. Security, privacy, auditability (S16)
17. User training needs (S17)
18. Direct use by third parties (S18)
19. Documentation (S19)

Putnam uses the fewest cost drivers other than size. Putnam's drivers are:

- Special skills factor (B), which is a measure of the additional labor required as a result of growing team size (management, administration, financial, QA, etc.) (P1)
- Manpower build-up parameter (MBP), which is a measure of the rate at which staff can be added to a project (P2)
- Productivity parameter (PP), which is a macro measure determined by backfire analysis of all other factors which impact productivity (P3)

Jones (1986) has done extensive research in the field of software metrics and productivity. He lists 20 factors, which, based on his research, have shown quantifiable impact on software cost based on historical data:

1. The programming languages used (J1)
2. Program size (J2)
3. The experience of the programmers and design personnel (J3)
4. The novelty of the requirements (J4)
5. The complexity of the program and its data (J5)
6. The use of structured programming methods (J6)

7. Program class or the distribution method (J7)
8. Program type or the application area (J8)
9. Tools and environmental conditions (J9)
10. Enhancing existing programs and systems (J10)
11. Maintaining existing programs and systems (J11)
12. Reusing existing modules and standard designs (J12)
13. Program generators (J13)
14. Fourth-generation languages (J14)
15. Geographic separation of development locations (J15)
16. Defect potentials and removal methods (J16)
17. Documentation (J17)
18. Prototyping before main development begins (J18)
19. Project teams and organization structures (J19)
20. Morale and compensation of staff (J20)

Additionally, Jones (1986) lists 25 other significant factors that impact cost but for which there is insufficient data to produce a true quantitative assessment:

1. Schedule and resource constraints (J21)
2. Unpaid overtime (J22)
3. Staff size (J23)
4. Total enterprise size (J24)
5. Attrition during development (J25)
6. Hiring and relocation during development (J26)
7. Business systems and/or strategic planning (J27)
8. User participation in requirements and design (J28)
9. End user development (J29)
10. Information centers (J30)
11. Development centers (J31)
12. Staff training and education (J32)
13. Standards and formal development methods (J33)
14. Cancelled projects and software disasters (J34)
15. Project redirections and restarts (J35)
16. Project transfers from city to city (J36)
17. Response time and computer facilities (J37)
18. Physical facilities and office space (J38)
19. Acquiring and modifying purchased software (J39)
20. Internal politics and power struggles (J40)
21. Legal and statutory constraints (J41)
22. U.S. export license requirements (J42)
23. Enterprise policies and practices (J43)
24. Measuring productivity and quality (J44)
25. Productivity and quality improvement steps (J45)

Examination of the factors listed by the different researchers in this file shows a great deal of overlap and duplication, serving to support the findings of each researcher. Using this overlap and duplication, the factors, as identified independently by different researchers, can be categorized into four major areas that impact the cost and duration of a software project. These four areas are product, project, organizational (group human factors), and personnel (individual human factors). Within each of these major areas the related factors can be further categorized into groups, eliminating the duplication and overlap between the researchers yet accounting for all factors that impact cost and duration within a software development project.

Software Product Factors

Prd-1   High reliability requirements (B1,B6,J16)
Prd-2   Code size (B2,J2)
Prd-3   Database size (B5)
Prd-4   Product complexity (B3,B18,S9,J4,J5)
Prd-5   High performance requirements (B4,S3,S5,J37)
Prd-6   Distributed function (B16,B24,S1,S2,S13,J7,J8)
Prd-7   Large number of users (S4)
Prd-8   On-line data entry (S6,S8)
Prd-9   Requirements of other applications (S10,S14,S15,J10,J11,J12)
Prd-10   Ease of use requirements (S11,S12,S7,S18)
Prd-11   Security, privacy, auditability (B25,S16)
Prd-12   The programming languages used (B17,J1,J14)
Prd-13   Use of COTS[4] software (J39)
Prd-14   Legal and statutory constraints (J41,J42)

Organization Factors

Org-1   Modern programming practices (B13,J6,J18,J33)
Org-2   Use of software tools (B14,J9,J13)
Org-3   Strong project management (B21,J19)
Org-4   Supportive human resource polices (B20,J20,J22,J25,J26)
Org-5   Quality of physical facilities and office space (J15,J30,J31,J38)
Org-6   Supportive enterprise policies, practices, and politics (J27,J40,J43)
Org-7   Productivity and quality improvement steps (J44,J45)

Project Factors

Prj-1   Schedule constraints (B7,B15,J21)
Prj-2   User Training Needs (S17)
Prj-3   Documentation (B23,S19,J17)
Prj-4   Available labor pool (P1,P2,J23)
Prj-5   Outside management controls, redirections, and restarts (B19,J34,J35,J36)

Personnel Factors

Per-1   The experience of the programmers and design personnel (B8,B9,B10,B11,B12,J3)
Per-2   End user development (J22,J28,J29)
Per-3   Staff training and education (J32)

An examination of these factors from a knowledge management perspective shows many, if not most, of these critical contributing factors are either knowledge or problem solving skills, products with a high knowledge content, knowledge work tools, or infrastructure needed to support knowledge work.

Other Factors

The factors identified in existing techniques, as well as the models they have been drawn from, deal with estimating the effort required to complete a software project. However, the required end result of estimation is the projected cost of the project, not just the effort involved. While an estimate of effort is necessary to forecast the staffing requirements for a project, management of the project, from go/no-go decision through monitoring ongoing performance requires a cost budget in dollars. However, the assumption in the existing software estimating models is that the effort expended is a direct reflection

---

[4]COTS = Commercial off the shelf.

of the cost and can therefore be substituted as a measure of cost. But there are other factors that impact the cost of a software project beyond the effort expended. In many cases these factors impact the effort expended, so there is a feedback loop, or mechanism, at work which must be accounted for in a full software cost model. For example, when labor rates are applied to the estimated effort, a labor cost can be calculated. But the labor rates are dependent on the skill levels used in the project. Higher skill levels produce higher productivity, which in turn reduces the required effort. But higher skill levels also command higher labor rates, which increase the cost. Where is the trade-off between higher skill level and higher labor rate? This is a feedback loop common to all knowledge work, but which is frequently overlooked when projecting costs.

Consequently, there are many factors, which have not earlier been specified, that impact the cost of software. Most of these factors are management and organizational factors, or factors directly related to the economic cost of software as opposed to the effort or labor cost.

Many of the factors that have been identified in this research are also still amalgams of lower-level contributing factors. Further research and development into contributing knowledge factors in knowledge work, and in the dynamics of their interactions, will increase the granularity of the model. While this will increase the complexity of the model, it will also continue to improve the accuracy. The philosophical stance taken with this modeling approach is that the problem solving method must be as complex as necessary to accurately solve the problem, rather than being by nature easy to use. However, tools developed that implement the method should be designed to be as easy to use as possible, hiding complexity where possible and providing guides and assistance to users elsewhere. An advantage to the wide availability of inexpensive desktop computers and graphic user environments is the ability to use technology to develop just such tools.

Dynamic Interrelationships between Knowledge Factors

The PNR model is illustrative in examining interrelationships between the many knowledge factors involved in software development, since it has undergone the most extensive testing against a wide variety of projects. However, key elements of both COCOMO and function point analysis also play a part in this study.

Cause and Effect Mapping of Factors in Software Development

Figure 3 is a cause-effect network showing the relationship between the factors in the PNR model discussed in previous sections. A network mapping such as Figure 3 has several advantages over the formulaic depiction of the PNR model. The relationship between the factors that make up the model are more obvious and easily understood in the network mapping than in the algebraic formulas. Algebraic formulas rely on the ability to understand abstract mathematical relations. The network mapping, as illustrated in Figure 3, provides a means of explicitly displaying the relations, allowing an observer to use the visual and spatial powers of the mind to augment the abstract reasoning abilities.

Another advantage of the network mapping method illustrated in Figure 3 is the ability to illustrate the cause/effect relation between the factors. Each factor is depicted as a node in the network. The relation between the node is depicted as a directed link between the nodes, with a "+" or "−" symbol to indicate the type of link. The direction of the link is the trace from cause to effect. A "+" link shows a positive relation. That is, as the value of the cause node (factor) increases, the value of the effect node (factor) increases, and as the value of the cause node decreases the value of the effect node decreases. A "−" link shows a negative relation in which an increase in the value of the cause node results in a decrease in the value of the effect node. For example, it is clear in Figure 3 that an increase in the size (S) of a software project will cause an increase in the duration of the project (t_d), which in turn causes an increase in the project effort (E). It is also clear that an increase in project productivity (PP) will cause a decrease in project duration (t_d), which in turn causes a decrease in the project effort (E). These relations are both well known as well as intuitive to those experienced with software development projects. Yet even to the experienced practitioner the relation between factors illustrated in Figure 3 is clearer and more intuitively understood than the same relation described in the formula:

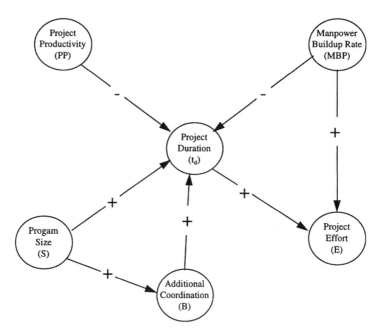

FIGURE 3   Cause–effect mapping of basic PNR model.

$$time = \left(\frac{1}{MBP} * \left[\frac{size * B^{\frac{1}{3}}}{PP} \right]^3 \right)^{\frac{1}{7}}$$

Adding Additional Factors to the Network Map

As was shown in the previous section, there are many factors that affect the size and duration of a software project and are not accounted for in current algebraic estimation formulae.

Figure 3 is a network mapping of the basic PNR model. However, as has been noted this model is incomplete in that many factors are not represented, or have been lumped together, or hidden, in a macro-parameter such as PP. A key factor in any software estimation is the labor rates applied and the relation of the labor rates to effort in determining the actual cost of the software. In fact, it is the dollar cost of the software that is actually of most interest to decision makers, not just the effort estimate. Adding the labor rate factor to the model is illustrated in Figure 4.

A primary determinant of the prevailing labor rate in a given software project is the average skill level of the personnel assigned to the project. The higher the skill level employed, the higher the labor rate. But high skill levels also increase the overall productivity of the project staff, which in turn decreases the duration and effort of the project. In a similar manner the procurement of productivity enhancing tools for a project increases the productivity of the staff, thereby reducing the duration and effort required, but directly adds to the dollar cost of the project. These effects are illustrated in Figure 5.

Some factors do not cause a definite effect on other factors, but do influence them. An example of an influencing effect rather than a causal effect is program complexity. As illustrated in Figure 6, program complexity directly causes a decrease in project productivity, since more labor must be expended to solve the more difficult problem set for the same amount of code or functionality. It can also be argued that increased program complexity tends to drive up the project skill levels required, which in turn drives up the labor rates and cost. But this effect tends only to influence the project skills rather than drive them. Thus a knowledge work estimating model must be able to account for both causal effects of one factor on another as well as the less distinct and less easily measured influencing effect.

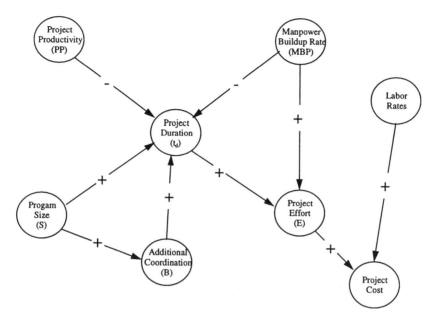

FIGURE 4 Cause–effect mapping of basic PNR model with labor rates and dollar cost added.

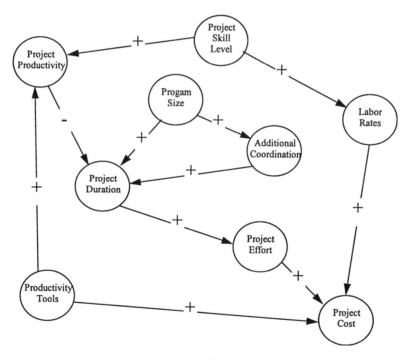

FIGURE 5 Dual impact of skill level and productivity tools.

Figure 7 shows the influencing effect labor rates have on staff resignation rates and the rate at which a company can hire new personnel. Low labor rates, while directly holding down the cost of a project tend to increase resignation rates and decrease the rate at which new personnel are hired. These factors in turn directly cause a decrease in available staff, which in turn limits the rate at which personnel can be added to a project. As can be seen, the effects of a single factor tend to ripple through the development process, often with unwanted impact on duration and/or cost of the project. This is a key reason why a

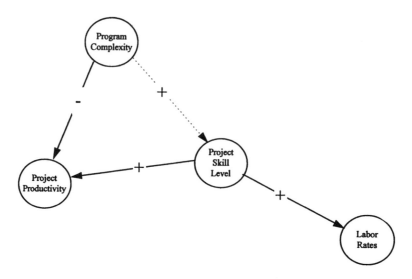

FIGURE 6   The effect of program complexity on productivity and skill levels.

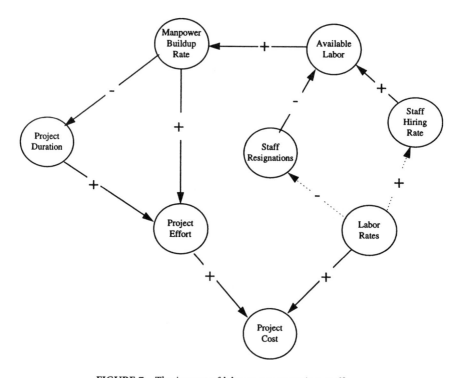

FIGURE 7   The impact of labor rates on project staffing.

knowledge work management model must account for all factors in order to avoid cost and schedule overruns due to unforeseen impacts of management decisions and policies. A network mapping of the process, such as shown in Figure 7, clearly shows the impact of project management on software productivity and ultimately software cost and duration.

Another effect that must be included in an accurate estimating model is the impact of cycles within the development process. Figure 8 illustrates a potential cycle of cause and effect. As shown, as the overall cost of a software project rises, there is a rise in management concern and consequent direct involvement in the development process. This involvement could take the form of more detailed reporting, tighter

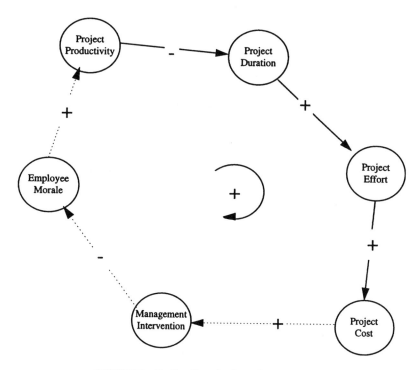

FIGURE 8 Cyclic effects in the software process.

controls, questioning of design decisions, etc. The effect of increasing management oversight is to increase the stress felt by employees, which in turn results in a decrease in employee morale. Lowered morale in turn impacts productivity, which increases the duration and effort, which in turn increase the cost of the project. The cycle then repeats. This is an example of a potential vicious cycle with a negative impact on the cost and schedule of the software project. However, efforts through total quality management (TQM) and quality and productivity enhancement (QPE) programs to reverse the impacts can turn the vicious cycle into a virtuous cycle, with management decisions increasing employee morale, which in turn increases productivity and decrease costs. Because knowledge work especially is dependent on highly skilled staff, factors that impact the effectiveness and continued availability of that professional staff must be clearly accounted for in management models. A network model that depicts these cycles clearly, and allows for their impact within the cost estimation and development process, makes a powerful tool when trying to forecast the potential improvements a given TQM or QPE initiative might have against the costs of implementing such a program.

There are many more factors that either directly or indirectly impact knowledge work activities such as software development. Figure 9 depicts a more complete network model for software development. This figure is illustrative of the complexity underlying knowledge work and the many different skill sets and knowledge types that must be considered when managing a knowledge work process such as software development.

As shown, a number of additional factors have been identified in the course of developing the network model. These additional factors are:

1. Project cost
2. Labor rates
3. Projected inflation rate
4. New staff hiring rate
5. Existing staff departure rate
6. COTS integration code

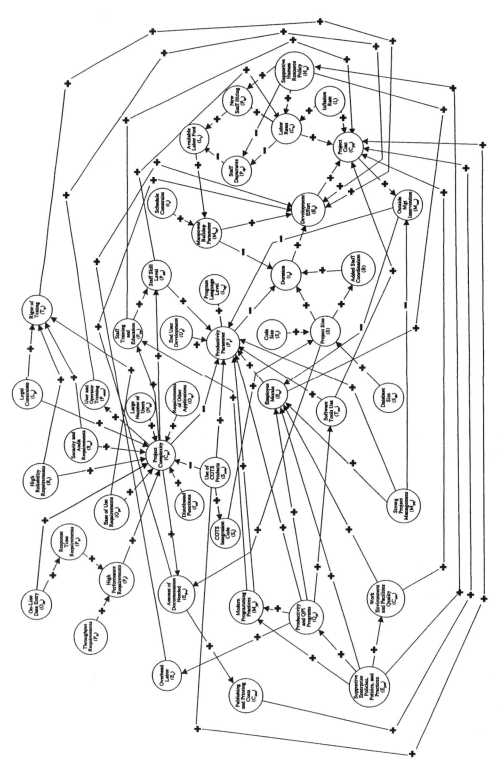

FIGURE 9   Software development networked factors model.

7. Documentation publishing and printing costs
8. Rigor of testing required
9. Overhead labor
10. Throughput requirements
11. Response time requirements
12. Employee morale

Most of these factors are intermediate factors rather than direct inputs or direct outputs. Project cost is a direct output, and is in fact the desired end result of the model. The rest are a combination of input from a model user and algorithmic computation to produce an intermediate result. However, these intermediate results are as important as end results in monitoring the software development and management process since they identify points of metric collection.

A detailed model of this type points out glaringly the lack of metrics being collected to adequately measure and monitor the business of developing software specifically and knowledge work in general. Metrics collection tends to focus on the size of projects developed, either in lines of code or function points, and the number of labor hours expended. However, these metrics represent only two of 47 potential monitoring points.

This lack of measurement is a primary reason why current QPE initiatives often go astray. They are based on only a portion of potential insight into the software organization and process. Consequently, results, for good or ill, which an QPE initiative produces, may not be measured until it is reflected, late, in another measure going off scale.

Metrics for Knowledge Work

Depicting the relationships between factors is an important step in understanding knowledge work processes, but to develop this approach into a practical tool for forecasting cost and schedule as a means of quantifying the links, or relations, between nodes as reflected in the PNR and other estimating equations must be developed.

Quantification

Proceeding from concept to quantification is a primary requirement for any estimation model. The network mapping used in the previous section conceptually makes very clear the relationships between the many different factors that affect the cost and duration of a software project. The same techniques would also work with other forms of knowledge work. The network model is also easily extended by adding new nodes as additional factors are identified and drawing in the cause–effect links. However, the next step must be to define the quantitative relation between the nodes in order to produce a useful predictive model. An advantage to the network mapping concept is the ability to break the quantification into smaller portions and analyze each node in turn. Each node can use one or more different approaches to quantification, from algebraic formulas for those nodes for which a great deal of knowledge is had, to estimated constant values for those nodes for which little experimental knowledge is available and estimators must rely for the time being on estimates from experience (the infamous but ubiquitous SWAG[5] technique). Later, as more knowledge is gained, less precise quantification techniques can be replaced with more accurate formulas.

The network mapping allows each node to be analyzed with respect only to the other nodes to which it is linked in the mapping. The value of each node is a function of the values of its input or cause nodes. For example, the development duration (td) is a function of the productivity parameter (Pp), the project size (S), the special skills factor (B), and the manpower buildup parameter (Mbp), all of which are nodes linked as input or cause nodes to the duration node in the network map. This relation can be expressed symbolically as:

---

[5]SWAG = Scientific Wild A_s Guess.

$$td = f(Pp,S,B,Mbp)$$

The duration node is in turn a cause node for the effort node. The value of the effort node can be expressed as a function of duration (td), the manpower buildup parameter (Mbp), the labor required for documentation (Edoc), and overhead labor (Eo). This relation can be expressed symbolically as:

$$Ed = f(td,Mbp,Edoc,Eo)$$

Relations for all other nodes in the network map can be developed in the same manner by examining the map and associating each node with the input or cause nodes to which it is linked. Table 4 lists the functional relation statements for all nodes as shown in the complete network model of Figure 9. As seen in Table 4, and as is observable from the network map, some nodes have no cause nodes and consequently have no functional relationship with other nodes. These nodes have outputs only to other nodes, and consequently form the inputs to the model. All other nodes that have both inputs and outputs are result nodes from which estimates and measurements can be taken, and for which comparative metrics should be collected to calibrate the model to a given environment as well as monitor activity and progress in the software development process and the software organization. The same process can be applied to other forms of knowledge work to identify what metrics should be considered to give management full insight to the knowledge work process being managed.

With the functional relationships for each node defined from the network mapping, each node can be analyzed to determine the best approach to computing a value for that node based on the defined inputs and relations. Three methods can be used on the identified qualitative relation definitions:

1. Known quantitative relations from published research and existing models
2. Statistical and experimental research in software metrics
3. Estimation based on empirical observations or experience

For software development, known quantitative relations from published research is available primarily from the PNR and COCOMO estimating techniques. Both methods are based on statistical analysis of project databases and provide formulas that relate the size of a project and other parameters to the duration and effort. These formulas have been shown through practice to have significant value in constrained circumstances. Therefore, they can be used in those nodes in which the defined functional relationships between nodes match the constraints defined for the formulas. Other types of knowledge work also likely will have an existing corpus of research from which an initial set of quantitative relations can be drawn.

The network mapping shown in Figure 9 started with the formulas from the PNR method, so these formulas can be used to quantitatively define those nodes that match the constraints of the PNR formulas. For example, the duration node, defined from the network map as a function:

$$td = f(Pp,S,B,Mbp)$$

can be quantitatively defined using the established PNR formula:

$$t_d = \left(\frac{1}{M_{bp}} * \left[\frac{S * B^{\frac{1}{3}}}{P_p} \right]^3 \right)^{\frac{1}{7}}$$

Similarly, the special coordination skills factor (B), shown earlier in table form in Table 1, can be quantitatively defined from PNR as an algorithm based on size (S), as follows:

Table 4 Functional Relationships between Network Nodes

Factor Name	Symbol	Function of:
Product Factors		
High reliability requirements	R_l	Input
Code size	S_c	Input
Database size	S_{db}	Input
Product complexity	C_{px}	$f(S_{cots}, f_{nd}, P_f, O_{ps}, O_{de}, R_l, S_{ec}, L_{gl}, N_{us}, O_{ap})$
High-performance requirements	P_f	Input
Distributed function	f_{nd}	Input
Large number of users	N_{us}	Input
On-line data entry	O_{de}	Input
Requirements of other applications	O_{ap}	Input
Ease of use requirements	O_{ps}	Input
Security, privacy, auditability	S_{ec}	Input
The level of the programming languages used	L_{ng}	Input
Use of COTS software	S_{cots}	Input
Legal and statutory constraints	L_{gl}	Input
Development Organization Factors		
Modern programming practices	M_{pp}	$f(Q_{pi}, E_{pol})$
Use of software tools and supporting infrastructure	T_{sw}	$f(Q_{pi})$
Strong project teams and organization structures	M_{prj}	Input
Supportive human resource polices	H_{mn}	Input
Quality of work environment, physical facilities, and office space	C_{env}	$f(E_{pol})$
Supportive enterprise policies, practices, and politics	E_{pol}	Input
Productivity and quality improvement steps	Q_{pi}	$f(E_{pol})$
Project Factors		
Schedule constraints	t_s	Input
User training needs	P_{ustrn}	$f(C_{px})$
Documentation	E_{doc}	$f(C_{px}, S)$
Available labor pool	L_a	$f(P_{hr}, P_{dp})$
Outside management controls, redirections and restarts	M_{out}	$f(C_{prj}, M_{prj}, E_{pol})$
Development Personnel Factors		
The experience/skill level of the programmers and design personnel	P_{skl}	$f(P_{trn}, C_{px})$
Enduser development	U_d	Input
Staff training and education	P_{trn}	$f(M_{pp}, C_{px})$
Standard PNR Factors		
Duration	t_d	$f(P_p, S, B, M_{bp})$
Project size	S	$f(S_c, S_{db}, S_i)$
Special coordination skills factor	B	$f(S)$
Productivity parameter	P_p	$f(T_{sw}, M_{prj}, M_{out}, E_m, S_{cots}, Q_{pi}, C_{px}, U_d, P_{skl}, L_{ng})$
Manpower buildup factor	M_{bp}	$f(t_s, L_a)$
Development effort	E_d	$f(t_d, M_{bp}, E_{doc}, E_o)$
Additional Factors Identified in the Network Model Process		
Project cost	C_{prj}	$f(E_d, P_{trn}, C_{lr}, I_r, C_{doc}, S_{cots}, T_{sw}, C_{env})$
Labor rates	C_{lr}	$f(I_r, P_{skl}, H_{mn})$
Projected inflation rate	I_r	Input
New staff hiring rate	P_{hr}	$f(H_{mn}, C_{lr})$
Existing staff departure rate	P_{dp}	$f(H_{mn}, C_{lr})$
COTS integration code	S_i	$f(S_{cots})$
Documentation publishing and printing costs	C_{doc}	$f(E_{doc})$
Rigor of testing required	T_{st}	$f(R_l, S_{ec}, L_{gl})$
Overhead labor	E_o	$f(Q_{pi})$
Throughput requirements	P_{Ft}	Input
Response time requirements	P_{Fr}	$f(O_{de})$
Employee morale	E_m	$f(Q_{pi}, E_{pol}, C_{env}, H_{mn}, M_{prj}, M_{out})$

```
Select SIZE;
   Case SIZE <= 15000;
      B = 0.16;
   Case SIZE > 15000 and SIZE <= 20000;
      B = 0.18;
   Case SIZE > 20000 and SIZE <= 30000;
      B = 0.28;
   Case SIZE > 30000 and SIZE <= 40000;
      B = 0.34;
   Case SIZE > 40000 and SIZE <= 50000;
      B = 0.37;
   Case SIZE > 50000;
      B = 0.39;
```

In similar manner the COCOMO model provides formulas and algorithms for relating the productivity parameter to a number of its input nodes. By defining an algorithm based on past research for each node, the productivity parameter could be quantitatively defined as a geometric series of values of the contributing nodes.

However, not all of the nodes that contribute to productivity have been quantitatively defined in past research. Many of the nodes have no established formulas from past published research, but are quantifiable with the appropriate statistical research. For example, the amount of documentation produced as a function of project size and complexity could be determined based on analysis of projects of varying size and complexity. However, while size metrics are routinely collected in software organizations, complexity and documentation metrics are not. This is a shortcoming throughout the software industry and the rest of the knowledge work field. Metrics research has focused on only the most obvious contributors to software duration and cost. Many contributing parameters have large impacts on the duration and cost of a software project, but metrics have never been collected for them because they have never been identified with software development.

However, one of the values of the network mapping model is to make clear the relationship between the many contributing parameters in knowledge work cost, and pointing out the importance of collecting metrics in all applicable areas. The end result of this dearth of metrics collection is we understand far less about a particular knowledge work process and knowledge organization than we otherwise might. This argument can be made for other areas of knowledge work. Software development is not unique in this respect.

In some cases the nodes identified in the network mapping model are not readily quantifiable. For example, employee morale is an important qualitative contributor to software productivity, and experience and intuition say it plays a key role. However, there is no identifiable way to measure employee morale quantitatively. Other factors are similarly difficult to measure. Hence even if an adequately thorough metrics program was in place, there would be no way to capture metrics for these nodes. One potential solution is the use of reasoning techniques developed in the artificial intelligence field for dealing with uncertain information. Heuristic algorithms, fuzzy logic, and rough set theory all are candidates for use in quantifying nodes that are not readily measured directly. These techniques can be calibrated using a backfire technique with other quantifiable nodes held constant with known measured values. At some future date, once metrics have progressed, these fuzzy techniques can be replaced with newly developed measures for these areas.

Conclusions

Software development is a strong example of knowledge work in the modern economy. Difficulties experienced by software managers are often reflections of similar difficulties experienced by managers in other knowledge work fields. One area of common difficulty in knowledge work is cost and schedule estimation. Existing estimation methods have proven utility based on industry experience. Some, such as the PNR formulas and COCOMO, for software, have wide acceptance and use. However, these methods have serious shortcomings, and often produce estimates that turn out be quite wide of the actual budget and schedule experienced on a software project. This is particularly true in organizations without effective metrics programs. However, even in mature software organizations with extensive experience in metrics collection and analysis, existing estimation models often produce results that vary from actual costs by 40%.

The reason for these wide variances is that the existing models account for only a portion of the factors that impact the cost and duration of a knowledge work project. In fact, most software models produce estimates only of the labor involved in the production of code, and have no capability to account for the many elements of cost outside this factor. Additionally these methods do not take into account the many impacts on productivity. Most factors are simply encased in an overall productivity factor that serves as a macro measure of everything in the organization that impacts development productivity. As a result, the organization has no real insight into the details of the knowledge work process, or into the workings of the organization.

A more robust approach to knowledge work estimation is to identify each individual factor in the work process, or which has an impact or influence on the process, and map the cause and effect relations between these factors. By presenting the process model as a directed graph or network, the model can be easily extended as new factors are identified, relations between factors better understood, or identified factors broken down further into finer granularity. This mapping provides significantly improved insight into the estimation, development, and management process, and clearly illustrates the many interdependencies between product, process, personnel, and organization.

This approach also provides a foundation for deriving quantitative values for these factors by allowing each node to be considered independently of the other nodes to which there is no direct link. Functional relationships are easily described from the network mapping. The relationships can then be analyzed either through past research, statistical analysis, or through estimation techniques, to provide a means to assign quantitative values to each node. The nodes can then be summed to produce an accurate estimate of project duration and effort.

Each node also defines a metric that should be collected in the course of development in order to calibrate the model for future use as well as to measure progress during the project development. These measures can then be used to estimate the effect of proposed TQM and QPE plans prior to implementation, as well as estimate the cost of implementation. A detailed process and management model such as this can significantly improve the impact of TQM programs by providing accurate estimates of the cost effectiveness of proposed programs. For example, a TQM program designed to improve productivity by 20%, but which increases costs 40%, could be seen ahead of time to not be cost effective. On the other hand, the metrics for each node in the network can be examined during the progress of software projects to identify those that are most adding to cost and/or schedule, with TQM programs designed to effectively improve those factors at minimum cost.

While this improved approach to measuring and estimating software, and by extension other forms of knowledge work, holds much promise as a technique for estimating costs and duration for software, there is still much work to be done. The process of modeling the process and organization has been developed here, but the process of quantifying each node is still incomplete. Data for quantitative studies for each of the factors in the model is scarce, since most organizations do not collect more than size and labor metrics, although much insight could be gained from detailed analysis of accounting and time records. Unfortunately, most of this data is proprietary and confidential, and consequently not available for public study. Efforts will continue, however, to devise an approach to compute values for each of the nodes in the model. These efforts must also include a means of keeping node quantification up to date with changes to the network as the model evolves, as well as with changing technology in the industry.

Areas for Future Research

There are two main areas for continued future research in developing this approach to measuring knowledge work. The first, and perhaps the most important from the aspect of industrial performance and support for future modeling, is improved metrics collection. Most of the factors identified in this paper have no statistical background available. Hence quantification of the factors is difficult at best, and frequently can be done only by empirical estimates. Reasoning techniques for uncertain information provide some assistance, but without metrics collected from experience there is no good way to calibrate the reasoning rules. Consequently, much improved metrics collection techniques that provide the needed information without injecting an additional burden on the knowledge work processes are in great demand. In some cases the definition of what metrics to collect must first be researched. For example, how can employee morale be quantified such that a meaningful metric can be collected in the workplace?

The second area for continued research is in defining formulas and/or quantification algorithms for the different nodes. As metrics come available a great deal of statistical research will need to be done to produce a means of calculating accurate, meaningful values for each of the factors identified in the network. Further, these formulas need to be extensible themselves in order to facilitate changes to the model over time. In this way the model can continue to evolve as understanding of the knowledge work processes continues to improve, as well as the technology that drives the process continues to evolve.

References

Boehm, Barry, *Improving Software Productivity*, Computer, September 1987.

Boehm, Barry, *Software Engineering Economics*, Prentice-Hall, Englewood Cliffs, NJ, 1981.

Boehm, Barry, and DeMarco, Tom, *Software Risk Management*, IEEE-Software. v. 14 May/June 1997 pp. 17-89.

Comptroller General, General Accounting Office, Report to the Congress of the United States, *Contracting for Computer Software Development — Serious Problems Require Management Attention to Avoid Wasting Additional Millions*, November 9, 1979.

Daly, Edmund B., *Management of Software Development*, IEEE Transactions on Software Engineering, May 1977, Vol. SE-3.

Garmus, David, and Herron, David, *Measuring the Software Process : A Practical Guide to Functional Measurements*, Yourdon Press, Prentice-Hall, Englewood Cliffs, NJ, 1996.

Jones, Capers, and Jones, Capers T., *Applied Software Measurement : Assuring Productivity and Quality*, Second Edition, McGraw-Hill, New York, NY, 1997.

Jones, Capers, *Programming Productivity*, McGraw-Hill, New York, NY, 1986.

Jones, Capers, *Viewpoint: 1998 — A Year of Resource Conflicts and Personnel Shortage*, IEEE-Spectrum. v. 35 Jan. 1998 p. 51.

Kosko, Bart, *Fuzzy Thinking, The New Science of Fuzzy Logic*, Hyperion Books, New York, NY, 1993.

Naisbitt, John, *Megatrends — Ten New Directions Transforming Our Lives*, Warner Books, New York, NY, 1984.

Norden, Peter V., *Useful Tools for Project Management*, Operations Research in Research and Development, edited by B. V. Dean, John Wiley and Sons, 1963.

Putnam, Lawrence H., and Myers, Ware, *How Solved Is the Cost Estimation Problem?*, IEEE-Software. v. 14 Nov./Dec. 1997 pp. 105-107.

Putnam, Lawrence H., and Myers, Ware, *Measures for Excellence, Reliable Software on Time, within Budget*, Yourdon Press, Englewood Cliffs, NJ, 1992.

Study for the United Kingdom Department of Trade and Industry, PA Computers and Telecommunications, London, 1984.

Stewart, Thomas A., *Intellectual Capital – The New Wealth of Organizations*, Doubleday, New York, NY, 1997.

Symons, Charles R., *Software Sizing and Estimating, Mk II FPA*, John Wiley and Sons, Chichester, England, 1991.

Ware, Myers, *Measuring Assets in the Information Age*, IEEE-Software. v. 14 Jan./Feb. 1997 pp. 89-90.

Ware, Myers, *Why Software Developers Refuse to Improve*, Computer. v. 31 no. 4 Apr. 1998 p. 112.

Woodward, Chris, *Trends in Software Development Among PEP (Productivity Enhancement Program) Members*, PEP Paper 12, Butler Cox PLC, London, England, December 1989.

Section III
Knowledge Management:
Some Elements

7

Knowledge Selection: Concepts, Issues, and Technologies

Clyde W. Holsapple
University of Kentucky

K. D. Joshi
Washington State University

In order for the knowledge to be used it must be selected out of various masses. Knowledge that cannot be selected is knowledge lost in vain.

V. Busch

7.1 Introduction

The view that knowledge is a valuable organizational resource has become widely recognized and accepted in the business community. One consequence is the increase in organizations' efforts to deliberately manage knowledge in a systematic manner. The conduct of knowledge management (KM) in an organization involves the exercise of participants' skills in executing knowledge manipulation activities that operate on the organization's knowledge resources. The conduct of KM is influenced by a variety of factors that can facilitate or impede its effectiveness. Four basic knowledge manipulation activities that appear to be common across diverse organizations are: **acquiring**, **selecting**, **using**, and **internalizing** knowledge (Holsapple and Joshi, 1998). This chapter focuses on knowledge selection activity and discusses its relationships with other knowledge manipulation activities, knowledge resources, and KM influences. Also, technologies for knowledge selection are surveyed.

Within the realm of knowledge management, the role of knowledge selection is extremely important. An organization may have vast knowledge resources, but it is of little significance if the knowledge relevant to some need is not easily and readily available at the desired time, in the desired place, and in a usable form. Managers "… fear the knowledge in their organization is going to waste simply because hardly anyone knows it exists" (Hibbard, 1997). A Reuters survey of 1300 professionals and managers revealed that 38% waste a "substantial" amount of time trying to "locate" the right information, 47% say "collection" of material distracts them from their main jobs, 25% require an "enormous amount" of information, 65% require a "lot" of information, 31% receive a "huge amount" of unsolicited information, 49% feel they are unable to handle volumes of information received, and 94% do not believe that situation will improve (Excalibur® Technologies, 1997). In one way or another, all of these findings are concerned with knowledge selection. Steve Kaye, president of KnowledgeX, contends that "The average business person spends 50% of their time in looking for information, 30% trying to make that information look pretty, and only 10 to 20% actually reading and analyzing data" (McCune, 1998).

In coping with knowledge selection issues in an organization, it can be helpful to have a thorough understanding of the nature of knowledge selection, a recognition of obstacles to effective knowledge selection, and familiarity with technologies available to support this activity. This chapter addresses these topics. Appreciating the nature of knowledge selection involves comprehending its role within the conduct of KM: What organizational participants perform knowledge selection? What tasks are involved with knowledge selection? What relationships does knowledge selection have with other knowledge manipulation activities and with various knowledge resources? What factors facilitate and constrain the performance of knowledge selection?

There exist a variety of obstacles that can impede knowledge selection or that may be converted into opportunities by effective knowledge selection. Some of these are the volume of knowledge available, how adequately to internalize knowledge for future use, and the diversity of knowledge resources and their attributes. The extent of knowledge available is increasing rapidly. For instance, Ford's knowledge base has the equivalent of more than 30,000 paper pages and selecting relevant knowledge efficiently is an extremely critical function (O'Leary, 1998). When knowledge is acquired by or generated in an organization, that piece of knowledge needs to be internalized into an organization's knowledge resources in an appropriate manner in order to make future selection possible or effective. Knowledge is not monolithic. An organization's knowledge resources can include employee knowledge, knowledge in computing systems, artifacts (e.g., manuals, videos, paper documents), infrastructure, and culture. Each knowledge resource can be characterized in terms of a variety of attributes such as degree of importance, location, mode (tacit vs. explicit), type (descriptive vs. procedural vs. reasoning), and age. The appropriateness of a methodology for selecting knowledge will depend on the knowledge resource involved and its attributes. Thus, the diversity of knowledge resources poses a challenge for effective execution of the selection activity.

There is a growing array of technology that aims to perform or support knowledge selection. Determining what technology can help satisfy an organization's knowledge selection needs can be a challenging task for today's managers. Understanding the functionalities offered by knowledge selection technologies can be helpful in identifying products that meet an organization's knowledge management needs.

This chapter's treatment of knowledge selection begins with a brief conceptualization of knowledge management. This provides a context for the subsequent examination of the nature of knowledge selection in terms of its subactivities and their functionalities. We then identify and discuss a variety of issues that deserve the attention of those who manage or research knowledge selection. Finally, we survey commercially available knowledge selection technologies.

7.2   Knowledge Selection — A Knowledge Manipulation Activity

The conduct of knowledge management involves various patterns of knowledge manipulation activities operating on knowledge resources. Knowledge management conduct occurs under the guidance and constraints of various influences. In order to fully understand the knowledge selecting activity in context, it is helpful to have an overview of its relationships with other knowledge manipulation activities, the nature of its interactions with knowledge resources, and the kinds of influences on it. This overview is based on a KM framework created through an initial synthesis of concepts from the KM literature and subsequent refinements via a Delphi process (Holsapple and Joshi, 1998). Following this contextual overview, we examine what can happen within an instance of knowledge selection.

An Overview of Knowledge Management

An organization has a variety of knowledge resources on which knowledge manipulation activities can operate. These knowledge resources can be classified broadly as content resources and schematic resources. A main distinction between the two is that the existence of content resources does not depend on the existence of the organization, whereas a schematic resource has no existence aside from that of the organization. Content resources include participants' knowledge and artifacts. An organization's participants include employees, customers, suppliers, and computer systems with the ability to represent and process knowledge. A knowledge artifact is an object that conveys or holds usable representations of knowledge (e.g., file cabinets contents, memos, videos, manuals, patents, products), but has no ability to process those representations. Schematic resources include infrastructure, culture, purpose, and strategy. By infrastructure, we mean formal roles that an organization's participants can play, relationships (e.g., authority, communication relationships) among those roles, and regulations (i.e., procedures and rules governing roles, relationships, and their use by participants) (Holsapple and Luo, 1996). Schematic resources are represented and conveyed in the form of an organization's behaviors. Our examination of knowledge selection is primarily with respect to manipulating content resources.

Factors that influence the conduct of KM can be classified as managerial influences, resource influences, and environmental influences. Managerial influences are concerned with administering the conduct of KM in an organization. They include four kinds of factors: leading, coordinating, controlling, and measuring the conduct of KM. Resource influences are of four types: financial, material, human, and knowledge. These organizational resources both limit and enable the conduct of KM. The environmental influences are factors external to the organization that can affect how KM is conducted within it (e.g., competitors, technology, markets, regulators).

Activities that an organization's participants perform when manipulating knowledge resources can be viewed from multiple levels. At a high level, there are such activities as experimentation and decision making. At a more basic level, there is the activity of selecting knowledge along with its companion knowledge manipulation activities of acquiring knowledge, internalizing knowledge, and using knowledge. The latter refers to the activities of generating knowledge and externalizing knowledge.

Arrows in Figure 1 indicate major knowledge flows among knowledge manipulation activities. For instance, execution of the selection activity entails a knowledge flow into that activity from the organization's knowledge resources and a consequent flow to either the acquiring, internalizing, and/or using activity. The internalizing activity produces a knowledge flow that impacts the state of the organization's knowledge resources; this, in turn, impacts subsequent knowledge selection.

Aside from the main knowledge flows, activities can send and receive ancillary messages (Figure 2). An example is the generating activity requesting a knowledge flow from the selection activity. Requests for selection can range from procedural (indicating how the selection is to be accomplished) to nonprocedural (indicating what knowledge is desired). They can range from explicit (e.g., selection of last year's sales from a database) to fuzzy (e.g., where the knowledge seeker is not sure what knowledge should be selected). Some selection requests may require fast responses; others tolerate performance of a selection activity in the background. They can range from one-time, on-demand requests (e.g., pull requests) to standing requests that require continual monitoring (e.g., push requests). Another example of an ancillary message is a feedback message (e.g., an activity comments on the value of a knowledge flow that is received).

Acquiring knowledge refers to the activity of identifying knowledge in the organization's environment and transforming it into a representation that can be *internalized*, and/or *used* within an organization.

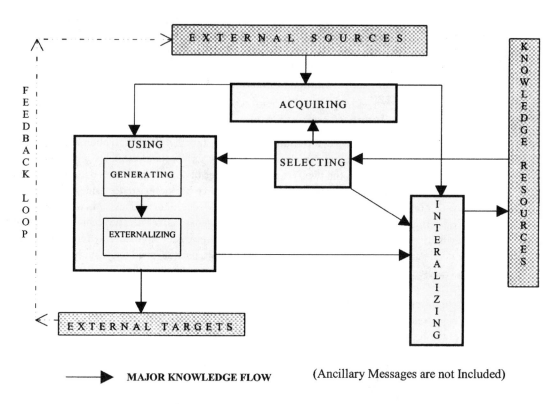

FIGURE 1   Major knowledge manipulation activities (Holsapple and Joshi, 1998).

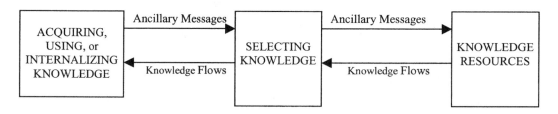

FIGURE 2   Knowledge selecting activity and relationships.

Selecting knowledge refers to the activity of identifying needed knowledge within an organization's existing *knowledge resources* and providing it in an appropriate representation to an activity that needs it (i.e., to an *acquiring, using,* or *internalizing* activity). This activity is analogous to acquisition, the main distinction being that it manipulates resources already existing in the organization rather than those in the environment. The distinction is an important one in that the two activities can require different skills, levels of effort, and costs. **Internalizing** is an activity that alters an organization's *knowledge resources* based on *acquired, selected,* or *generated knowledge.* It receives knowledge flows from these activities and produces knowledge flows that impact the organization's state of knowledge. **Using** knowledge is the activity of applying existing knowledge to *generate* new knowledge and/or produce an *externalization* of knowledge.

Knowledge selection plays a pivotal role in the conduct of KM in an organization. It is through this activity that the other activities interact with the existing knowledge resources. It acts as an interface between knowledge resources and other knowledge manipulation activities. Knowledge selection provides raw materials to the other knowledge manipulation activities. For instance, when the organization is in the process of acquiring knowledge, it may select knowledge from internal knowledge resources to evaluate the quality of this knowledge; or, while internalizing the knowledge, it may select knowledge to identify the best targets for depositing the knowledge; or, while generating a demand forecast, the using activity can request a selection of the past years' sales. On the other hand, what the selecting activity can do is impacted by these other manipulation activities. It can only select what has been internalized, and an organization internalizes the knowledge that has been generated and/or acquired.

Knowledge selection can be performed by an individual participant in an organization or may be carried out by some configuration of multiple participants. For instance, an individual's knowledge manipulation skills can be applied to selecting knowledge or the skills of multiple participants may jointly be brought to bear on selecting that knowledge.

The Knowledge Selection Activity

As noted above, knowledge selection refers to the activity of identifying needed knowledge within an organization's existing *knowledge resources* and providing it in an appropriate representation to an activity that needs it. Knowledge selection is triggered either by a request received from a knowledge seeker (pull) or through automatic alert criteria (push) that have been prespecified. Knowledge selection involves several subactivities, which we refer to as identifying, capturing, organizing, and transferring. These four subactivities are not necessarily followed in a strict sequence, but rather there can be overlaps and iterations. In order to perform a subactivity, a participant needs certain functional skills such as those outlined in Figure 3.

Identifying appropriate knowledge within an organization's existing resources involves such functionalities as locating, accessing, valuing, and/or filtering. When the knowledge selection is pulled by a knowledge seeker with a clearly formulated request, identification primarily involves locating the resources from which the knowledge is to be captured. If alternative candidates are located, then determining which is appropriate can involve evaluations such as calculating transaction costs and assessing knowledge quality. Valuation may depend on accessing (e.g., interviewing, sampling) candidate resources — not to capture the knowledge they hold, but to learn about the parameters of selecting knowledge from them. If a knowledge seeker's request is unclear, then the identifying functionality interacts with the seeker for clarification. This may be an iterative process that proceeds until it can identify the knowledge being sought. Such interaction is a basis for filtering, wherein the identification progresses by eliminating undesired candidates.

When knowledge selection is pushed, the identification functionality is more proactive. It is cognizant of standing requests or profiles that embody predetermined criteria of interest to a knowledge seeker. Its locating functionality reviews changes in available knowledge resources with respect to its profiles. The review may be continual, periodic, or intermittent. In this way, candidate resources for satisfying knowledge needs are located as they arise. The access, value, and filter functionalities can then operate on these candidates.

Capturing knowledge involves the functionalities of retrieving and/or collecting/gathering knowledge from the organization's knowledge resources. Retrieval refers to extraction of knowledge from identified knowledge resources (e.g., from a document, a computer system, an employee). In cases where knowledge to be selected resides in multiple identified resources, then capture involves coordinated collection or gathering from a variety of resources. The sequencing and timing of retrievals within a collection effort can have a material effect on what knowledge is captured.

There are variations for implementing capture functionalities depending on the type of knowledge resource involved and its attributes. For instance, capturing knowledge from an employee is implemented differently than capturing it from a computer system or an artifact. Moreover, implementations can differ from one employee to another, one computer system to another, and so forth. Even for a single type of knowledge resource, implementation can vary according to the attributes of the knowledge to be captured (e.g., tacit vs. explicit, text vs. video, new vs. old, descriptive vs. procedural). Once knowledge has been captured, it may be necessary or useful to organize it prior to transfer to the knowledge seeker.

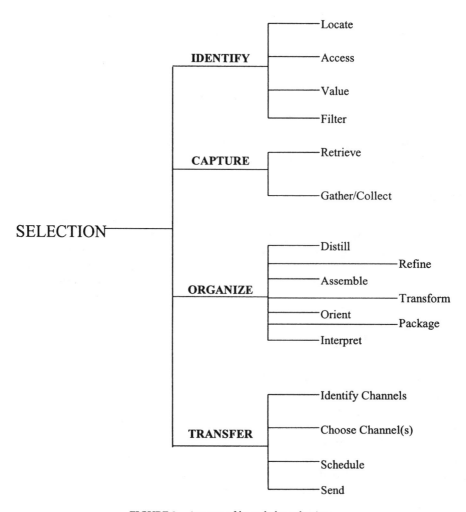

FIGURE 3   Aspects of knowledge selection.

Organizing captured knowledge involves such functions as interpreting, distilling, refining, assembling, transforming, orienting, and/or packaging it into representations appropriate for subsequent knowledge manipulation activities. Distilling, refining, assembling, and transforming are concerned with revamping the internal organization (i.e., content) of captured knowledge. Orienting and packaging are concerned with rearranging the outward organization (i.e., appearance) of captured knowledge. All of these functionalities depend on an interpretation ability.

In the absence of an organizing subactivity, captured knowledge is transferred verbatim to the knowledge seeker. In some circumstances this may be acceptable. In others, the captured knowledge may be too extensive, raw, fragmented, or unclear for the knowledge seeker; or its representation may be unpalatable or inconvenient for the seeker. The organizing subactivity customizes knowledge representations to suit a knowledge seeker. The seeker's organizational preferences may be part of a knowledge request or the seeker's profile.

An instance of knowledge selection includes one or more transfers of captured/organized knowledge to one or more seekers. This knowledge flow can support acquiring, internalizing, or using activities performed by knowledge seekers. The **transfer** subactivity involves such functionalities as channel identification and choice, scheduling, and sending. The prospective channels (medium) for knowledge flow are identified. Channels can be of various forms (e.g., cables, wireless, e-mail, telephone, face-to-face). One or more are chosen (e.g., the most efficient and reliable channel) to deliver knowledge to the knowledge seeker(s). Channel choice also depends on the abilities of the selector and preferences of the seeker as stated in a request or specified in a profile. If there are competing knowledge flows that need to be delivered, transfer may also have a scheduling functionality that determines when to send a flow based on criteria such as importance, magnitude, and urgency. Scheduling may also determine which paths of a channel will be used in routing knowledge flows.

7.3 Issues Related to Knowledge Selection

Having sketched out the context and nature of knowledge selection, we now examine several issues associated with the selection subactivities. We also consider issues related to the interaction of knowledge selection with other knowledge manipulation activities, with influences on knowledge management conduct, and with knowledge resources. There are many such issues. Those highlighted here are representative rather than exhaustive. Exploring such issues can help deepen an understanding of knowledge selection, suggest areas where additional research is warranted, and offer a useful background for studying (or creating) knowledge selection technologies.

Pull vs. Push Strategies

As we have seen, push and pull are distinct strategies for instigating knowledge selection. In the pull case, a knowledge flow from the selection activity to other knowledge manipulation activities is triggered by a knowledge seeker's request. In the push case, a knowledge flow is triggered automatically by the selection activity without an explicit request from a knowledge seeker. Each strategy has its advocates. Manville and Foote (1997) are of the opinion that "people networks leverage knowledge through organizational pull rather than centralized push … companies that push information at their people may actually cause information overload, blocking them from developing their own networks." On the other hand, according to Dennis Fennessey, national director of client service technology at Deloitte & Touche, "right now in their organization it is 80% pull; tomorrow it will be 80% push, … push system saves time by matching available information with individual interest profiles" (Graef, 1998).

The pull strategy requires that the knowledge request is well framed and articulated (Manville & Foote, 1997). This means that either the knowledge seeker knows exactly what is needed and how to formulate a request or a sophisticated set of functionalities is embedded in the identify subactivity that will help the knowledge seeker articulate the request via interactions. According to Paul McNabb, founder and publisher of Cambridge Information Network, "one knowledge management solution that is badly needed is a tool that helps users determine what they need to know in the first place … The biggest issue in gathering knowledge is simply knowing what questions to ask. What the industry needs is a collaborative filtering software that looks at historical information to help the user determine what information they need to know" (Dash, 1998).

There can be situations where a knowledge worker is unaware of a knowledge need or of the existence of knowledge that can be helpful. In such cases, a push strategy alerting the knowledge worker about the existence of that knowledge would be beneficial. A simple example of this is the automatic help facility in Microsoft Word. The software recognizes when a letter is being composed and alerts the user about availability of letter-formatting guidance. However, an implementation of the push strategy that does not make knowledge available at the time it is needed may lead to information overload or simply be annoying. There are many factors that impact push system performance: push timing, push frequency and interval, push criteria, and push content. A quasi-push strategy can be appropriate: "good" push timing, frequency, and interval are determined and the selection activity makes the knowledge seeker aware of the existence of potential helpful knowledge rather than immediately pushing the whole content.

Push and pull strategies for initiating knowledge selection can, of course, coexist. A key issue, aside from implementing each, is determining which should be used under varying selection circumstances. Although research is needed to address this issue, such a determination would appear to depend on such factors as their respective costs, capabilities of the knowledge selector, preferences of the knowledge seeker, nature of the task confronting the seeker, and quality of the knowledge selected.

Locating Issues

Locating the knowledge that a knowledge seeker desires involves searching an organization's knowledge resources. The nature of a search to locate requested knowledge depends on who is conducting the search and what knowledge resources are being investigated. A search can be conducted by a human participant (an individual or multiple individuals), a material participant (a computer system), or a hybrid of the two. Any of the organization's knowledge resources can be searched, the most common being content resources: human participant knowledge, knowledge stored in computer systems, and knowledge represented in artifacts. The location effort can be wide-ranging, spanning many of an organization's repositories and having diverse knowledge representation schemes, or it may be narrowly confined to a single repository. A knowledge selector's locating chore may be limited by time constraints, cost ceilings, and security restrictions on the seeker. It can be facilitated by maps, directories, indices, and its past locating experiences.

Capture and Organizing Issues

When a variety of resources with diverse attributes have been identified in selection, the knowledge capture may be performed by a single selector (e.g., the participant who did the identification) or distributed to multiple participants. In the former case, the participant doing the capture may need an array of collection skills to cope with the different sources. In the latter case, there is an issue of coordination: assigning appropriate participants to the capture tasks, sequencing their work, dealing with possible passage of knowledge among them (e.g., one's capture task may require results of another's capture task), and gathering the results of the distributed capture in preparation for organizing them for the knowledge seeker.

Selection's organizing subactivity exists for the purpose of customizing requested knowledge to suit seeker needs. Failure to customize captured knowledge for seekers implies that it is either incapable of differentiating its audience or it is simply insensitive to their needs. For instance, if the same set of report formats is used to display knowledge to a CEO, a director of research, and a customer, then it is implicitly assumed that they all want to see the selected knowledge arranged in a same way (Maurer, 1998). For knowledge selection to be most effective, it should be able to organize knowledge to suit each knowledge seeker's needs. In doing so, there is not only the issue of presenting captured knowledge in suitable ways, but also of adjusting its content to fit each seeker's needs. The adjustments can range from eliminating unneeded portions of captured knowledge to modifying it (e.g., producing a summary from details). Implementation of the organizing subactivity can thus be seen as a problem of mapping from one representation to another, where the mapping functions are defined in terms of a seeker's needs and preferences.

Selection Skills

As Figure 3 suggests, multiple skill levels are required to execute knowledge selection: functionality skills (e.g., locating, retrieving, interpreting, scheduling), subactivity skills (e.g., skill at capturing knowledge), and overall selection skills (e.g., skills of coordinating subactivities, selection control, and audit skills). For a particular knowledge selection effort, all of these skills may reside in a single selector. Alternatively, the skills may be distributed across multiple participants that comprise a selector. Inadequate selection skills (at any of the levels) can lead to underutilization of existing knowledge resources. Therefore, management should pay significant attention to cultivating participants' selection skills.

Interplay of Selection and Internalizing Activities

The link between internalizing and selecting is crucial. First, only knowledge that has already been internalized can be selected. Second, poor internalization leads to poor knowledge selection even in the presence of great selection skills. During internalization, it is vital to distribute and structure organizational knowledge in a manner that allows for effective and efficient knowledge selection. That is, which knowledge resources will have their knowledge states altered (e.g., by addition, alteration, deletion), and how so? The answer to this question will affect how well the organization is poised to accomplish knowledge selection.

Knowledge structuring depends on the type of knowledge resource(s) targeted for internalization. For instance, if the target is a computer system, then special attention needs to be paid to such issues as indexing, categorizing, classifying, filtering, and linking. On the other hand, if internalization is targeted at human participants, then there is the issue of how humans can internalize (e.g., via training, observation, experience, socialization) knowledge so as to be poised for facile selection. If internalization involves distribution toward multiple knowledge repositories, there may be a greater internalization cost, but also a more flexible selection process (e.g., increasing the possibility of rapidly locating knowledge).

To foster intelligent and customized knowledge selection, it is important to internalize knowledge about knowledge (i.e., meta-knowledge). Meta-knowledge (e.g., tags that characterize knowledge content, its authors, its inception, its expiration) allows knowledge selection based on context as well as content. For instance, engineers at Hughes Space and Communication Co. often exchange designs, but design knowledge (content knowledge) is not sufficient, the context of a design (i.e., why this design was chosen and what knowledge was used to select this design) is also crucial (Hibbard, 1997).

Selection vs. Acquisition

In many respects knowledge selection and knowledge acquisition (Figure 1) are similar. These two activities have comparable subactivities. The primary difference between them is that the acquiring activity locates and captures knowledge from outside the organization, whereas the selecting activity deals with organizational knowledge resources. An organization tends to have more control over selecting than acquiring because it has more control over the distribution and structuring of knowledge deposited in organizational knowledge resources than over the state of external knowledge sources. Not only is the level of control different, but the quality and cost of selected knowledge can also vary from those of acquired knowledge. A couple of interesting issues are determining criteria for deciding when to do acquisition instead of (or in addition to) selection and investigating what selection skills or mechanisms can be applied to acquisition (and vice versa).

Just-in-Time Selection

Ideally, the result of knowledge selection should be made available to a seeker just in time for application to the seeker's task. The seeker should not have to wait for the needed knowledge, nor should the seeker be forced to keep too large an inventory of selection results (for subsequent personal reselection when needed). In the pull case, a lengthy or numerous selection effort may distract the seeker from the task at hand. If the knowledge is pushed too far ahead of time, the seeker may not remember to apply that knowledge. If it is pushed after the time it is needed, then it is rendered useless. Thus, an important issue for optimal application of knowledge is how to achieve just-in-time knowledge selection.

Influences on Knowledge Selection

Organizational resources (i.e., financial, human, knowledge, and material) influence the conduct of KM in general and the performance of knowledge selection in particular. Financial resources put a ceiling on knowledge selection expenditures. Similarly, human participants' knowledge handling skills both constrain and facilitate selection. Also, material resources that are available can affect knowledge selection (e.g., selection capabilities of computer systems). The six types of knowledge resources influence selection, not only because selection must operate on them, but also because they (e.g., culture, infrastructure, participant knowledge) can condition when and how knowledge selection happens. Appreciating resource influences on knowledge selection is important in managers' efforts to ensure that selection happens as desired and in researchers' efforts to study how selection does or should occur.

Managerial influences on knowledge selection involve a host of issues in the categories of leadership, control, measurement, and coordination. Leadership is concerned with being a motivating catalyst for effective KM, so that participants have the knowledge they need to perform their tasks well. In the case of knowledge selection, the key leadership issue is one of creating conditions that foster cooperation among participants so there is a willingness of each to do conscientious knowledge selection for others.

Control is concerned with ensuring that knowledge being manipulated is of sufficient quality and quantity, and that it is well protected. How knowledge resources are controlled dictates the validity and utility of knowledge that is selected. Another control issue influencing selection is maintaining the security of knowledge resources (i.e., avoiding unauthorized exposure). Regular selection audits using legal, social, or technological means can aid in controlling selection for security reasons.

Measurement is concerned generally with appraising knowledge management. In particular, measuring selection performance gives managers a basis for evaluating it and formulating ways to improve it as needed. Some factors that can be used to measure selection performance are selection successes, time, ease, cost, impact on other knowledge manipulation activities, and flexibility.

Coordination refers to managing dependencies among knowledge manipulation activities. In connection with selecting knowledge, activities within (subactivities) and across (other knowledge manipulation activities) need to be well coordinated. This involves identifying knowledge manipulation activities and selection of subactivities that are to impinge on an instance of knowledge selection, sequencing their execution, and determining participants that will execute selection and its subactivities. An example of a coordination approach is Beckman's (1997) eight-stage process for sequencing knowledge manipulation activities.

Another class of influences on knowledge selection comes from an organization's environment. Environmental influences include factors such as competitors' actions, market dynamics, government regulations, and technology advances. A later section of this chapter is devoted to selection technologies available to organizations.

Selection from Schematic Knowledge Resources

The concept of selecting from a content knowledge resource is fairly straightforward, be it push or pull in orientation. The seeker makes a request (direct/on-demand/pull or implied/standing/push) of a participant to supply desired knowledge, or the seeker observes artifacts. In the case of selection from schematic resources, requests do not come into play, nor is there an object to observe. Instead, selection occurs by observing actions, behaviors, and outcomes. As an example of knowledge selection from a schematic resource consider the entry of a new employee into an organization. He or she selects knowledge from the organization's cultural resource by observing the actions and interactions of current employees. If it is observed that employees who hoard knowledge are those who get promotions and bonuses, then the new employee has determined from the culture that knowledge hoarding is valued more than knowledge sharing.

Issues related to selection from schematic resources include guarding against bias by the seeker who observes, realizing the limitations of inductive use of observed knowledge, and substituting selection from a content resource (e.g., an artifact describing culture) for observation of a schematic resource. Selection directly from schematic resources (especially in a complex organization) can be difficult and time-consuming; as a schematic resource develops, parallel developments of a corresponding artifact codifying that knowledge can prove beneficial. Indeed, organizations often codify infrastructure, purpose, and strategy knowledge as artifacts (e.g., mission statements, policy manuals, organizational charts). Selection from artifacts can be more timely, facile, uniform, and formal.

Attribute Effects

Attributes of knowledge being selected can dictate what mechanisms or methods need to be employed in implementing the subactivities' functionalities. The ease of selection can depend on these knowledge attributes. For instance, knowledge can be tacit or explicit. Generally, tacit knowledge is harder to select than explicit knowledge. A human selector is usually more skilled at selecting tacit knowledge, and a computer selector is generally very efficient at selecting explicit knowledge. Knowledge can be represented in different forms: text, image, audio, or video. Skills needed to select from textual knowledge are different than skills needed to select from audio or video knowledge. Organizations need to be equipped to handle knowledge selection in the face of diverse attributes. Issues include identification of major knowledge attributes, of skills helpful in selection for each, and of methods for developing those skills in an organization's participants.

7.4   Knowledge Selection Technologies

Over the past few years there has been a rapid growth in technologies that their vendors characterize as knowledge management software. These technologies fit into various portions of the knowledge management space spanned by the previously described influences, manipulation activities, and knowledge resources. It is fair to say that no single product offering satisfies all of an organization's KM needs. Indeed, many (if not most) of those needs are outside the realm of technology. Nevertheless, KM technology is increasingly being adopted as an important ingredient in addressing knowledge management needs. As a basis for doing so, there are a couple of prerequisites. First, it is important to develop an appreciation of the conduct of KM in an organization, including current practices and desired practices. Second, an awareness of candidate KM technologies is valuable.

The framework and issues already described can contribute to organizing a consideration of the first prerequisite. As one starting point for meeting the second, we now survey features of representative commercial KM products that have an emphasis on performing or supporting an organization's knowledge selection activities. In one way or another, each of these products is a technology for selecting knowledge from a repository and providing it to a knowledge seeker. It is a knowledge provider that may select from its own repository or from knowledge repositories that have been created and are maintained independently.

A single entity may be both the seeker and provider of some knowledge. For instance, in the course of solving a problem, a person may reach a point of recognizing that further knowledge is required; rather than outsourcing the selection activity, that same person may function as the provider by identifying, capturing, and organizing the knowledge sought. Here the focus is on the situation where seeker and provider are distinct: a computer-based provider and usually a human-seeker. In such situations, a knowledge seeker tends to be skilled in the use of sought knowledge, but does not have the time, inclination, or ability to do the selection of it. Conversely, a knowledge provider tends to be skilled at knowledge selection, but is not in how to use it.

Clearly, it is important to have a good fit between a provider and a seeker. Ideally, the provider should be able to select knowledge on any subject of interest to the seeker, from any relevant repository in an organization, and regardless of its storage format. Ideally, it should minimize the knowledge seeker's effort by making it easy for the seeker to indicate what knowledge is sought (on a push or pull basis) and by presenting selection results in a manner customized to the seeker's preferences. The provider technology may include various features based on artificial intelligence such as understanding the context of an instance of knowledge selection, learning from a seeker's previous selection requests, and offering advice about other knowledge selection that may be beneficial.

Criteria for evaluating a knowledge provider's functionality include how it copes with the four selection subactivities: identifying, capturing, organizing, and transferring. With respect to identifying, for example, how easy it is for a seeker to express what is to be selected, what repositories make up the domain for locating and accessing knowledge, what volumes of knowledge can be searched in a timely way? What repository types are amenable to knowledge capture by the provider, what are the limits on the volume of knowledge that can be captured, and how timely is the capture? What degree of customization is allowed by provider's organizing feature and how easy it is to specify the customization? Can the provider transfer knowledge to a seeker on a push and/or pull basis, can it time that transferral to suit a seeker's needs, can it transfer captured knowledge to multiple seekers (perhaps organized into their own customized presentations)? Aside from such functional criteria, there are other issues to consider such as the

TABLE 1 Representative Vendors of Knowledge Selection Technologies

Vendor	Products	Orientation
Dataware Technologies* www.dataware.com	Knowledge Management Suite	Cross-repository selection
Excalibur Technologies Corporation* www.excalibur.com	Excalibur RetrievalWare*, Excalibur Screening Room™	Cross-repository selection
Folio* www.folio.com	Folio Views*, Folio Builder*	Selection from a native repository
Fulcrum* www.fulcrum.com	Knowledge Network™	Cross-repository selection
GrapeVINE* www.grapevine.com	grapeVINE*	Selection add-on
Hyperknowledge* www.hyperknowledge.com	Hyperknowledge*	Selection from a native repository
Inxight* www.insight.com	LinguistX™ and VizControls™	Selection modules
KnowledgeX* www.knowledgex.com	KnowledgeX*	Selection from a native repository
Verity* www.verity.com	SEARCH'97™	Cross-repository selection

initial and ongoing costs of a provider technology, the degree of vendor support, and compatibility with an organization's existing human computer systems.

Currently the knowledge management software industry is in an early stage. Perspectives of vendors on what KM technology should include are diverse and the terminology they use to describe their products is similarly quite varied. The absence of a common KM vocabulary makes it difficult to readily compare products and their functionalities. Here we survey a representative set of knowledge management technologies that perform or support knowledge selection. Each product description is related to the previously presented concepts and issues. Practitioners looking to assess various knowledge selection technologies could use the subactivities and functionalities as a starting point for understanding and comparing them.

The KM products surveyed in this chapter are identified in Table 1, along with vendor information. This is not an exhaustive list of KM technologies, and it should be kept in mind that new KM vendors and products will continue to appear. Also, capabilities of the products surveyed here will evolve and expand. The following product characterizations are based on product descriptions found in vendor brochures and Web pages, vendors' conference presentations, and witnessing product demonstrations. Moreover, vendors were invited to characterize their products with respect to the selection subactivities and functionalities. Feedback from those that responded to this questionnaire is incorporated into the descriptions.

Knowledge Selection from a Native Repository

Traditional technologies can be seen from a KM perspective, even though they typically are not marketed or described as KM technologies (Holsapple and Whinston, 1988). For instance, an essential feature of database management software is to accept knowledge seeker requests to select subsets from what has been stored. In the database query case, the emphasis is on conditional selection of descriptive knowledge (i.e., data, information) from files structured according to a vendor's conventions (e.g., DB2, Oracle, Access). Similarly, some emerging knowledge management products provide facilities for creating special native repositories of knowledge and for subsequent selection from them. For instance, the special repository could be a hyperknowledge structure of various kinds of nodes with labeled associative connections (Chang, Holsapple, Whinston, 1994). The labels indicate semantics of relationships among nodes and the hyperknowledge structure represents procedural and reasoning knowledge, as well as descriptive knowledge (Holsapple, 1995)

An example of a product of this type is Hyperknowledge* from a vendor with the same name. It provides facilities to interactively build the content of a hyperknowledge repository, with a notable emphasis on storing procedural and reasoning knowledge involving WHAT (is the objective), WHY (is it an objective) and HOW (can it be achieved) connections among embedded document nodes. Known as a knowledge base, this repository can also accommodate descriptive knowledge and other connections such as WHO, WHERE, WHEN, and HOW MUCH. These can link to repositories outside the native knowledge base including word processor documents, presentations, Web pages, video/sound clips, and various application files.

Using Hyperknowledge question-answer facilities, an organization has a way to internalize representation of schematic knowledge resources (i.e., aspects of purpose, strategy, infrastructure, and culture) in a computer-based participant rather than in artifacts. This participant performs knowledge selection from its native knowledge base repository and linked external items. A knowledge seeker's request can be a query or take the form of navigation through a displayed map of knowledge base contents. The map shows the linked nodes, allowing a seeker to get to the desired level of detail and to open documents of interest. By navigating the labeled links, a seeker remains aware of the context of selected knowledge. Selection requests are subject to a seeker's access privileges.

Another example of technology for selection from a native repository can be seen in Folio Views* and Folio Builder*. Folio Builder internalizes knowledge from multiple sources into a repository from which Folio Views selects knowledge in response to a seeker's requests. Folio Builder helps automate the gathering of knowledge from multiple sources and documents having diverse formats (e.g., MS Word, RTF, HTML). It is also capable of capturing multimedia objects (e.g., graphic, sound, video objects). Results are internalized in a repository known as a Folio infobase, where they are structured, indexed, and compressed. Every word in an infobase is indexed. An infobase administrator establishes security and rights parameters for an infobase. The administrator also maintains infobase contents and distributes the infobase as needed.

Folio Views performs knowledge selection on infobases as native repositories. It provides a variety of text search features including Boolean searches, phrase searching, proximity searching, word stemming, natural language processing, relevance ranking, wildcard searches, term-proximity, and field searches. It supports document-level searches as well as restrictive searches within a level, field, group, highlighted text, or notes. It allows knowledge seekers to build custom queries using templates or to respond to fill-in-the-blank query dialog forms created by an infobase administrator. Folio Views can organize captured knowledge in various ways. For instance, it can create a dynamically linked table of contents for captured knowledge. This helps a seeker navigate selection results in an organized fashion. For rapid visual identification, Folio Views allows introduction of personal annotations such as searchable highlighters, sticky notes, and bookmarks. Users can also customize and extend the standard interface it provides for viewing captured knowledge.

As with Hyperknowledge and Folio, KnowledgeX tightly couples internalization and selection activities. First, it does knowledge acquisition from a variety of external sources (e.g., the Internet, content providers, databases, news media, magazine articles, annual reports, press releases, graphic images, SEC documents) and selection from internal sources (e.g., internal databases). These acquisition and selection results are internalized into a centralized repository of interrelated knowledge categories. KnowledgeX allows users to add or update contents of the repository based on their own knowledge.

Second, KnowledgeX reveals relationships among the categories in graphical displays that provide intuitive and visually informative cues for knowledge seekers to search and browse the native knowledge repository. This knowledge selection can be done at detailed levels or at a macro level by dragging and dropping the appropriate icons (e.g., discover all of the individuals associated with Acme Corp. by dropping an "all individuals" icon onto the Acme Corp. icon). Users can indicate what knowledge they seek in terms of natural language descriptions involving topic categories and relationships. Captured knowledge can be organized according to report templates and transferred via e-mail or posting on an intranet server.

Third, KnowledgeX alerts a knowledge seeker who has subscribed about changes to topics or relationships among topics. Subscriptions can be established that range from an object level to the level of global changes for the entire knowledge repository. A subscription can be targeted to an individual or a distribution list. Alerts are pushed via e-mail.

Cross-Repository Knowledge Selection

An alternative or complement to creating a special repository on which knowledge selection software operates is to rely on existing repositories. In this case, the selection software operates on diverse, distributed repositories whose files are stored in a variety of formats. The rationale is that organizations have many such repositories that have been created and have evolved over the years to satisfy specific needs. While they continue to satisfy those needs, it would be beneficial for a seeker to be able to tap them for his or her own knowledge needs on an ad hoc or as-needed basis. But the problem is that there can be many of these repositories, difficulty in determining what knowledge is where, difficulty in coping with the diverse interface conventions of the various computer systems, and difficulty in organizing results from distinct repositories.

A solution is technology designed to integrate the "knowledge silos" in an organization, fostering lateral thinking (deBono, 1973). This integration tends to be virtual; the silos remain, but a seeker does not need to be concerned with them. The knowledge selection technology is designed to give a single interface for making requests, to handle identification and capture from multiple repositories, and to organize results of capture in a uniform manner without regard to originating silos.

The Knowledge Management Suite from Dataware® allows the construction of a knowledge warehouse to unify access to knowledge silos without replacing them. A knowledge warehouse is organized according to a hierarchical taxonomy of knowledge categories peculiar to a particular organization. Called a knowledge map, this taxonomy is used to categorize knowledge assets that will be subjected to selection. These assets can exist in knowledge silos having such diverse formats as those of Word, Excel, WordPerfect, HTML, PowerPoint, relational databases, Lotus Notes, and document management systems. Graphic, video clip, and sound clip formats are also allowed. Users can also contribute knowledge directly to the warehouse in any of the map's categories. A knowledge base administrator can specify limits on access to various areas of the warehouse or to particular types of knowledge. The administrator can also alter knowledge map categories as well as what assets are available in the categories.

From a knowledge seeker's viewpoint, the Knowledge Management Suite looks essentially like a Web browser. The knowledge map is portrayed as a folder, containing the highest level of an organization's knowledge categories as other folders, which contain the next hierarchical level of categories as still other folders, and so forth. Selection can be accomplished in a variety of ways including search by map category, natural language search, Boolean search, proximity search, term-weighted searching, search by object type (e.g., PowerPoint files), and search by context (e.g., author, date).

Relevant assets that are identified during knowledge selection are presented according to the knowledge map taxonomy. For any category, the seeker can view the title of each identified asset along with an evaluation of its relevance, its type (e.g., a Word document), size, and contextual characterization. When a user chooses one of the identified assets, it is captured for organization and transferral according to a range of options such as direct viewing in the browser window, execution of the computer system in whose repository it resides, production of a summary of its contents, preservation on disk, or transferral via e-mail to other knowledge workers.

In addition to actually performing knowledge selection from computer participants, the Knowledge Management Suite can support selection from human participants through a directory facility that identifies human repositories of knowledge. These are presented to the seeker in ranked order, along with contact information from the knowledge warehouse. The capture, organize, and transfer subactivities of knowledge selection may then be performed for any identified person by the seeker or his or her assistant.

The Fulcrum® Knowledge Network™ is a suite of products for knowledge selection from a corporate knowledge base comprised of distributed heterogeneous knowledge repositories including those with formats of Microsoft Office files, Lotus Notes, Web sites, databases, document management systems, Microsoft Exchange Server, and various file systems. A hierarchic folder-oriented knowledge map is constructed for linking to such repositories in an organization. This map can include both shared and private folders. The former are created and maintained by a Knowledge Network administrator for general use in the organization. In contrast, a private folder is tailored to a specific individual's knowledge needs. The map, indexes, and document attributes are held in the Knowledge Center for use by the Knowledge Server software in performing selection activities.

The Knowledge Server provides a seeker with a browser-style interface. Searches can be based on natural language queries, Boolean queries, properties, and profiles of documents, map navigation by source or subject, and content similarity ("find more like this"). For a given request, identified items from the geographically dispersed knowledge base are displayed in terms of a title, score, and summary. Items for which the seeker does not have adequate access privileges are not displayed. Summaries can be accessed from the identified items or generated by the Knowledge Server (up to a maximum number of lines or percentage of the total item's content). Proactive agents can perform push selection delivered via e-mail or the browser interface according to user-defined settings.

The SEARCH'97 product family from Verity® can do knowledge selection across such diverse repositories as Microsoft Office documents, Lotus Notes, HTML pages, e-mail, Microsoft Exchange folders, network file systems, and various databases. This cross-repository selection is based on "Verity collections." Each collection is comprised of indices for documents in various repositories, but related to some topic domain or subject category. SEARCH'97 provides administrators with facilities for building and maintaining collections; it also includes a spider for automatically indexing intranet contents and file systems. An administrator can manually assign a document to a collection category or this can be automatic, based on analysis of document content relative to a predefined or customized set of categories.

Through a browser interface, a knowledge seeker can submit a request against any Verity collection (or category in a collection) using natural language, keyword matching, concept-based search criteria, Boolean logic, fuzzy logic, field and proximity search criteria, query by example, or navigation of a hierarchic category map. Upon identifying documents that satisfy a request, SEARCH'97 displays a list of their titles ranked by a relevancy score. These can be arranged automatically into categories based on common themes. A seeker can click on any title to have the corresponding document captured and transformed into an HTML rendering for display via the browser. If desired, a summary of the captured document can be generated for display. Selection results can be transferred to both the seeker and designated third parties.

SEARCH'97 has an agent for push-style knowledge selection based on profiles for individual, group, or organization-wide knowledge needs. These profiles are essentially stored requests, processed according to a user-specified frequency. Results can be delivered via e-mail or custom Web pages.

The Excalibur RetrievalWare® product family performs cross-repository knowledge selection from such repositories as databases, text, e-mail, groupware files, intranets, and document management systems. Its FileRoom facility can be used to create repositories of scanned images from paper artifacts; they are organized into electronic folders within electronic cabinets. These, too, are included as a native component in cross-repository selection. A map of knowledge assets and corresponding indices into repositories form the basis for handling knowledge selection requests made through a browser interface.

In addition to conventional Boolean requests, a knowledge seeker can submit natural language requests. In processing such a request, RetrievalWare uses an expandable semantic net with over 2 million initial entries to perform concept analysis based on word meaning and relationships. Identified assets are presented in a ranked order, reflecting concept analysis results. A user can get information on why each of them is considered to be a "hit" in the concept search. When the seeker picks an identified asset, it is captured from its repository and presented according to the settings of preference parameters (e.g., full document, creation of a summary for presentation).

Excalibur Screening Room™ allows knowledge selection from analog and digital video assets. First, it analyzes and indexes contents of each pertinent video asset. This includes indexing based on closed captions. Selection requests can involve concept search, browsing, and finding images similar to previously found images. For instance, a training or maintenance video may be indexed to allow selection of any desired portion from it for display. Screening Room can also create video storyboards (i.e., video summaries) of captured video knowledge. The ability to select knowledge from video is a powerful substitute for in-person communication and direct observation, and can be useful in overcoming language barriers.

Knowledge Selection Add-ons

Another class of knowledge selection technologies is comprised of software designed as add-ons to established products such as groupware or workflow systems. Examples include the offerings of grape-VINE® for enhancing functionalities of groupware (e.g., Lotus Notes) that an organization may be using. Its components include knowledge charts, directories, and interest profiles. A knowledge chart is a hierarchy of keywords representing an organization's knowledge. This customized hierarchy contains terms and topics of sufficient value to the organization to be of interest for knowledge selection. Directories indicate where to look in attempting to locate desired knowledge. Interest profiles represent each user's knowledge needs, mapping them to four categories in the knowledge chart: routine, significant, action, and critical.

The version of grapeVINE® used with Notes monitors documents in multiple Notes databases using organization-specific criteria. Users receive alert messages (via Notes mail, grapeVINE® e-mail, and/or personalized Web pages) about any documents that match their personal interest profiles. Alert messages contain summary descriptions, links to the documents and any discussion related to them. Users can add ideas, assessments, and opinions to Notes databases and grapeVINE® selects this knowledge for those who need it. GrapeVINE® also supports selection based on browsing and keyword searches.

The grapeVINE® software provides a metrics system offering two capabilities related to knowledge sharing. First, there are periodic measures of the extent to which knowledge is being shared in the business, including demographics summarizing users' subjects of interest, their use of documents, and patterns of value assessment. Second, it monitors dynamics of knowledge selection such as trends in keyword selection and patterns in knowledge needs.

Knowledge Selection Modules

Some knowledge selection technologies are not offered as standalone software, but serve as a module that can be integrated with other vendors' products. Examples are two product families offered by Inxight®: VizControls™ user interface facilities and LinguistX™ natural language processing technology. Companies such as Microsoft, Oracle, Comshare, Verity, Infoseek, SPSS Inc., and Excite have integrated Inxight® technology into their products. This technology focuses solely on selection from text sources. VizControls is a visual display facility that helps organize presentations of captured text in various ways. For instance, results can be presented in the form of a hyperbolic tree that allows hierarchic navigation of the selected knowledge, or they can be arranged using two criteria (one on an x-axis and another on a y-axis).

LinguistX™ analyzes text to identify knowledge for capture. Its text analysis capabilities can determine what language (e.g., French, English) is being used. For a particular language, it then does contextual tagging and phrase extraction via word analysis and sentence analysis components. The word analysis component has features such as linguistic stemming and linguistic inflection. Linguistic stemming identifies the root forms of words (e.g., the root of the words thinking and thought is think) as a basis for locating all forms of a word. Linguistic inflection is the opposite of stemming; it generates all the variations of a root word. The sentence analysis component has features such as language identification, tokenization, and tagging. Tokenization breaks a stream of characters and punctuation into discrete words and sentences, recognizing special contractions and abbreviations. Part-of-speech tagging identifies the part of speech, tense, number, gender, and mood of each word in a sentence based on its context. For instance, it can recognize the difference between "ground" as a noun ("the ground beneath us"), "ground" as an adjective ("the ground spices"), and both the senses of ground being used as a verb ("to ground a wire" and "I ground the wheat"). Aside from doing identification, LinguistX™ also aids in the organizing subactivity by helping generate real-time document abstracts for captured text. This enables knowledge seekers to quickly get a sense of document content.

7.5   Conclusion

A subject of ongoing research, knowledge selection is far from being a mature topic. Nevertheless, it is a very significant topic. Because the knowledge selection activity plays a crucial role in the conduct of KM in an organization, practitioners need to pay careful attention to it. This chapter's coverage of concepts, issues, and technologies can benefit both researchers and practitioners. The conceptualization of the role of knowledge selection in an organization's conduct of KM suggests parameters for research and a checklist of considerations for practice. Issues that flow from the concepts suggest research projects to conduct and highlight insights and challenges for practitioners. The survey of selection technologies forms a starting point for research into more advanced technologies and sketches out representative options for practitioners. Current selection technologies have already gone a long way toward a realization of knowledge-based organizations (Holsapple and Whinston, 1987). Future developments in selection technologies will doubtless contribute to the efficiency, effectiveness, and pervasiveness of such organizations.

References

Beckman, T. A methodology for knowledge management. *Proceedings of the IASTED International Conference on AI and Soft Computing*, 1997.

Chang, A., Holsapple, C., and Whinston, A. A hyperknowledge framework of decision support systems. *Information Processing and Management*, 30(4), 473-498, 1994.

Dash, J. Emerging Tools and Technologies. *Software Magazine*. Business Intelligence (Feb. 1998). http://www.sentrytech.com/sm028f_bie.htm (March 30, 1998).

deBono, E. *Lateral Thinking*. New York: Harper and Row, 1973.

Excalibur® Technologies. Demonstration CD. 1997.

Graef, J. Managing knowledge can mean many things. *InternetWeek*, March 9, 1998.

Hibbard, J. Knowing what we know. *InformationWeek — Online*. News in Review. October 20, 1997. http//techweb.cmp.com/iw/653/53iukno.htm.

Holsapple, C. Knowledge management in decision making and decision support. *Knowledge and Policy*, 8(1), 5-22, 1995.

Holsapple, C. and Joshi, K. D. Knowledge management: a three fold framework. *Kentucky Initiative for Knowledge Management*, Research Paper No. 118, College of Business and Economics, University of Kentucky, 1998.

Holsapple, C. and Luo, W. A framework for studying computer support of organizational infrastructure. *Information and Management*, 31(1) 13-24, 1996.

Holsapple, C. and Whinston, A. Knowledge-based organizations. *The Information Society*, 5(2), 77-90, 1987.

Holsapple, C. and Whinston, A. *The Information Jungle. A Quasi-Novel Approach to Managing Corporate Knowledge*. Homewood, Illinois: Dow Jones Irwin, 1988.

Malloy, A. Supporting knowledge management: you have it now. *Computerworld*, Feb. 23, 1998.

Manville, B., and Foote N. Strategy as if knowledge mattered. April 8, 1997. http://www.fastcompany.com/02/stratsec.html (April 30, 1997).

Maurer, H. Web-based knowledge management. *Computer*, 31(3), 122-123, 1998.

McCune, J. Snooping on the net. *Tech Talk*. http://www.amanet.org/news/archive/techtalk/07tech.htm (March 30, 1998).

O'Leary, D. Enterprise knowledge management. *Computer*, 31(3), 54-61, 1998.

Reynolds, H. No action, no knowledge. Delphi Consulting Group, Inc. http://www.delphigroup.com/articles/1997/10031997MedInfReport.html (April 17, 1998)

Ruggles, R. Knowledge tools: using technology to manage knowledge better. (April 1997). Ernst and Young Business Innovation Center (Working Paper). http://www.businessinnovation.ey.com/mko/html/toolsrr.html (April 30, 1998).

8

Intellectual Capital and Knowledge Creation: Towards an Alternative Framework

James A. Sena
California Polytechnic State University

A.B. (Rami) Shani
California Polytechnic State University

8.1 Introduction

The Emergence of Knowledge Intensive Firms

What's happening today on the Netscape campus in Mountain View, California happens every day in fast-growing companies around the country. The scarcest commodity in business is not customers or technology or capital. It's people. More and more companies simply can't recruit great people fast enough. This talent shortage is their biggest obstacle to growth; solving it is their biggest strategic priority (Broden, 1997).

Netscape is one of many knowledge-intensive firms stepping to the forefront. These fast-growing firms (e.g., Cisco, Yahoo, etc.) are as rigorous about sourcing, selecting, and shaping new people as they are about designing new products and conquering new markets. Besides the difficulty in selecting the right people, an even greater challenge is to make these people productive contributors. At Cisco, employees start their first day with a fully functioning workspace and a full day of training in desktop tools — computers, telephones, voice mail. At MEMC Electronic Materials in Missouri, new engineers spend their first two days immersed in the company's history and operations. They then receive a workbook containing questions to be answered within two months about every part of the company — from manufacturing to computer operations, to purchasing, to employee benefits. In order to complete the workbook, they will need to talk to people all over the company.

Netscape has a theory about successful companies — great people build great products. Great products generate big profits, which provide resources to attract more great people — a self-sustaining cycle. The business runs itself. Tenaski (1995) relates that the nature of work has a changing landscape. Capital and labor-intensive firms are being replaced by knowledge-intensive firms. A firm's survival and success depend on producing newer, better, and more innovative products and services faster then ever before.

Knowledge work involves the creation of new understandings of nature, organizations, or markets and their application by a firm in valued technologies, products, or processes (Stebbins and Shani, 1995; Tenkasi and Boland, 1996). In firms such as Netscape or Microsoft, the emergence of knowledge work is their core business. Knowledge work is not limited to these firms; now virtually all organizations across all sectors are becoming more knowledge intensive (Drucker, 1994). A critical feature of knowledge work is that it requires multidisciplinary expertise and mutual learning in order to achieve a synthesis of technology and knowledge domains.

Knowledge-intensive firms are characterized by their emphasis on knowledge. Their key input is strategic and technical expertise that enables them to outperform their rivals. Knowledge work is a complex process requiring multidisciplinary expertise in order to achieve a complex synthesis of highly specialized state-of-the-art technologies and knowledge domains (Purser, Pasmore, and Tenkasi, 1992). In these knowledge-intensive firms competitive advantage and product success are a result of collaborative, ongoing learning.

Knowledge management presents significant organizational and technical challenges that require the integration of an effective human network with a wide range of technological opportunities (Lloyd, 1996). In order to achieve this task new skills, new mindsets and models, organizational commitment, and new ways of thinking are required.

Intellectual Capital as the Key to Long-term Success

In business today the massive surge of information has overwhelmed the traditional collection and analysis systems that firms use. Most information is lost or wasted. Critically important information is rarely acquired and even when acquired rarely turned into knowledge. Businesses have become quite sophisticated in turning data into information, but they are not nearly as good at turning information into knowledge. This is rooted in the failure of businesses to approach the problem of knowledge with the same rigorous system approach they use in creating their early data processing systems and later information systems. They have neglected to create intelligence systems and to turn information into knowledge. (Friedman et al., 1997)

One solution adopted by many firms (e.g., Coca-Cola, Monsanto, and General Electric) is the creation of a knowledge manager. Such a manager must organize, control, manipulate, and exploit all of the information that has been created inside the company and turn it into knowledge. According to Inkpen (1996), new organizational knowledge provides the basis for organizational renewal and sustainable competitive advantage. Organizational knowledge starts with individuals. This knowledge needs to be shared throughout the organization; otherwise it will have limited impact on organizational effectiveness. Thus, organizational knowledge creation represents a process whereby the knowledge held by individuals is amplified and internalized as part of an organization's knowledge base (Nonaka, 1994).

Lloyd (1996) discusses the knowledge value chain concept where ideas, know-how, skills, competencies, and other forms of intellectual capital can be transformed into intellectual assets with a measurable value to the business. He notes that the first step needs to be the visualization of intellectual capital as the interchange of human capital, organizational capital, and customer capital. Dow Chemical's vice president of R & D reflected on Dow's vision to "maximize the business value of Dow's intellectual assets" by "developing a management process that will help to maximize the creation of new value creating intellectual assets." While still evolving, Dow is satisfied that it has already demonstrated the ability to increase its earnings through the more effective creation and management of its intellectual assets.

Intellectual capital is becoming corporate America's most valuable asset and can be its sharpest competitive weapon (Roos and Roos, 1997). These assets — the knowledge of employees, customer and supplier relations, brand loyalty, market position, and knowledge — must be nurtured and leveraged. Peter Drucker (1994) tells us that the chief source of competitive advantage is the knowledge of the organization's members. He points out that there is a mutual dependence — the company needs to serve and nurture the knowledge worker while at the same time the knowledge worker needs the value-creating processes and infrastructure of the organization, as well as conversations with other knowledge workers to unleash and leverage their knowledge.

Edvinsson (1997), in describing Skandia's approach to measuring intellectual capital, relates that we are now in a knowledge era that requires a knowledge economy. A major portion of corporate investments now goes into knowledge upgrading or competence development leading to human capital. Other investments go into the development of information technologies devoted to value-added networks. He notes that an IT investment can initially lead to a short-term decrease in profits at the same time that the upgrading of IT enhances the value of the organization.

According to a survey of chief executive officers of large U.S. companies, Wiig (1994) emphasizes the fundamental role that knowledge and intellectual capital play within corporations today. These executives assert that knowledge is the "most important asset" and the foundation for success in the twenty-first century. Organizations have increasingly realized that knowledge and intellectual capital must be managed deliberately and systematically. Managers are coming to realize that a firm's viability depends on the competitive quality of its knowledge-based intellectual capital and assets; and the successful application of these assets to its operational activities (Wiig, 1997).

The dynamics of the intellectual capital role requires a new type of leadership capable of bringing about fast and fluid changes within an organization (Lloyd, 1996). At the Canadian Imperial Bank Corporation, Hubert St. Onge, the Director of Organizational Learning, describes the role of organizational learning in identifying, managing, and cultivating intellectual assets. He notes that individual expertise and learning must be transformed into explicit corporate knowledge. Human, structural, and customer capital need to be drawn together to strengthen business strategies and competitive advantage.

The Challenge of Knowledge Creation

Knowledge can be defined as undeniable facts and objective truths as well as an institutionalized, socially constructed enactment of reality (Furusten,1995). Knowledge is a set of justified beliefs (Marshall, Prusak, and Shpilberg, 1996). Davenport and Prusak (1998) provide a working definition of knowledge that we extend to include wisdom, the intellectual capital of organizations. Intellectual capital, or organizational wisdom, is the application of collective knowledge within the organization. This projection is depicted in Figure 1. Data is the first point of the progression. Objective facts about some event are entered into a record keeping or transaction processing system (e.g., a payment for a purchase). The data is evaluated based on typical validation criteria. Collectively, these transactions or data events are viewed in terms of data management. How fast is the entry and processing of transactions? How much does the system cost and how efficient is it? Qualitatively, the organization is concerned about the timeliness of the transaction events. Can the records be accessed when they are needed and do they make sense? Transaction processing systems (TPS) are the foundation for organizational operations. Sometimes, though, this storage of operational data can become overwhelming — more data is not always better than less. Data is essentially raw material for information.

Information is meant to change the way the receiver perceives something, to have an impact on a decision maker's judgment and behavior (Davenport and Prusak, 1998). The organization interprets, analyzes, and massages the data to produce reports or screen displays — anticipating what the business user or knowledge worker may want to see to perform their duties. The data is arranged and placed into data banks. Current data is frequently called operational data stores (ODS). Historical data is often

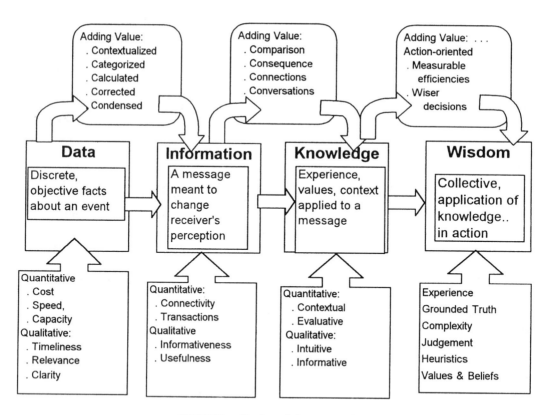

FIGURE 1. The knowledge progression.

summarized and aggregated, based on time intervals, and placed into a data bank called a data warehouse. The data warehouse is a nonvolatile, time-based repository in which knowledge workers can access, query, and analyze information in a variety of forms and arrangements to see trends over extended periods of time. What was the firm's financial position in the same quarter for the previous five years? What products in a particular segment were sold over those quarters?

Neither the ODS or the data warehouse would be particularly useful without some form of collective connectivity. Organizations today depend on their communication and network structure. Information moves around the organization through this formal (hard) network. It also moves about over the "soft" network — phone calls, face-to-face, written memos and e-mail. Frequently firms will use groupware products, such as Lotus Notes or Microsoft Exchange, to streamline both the hard and soft network. All of these data banks, infrastructures, and tools add to the value of data and information when properly used.

The "proper use" forum is knowledge. According to Davenport and Prusak, knowledge is "a fluid mix of framed experience, values, contextual information, and expert knowledge providing a framework for evaluating and incorporating new experiences and information. It originates and is applied in the minds of knowers. In organizations, it is often embedded not only in documents or repositories but also in organization routines, processes and norms." Use of the company's management information systems, the operational data stores, the data warehouses, and other institutional resources must be placed in the hands of individuals within the organization that are "knowers." Knowledge is applied within the organization based on experience, intuition, and judgment.

The application of knowledge by an individual is useful and important in an organization, but collective deployment is of greater value. Managers are much more likely to go to people they believe to possess knowledge and expertise than they are to look for information in databases. Recognizing knowledge as a corporate asset is understanding the need to manage and invest it with the same care paid to getting value from the more well known, tangible assets.

The concept of managerial knowledge (wisdom) can be conceived of as objective and universal techniques as well as common sense. One may argue that knowledge is created through the application of standardized work methods. However, managerial knowledge transcends the notion of standardized methods. It involves a process of transforming observations and interpretations of work-related events into some form of representation like what we have shown in Figure 1. By combining various representations managerial knowledge is created.

The management of knowledge is much more than the storage and manipulation of data and information. It is an attempt to recognize the human assets within the minds of individuals and leverage them as organizational assets that can be accessed and used by a broader set of individuals on whose decisions the firm depends. According to Nonaka and Takeuchi (1995), knowledge management requires a commitment to "create new, task-related knowledge, disseminate it throughout the organization and embody it in products, services and systems."

At the organizational level knowledge is generated from internal operations or from outside sources communicating with the corporate structure. Once created, knowledge is accessed when needed from sources inside and outside the firm. Knowledge can be transferred in a formal manner through training or in a less formal way through work-related experiences. Knowledge can be represented and conveyed in printed or displayed forms, reports, graphs and charts. At some time the validity of the knowledge has to be established. After validation, knowledge is internalized within the organizational framework in its processes, systems, business rules, and practices.

With the need to maintain a sustainable competitive advantage critical knowledge cannot reside passively in the minds of employees. It has to be accessed, synthesized, augmented, and deployed. The organization must learn rapidly and uniformly. Informal or tacit knowledge is no longer sufficient. Certain information technology innovations have come to the forefront to aid the firm in actively creating and utilizing knowledge. The data warehouse is a repository of time-based information expressed in summarized form. Information that exists throughout the organization is available and conveniently accessible to all managers and their staff.

The Need for an Alternative Paradigm

The discussion of the literature thus far reveals that while we have gained a significant insight into intellectual capital and knowledge creation during the past five years, organizations are far from fully utilizing their intellectual capital and the knowledge potential that exists within and outside the organizational boundaries. Our study suggests that this is a result of the lack of emphasis on the design for knowledge creation. In this chapter we explore knowledge creation from the perspective of organization design. Knowledge management is seen as a response to a "mess," that is, environmental conditions that call for the organization to develop methods for adaptation (Lillrank et al., 1998). It can be described as a set of design requirements and functions that an organization needs to fulfill for the adaptation to be successful. The organizational solutions for these requirements can be expressed in terms of design mechanisms and dimensions, alternative solutions, and line actions.

8.2   The Nature of Knowledge Creation

Knowledge work involves the creation of knowledge and its application to the firm as new or improved technologies, products, services, or processes.

Can Knowledge Creation be Managed?

Demarest (1997) defines commercial knowledge as "an explicitly developed and managed network of imperatives, patterns, rules and scripts, embodied in some aspect of the firm, and distributed throughout the firm, that creates marketplace performance." The test of a knowledge management system is the determination of whether it is a by-product of the firm's operations or an explicit objective of those operations.

In the business world, the use of "must" and other absolutes is commonplace. Whereas in the sciences there are demonstrable proofs, a firm has few substantive imperatives. Commercial knowledge, expressed in terms of strategies, goals, and operating plans are only expressions of a firm's need to enter a new market or achieve some level of economic growth. Commercial knowledge is useful only in proportion to its productivity.

Innovation begins with the construction of a new kind of knowledge within the firm. For innovation to flourish it has to be repeatable and reusable. Firms need to insure that their business processes have been aligned and fine tuned. According to Demarest (1997), this means that time-to-market and time-to-decide are the two most critical time metrics today. The entire process from front-end, market identification; product and service design; to the delivery of goods and services has to be managed. Knowledge management has a crucial role in this cycle and should be distributed and embodied through all parts of the value chain.

Knowledge management is a sequential set of activities related to observation, instrumentation, and optimization of the firm's knowledge economies. It is an officially sanctioned and formally valued set of activities within the firm that involve the construction or making of knowledge; the transformation of tacit knowledge into processes, practices, materials, and cultures; the distribution of knowledge throughout the firm's value chain and, the application and dissemination of knowledge to problems and opportunities — making knowledge work. Knowledge management also monitors, measures, and facilitates these activities.

Knowledge Management People

Creating Knowledge	KM Infrastructure	Applying Knowledge
• Publishers • Network members • Analysts • External sources	• Publishing coordinators • Knowledge organizers & architects • Knowledge managers • Staff in networks • Help desks • Information technology • Telephony	• Individuals • Network (led by knowledge manager) • Teams

The Interplay between Tacit and Explicit Knowledge

Based on observations and interviews at almost two dozen companies, Nonaka and Takeuchi, (1996) found that American and Japanese executives tend to hold fundamentally different attitudes about information and knowledge. Americans tend to put their faith in "explicit knowledge," or knowledge that is formal, unambiguous, systematic, falsifiable, and scientific. The Japanese are more inclined to value "tacit knowledge," or knowledge that is intuitive, bodily, interpretive, ambiguous, nonlinear, and difficult to reduce to a scientific equation. The two professors trace the differences between tacit and explicit knowledge back to the divergence in the Western and Japanese intellectual traditions — the rationalism of the West and the "oneness of body and mind" of the East.

Nonaka and Takeuchi challenge the concept and practice of benchmarking, in which companies keep a scoreboard on their competitor's business practices to stay a step or two ahead of them. They argue that this kind of practice leads to incremental improvement, not to true creativity or knowledge creation. In a Japanese company, knowledge is thought to be internally generated from basic principles laid out by top management, then improved by brainstorming from within the ranks and finally by some amount of feedback from external sources. Knowledge acquired by individuals becomes "organizational knowledge" shared among colleagues.

The generation of tacit knowledge is being targeted as a critical part of organizational knowledge in many firms. With its roots in the experience of individuals, tacit knowledge is difficult to process and hard to transfer. However, to understand the context of a piece of information most likely requires tacit knowledge. Through the use of computer-based training (CBT), simulations, the use of expert systems, and other model-based software tools tacit knowledge can be generated and transferred. Knowledge workers can gain experiences from which to develop their own tacit knowledge.

Explicit knowledge at the level of the individual is necessary but not sufficient. Generating organizational knowledge requires converting the tacit knowledge of the individual into explicit knowledge that is accessible to other organizational members. This most often is a social process whereby organizational members engage in a dialogue and gain new perspectives. In this mode of conversation, conflicts and interpretations are resolved — the premises of existing knowledge are questioned and new knowledge is generated.

To support the creation of organizational knowledge, a knowledge management culture needs to be established. Rewards and incentives need to be put in place to make knowledge workers aware of what behaviors and outcomes are desired by management. Management sends signals about what is important through its recruiting, priorities, promotions, and management styles.

The Role of Information Technology

According to Dash (1998), successful knowledge management involves more than merely deploying the newest and fastest IT products. Instead a business process needs to be created that enables employees to make better use of the information they have, and one that promotes a culture that encourages knowledge sharing. Today's widespread dependence on information technology has precipitated the need for more effective knowledge management. Improvements in information technology make it easier to collect, store, and distribute information. However, to be effective, knowledge workers need to be able to understand and act on that information. Knowledge management allows them to leverage their organization's resources to achieve their business goals.

Increasingly central to organizational management is an integrated data architecture that enables information about all facets of the organization as well as its partners and competition. Many computer firms provide specialized hardware and software, such as Compaq–Tandem Computers, who have created a combination of computer architecture, the online transaction processing (OLTP), distributed computer hardware, along with a set of data warehouse and online analytical processing (OLAP) software that can be tailored to specific industries. Marshall, Prusak, and Shpilberg (1996) in their paper about financial risk and the need for knowledge management describe a data warehouse specifically designed to inform management about portfolio risks, to which other systems have the responsibility of sending data. Architectures of this kind enable access to a wide variety of information that can be analyzed using a variety of techniques, such as slice-and-dice, roll-up, drill-down and pivoting.

There are other approaches to computer-information architecture as well, such as object-oriented technology, which encapsulates procedural knowledge with data. This approach allows the rapid definition of new objects or information types that belong to a class of related objects. The class system allows the information characteristics and associations to be passed down to the objects in its class hierarchy. An example would be the introduction of a new product wherein the associations of all related business unit activities would be inherent as it is created — such as the market niche, the customer profile of related product, projections of potential sales, and so forth.

To ensure that information technology becomes part of the firm's business plan it is essential to alter the way in which information technology is viewed within the organization. The key is for IT to be seen as the new engine for growth (Griffith, 1997). When IT is placed in the role of potential profit-maker, managers begin to view it in a far more positive way. One way to do this is to identify technology as a business proposition wherein the business units reap the profits as they bear the cost of development.

Information management is the application of sound management principles to information. It includes three components: data resources management, process management (i.e., business activities that collect or present information to knowledge workers), and information technology management. Underlying this organization is the belief that the value of data is optimized when data is so managed that it can be shared by many applications and knowledge workers, when processes are managed to maximize value-added activities and eliminate non-value-added activities, and when technology is exploited to enable just-in-time delivery of information (English, 1996).

Data resource management has three components — data management, database management, and data technology management. All of these components are directed inwardly without regard to the needs of the knowledge worker. Instead they have been designed to meet the needs for efficient operational processing. Each component is related to the control and use of corporate databases. With the advent of data warehouses, online analytical processing, and data mining new uses and new demands for corporate data have evolved. Knowledge must be represented in a form that is readily interpreted as useful information. Good presentation is essential if the information-knowledge is to be considered as valuable to the knowledge worker.

The choice of information to represent can be complex. In some firms, broad overviews of operations are described in "balanced scorecards" in which a few critical performance factors are updated daily to support managers and their staff in making decisions (Kaplan and Norton, 1996). These factors include customer measures, internal business measures, financial measures, strategic measures, and innovation and learning measures. These measures can be incorporated into executive information systems (EIS) that summarize information generated throughout the business and allow it to be viewed in terms of the critical performance variables established by a manager and his or her staff. The EIS operates similar to a data warehouse wherein the knowledge worker can summarize, aggregate, roll-up or drill-down to investigate key phenomena.

A firm's information model reflects the enterprise's knowledge base organized around the firm's fundamental resources, called subject areas, about which the enterprise must know information. The information model must be as understandable to the business executive as the organization chart, which models the organization's human resources, and the chart of accounts, which models financial resources in the form of sources of revenue and expenses.

The process and application models reflect the fundamental business value chains and are integrated through shared information that is commonly defined. The role of application development must evolve into a true process engineering and management function (Martin, 1994). Management principles must be applied to information technology (IT). A common IT architecture with interoperable technology components is requisite to control costs and add value. Information technology management increases the value of IT through maximum use with minimum support costs (English, 1996).

Successful companies rapidly create, disseminate and embed knowledge in new technologies and organizations (Itami, 1987). This knowledge forms the foundation for an infrastructure for new knowledge, assumptions, and controls. By establishing controls new business rules are defined that enable the consolidation and formalization of information gathering and dissemination to reveal knowledge. The deployment of groupware products and the introduction of the intranet provide channels to quickly provide information to all parts of the firm. Intelligent agents can be embedded on Web pages permitting the knowledge worker to examine information on a demand basis. The agents enable the rapid embedding of complex models into software.

Information management is no longer just the responsibility of the information services . It is not just a technical resource — instead information is a business resource used by business personnel (knowledge workers — data consumers), created by business personnel (data producers), and defined and guided by business personnel (data definers).

Knowledge workers use information as raw material in their work. Therefore, the reliance on data producers to create accurate and quality-based information is paramount. These same producers may be called on to capture facts that may not be needed in their jobs or business units but could be required by knowledge workers in downstream activities.

In companies that embrace IT as part of strategy the line between technological and business sides becomes blurred. There is a need for expertise on both sides. Some corporations have experimented with personnel blends to achieve an integrated strategy (Griffith, 1997). Teams made up of both business and IT employees are increasingly popular. Reebok, a sportswear manufacturer, uses blended teams for all major technology initiatives. Wisconsin Power has formed a "governance" committee composed of a mix of operational and technology employees to oversee their major IT decisions.

Rudy Ruggles, director of the Center for Business Innovation at Ernst & Young states that there are three basic issues that companies need to address with respect to knowledge management and the use of information technology (Dash, 1998):

1. The reuse of information at later points in time — storage of information in a repository using document management or data warehouse technologies; using "knowledge maps" to catalog information to help knowledge seekers trace know-how back to its source.
2. Collaborative techniques such as groupware and intranets are required to assist firms with widely dispersed knowledge workers that must work in teams.
3. Cultural and hierarchical barriers need to be overcome to facilitate knowledge sharing and use — through the use of decision support and expert systems to capture and share knowledge.

In an effort to address such problems many firms have created the "chief knowledge officer" position — someone who is responsible for corporate strategy who can work with a variety of constituents to oversee enterprise-wide knowledge management.

At Xerox, Dan Holtshouse, Xerox's Director of Corporate Strategy, says the imaging company internally developed a Web collaborative tool as a low-cost alternative to Lotus Notes. The tool allowed people with similar functions to share their best practices and includes bulletin boards, calendar tools, and file drawer space for sharing documents — enabling the users to create a "virtual workspace" (Dash, 1998). Within eighteen months more than 10,000 employees were using the tool in contrast with 500 when it was first employed.

Below is a table of common knowledge management technologies that have been adopted by knowledge intensive firms:

Knowledge Management Technology Adoption	
E-mail	100%
Internet	100%
Videoconferencing	100%
Project management systems	91%
Groupware	91%
Intranet	82%
Knowledge-based systems	82%
Customer management systems	73%
Skills inventory systems	64%
Yellow Pages for knowledge	44%

Source: Knowledge Management Consortium

In addition to these traditional technologies some companies are using newer types of software designed to meet the demands for tools that specifically address knowledge management issues. One such product is KnowledgeX, which is used to capture any information related to suppliers and competitors that comes via e-mail, word documents, or news feeds. The software builds a timeline of events, helping users understand how discrete actions are related.

A Forrester Research report, cited by Dash (1998), indicates that there are other offerings from software firms that help companies fill the need to capture qualitative information as opposed to the quantitative information found in databases and data warehouses. These software products allow employees to access company information located in a variety of sources — such as groupware applications, document management systems, e-mail, and new feeds — from a single point using the Web.

The Role of Organizational Learning Processes

The pressures to better utilize "human capital" as organizations are faced with tougher competition increased the interest in the phenomena of "organizational learning." Organizational learning is a system of principles, activities, processes, and structures that enable an organization to realize the potential inherent in its human capital's knowledge and experience (Shani & Mitki, 1998). According to Senge (1991), organizational learning incorporates all activities and processes taking place on the individual, team, and organizational levels. Schein (1993) notes that there are at least three distinctly different types of learning: knowledge acquisition and insights (cognitive learning), habits and skill learning, and emotional conditioning and learning anxiety. Recently two different kinds of organizational learning processes, *learning how* (organizational members engaging in processes to transfer and improve existing skills or routines and learning) and *learning why* (organizational members diagnosing causality), were identified (Edmondson and Moingeon, 1996). Organizations, by their very nature as social systems, are the environments in which learning takes place (Argyris and Schon, 1978; DiBella, Nevis, and Gould, 1996; Mitki, Shani, and Meiri, 1997). As such, the organization design plays a critical role in creating an environment that fosters knowledge creation and the development of human capital.

The Role of Work Design

Pasmore and Pursar (1993) tell us that resistance is more likely to arise as a roadblock when we intervene in knowledge systems. This intervention is intended to improve the rate of knowledge creation through team building and group dynamics. They recommend involving the knowledge worker directly in the design of the intervention. It is assumed that the knowledge worker is predisposed to knowing what it takes to think effectively. Knowledge workers "will engage in activities that they believe will make their work better for them — and passively or actively resist everything else."

Inputs to knowledge work include the knowledge workers, the training they receive, the experiences they have accumulated, and the knowledge that is readily available at the time work is undertaken. To determine what knowledge is available a search of the external environment may be necessary. The knowledge scan may include both a look outside the organization to see what is available and what will be needed as well as an inventory of what is currently available in the system — the difference between what we know and what we need to know creates the agenda for augmenting knowledge resources.

According to Purser et al. (1992), the actual mode of improvement ought to be determined by members of the systems as well as management. This sets the stage for identifying the resources that are available and the time frame for reaching the knowledge scan objectives. In any event some strategy toward acquiring knowledge must be put in place. A discussion about current and past projects could reveal variances in the way knowledge is handled. The behavior of the individuals in these projects sets the stage for what knowledge is given weights and which is discarded.

Planning for the knowledge scan involves a deliberation analysis process that examines:

- Who is involved in important knowledge processing
- What information is used by various knowledge workers
- What information is typically not available
- What the influence relationships are within the knowledge and decision making system
- What conflicts and barriers exist among the knowledge workers

The final step in the analysis of knowledge work is to examine factors that influence which knowledge is actually applied to the tasks undertaken. Considerations regarding costs, manufacturability, quality, safety, reliability, and compatibility need to be taken into account.

Taking all of these considerations together, a redesign of the organization may be needed for more effective knowledge work. Purser et al. (1992) note that the redesign may involve aligning knowledge with influence in decision making and developing structures that are flexible enough to support the nonroutine nature of knowledge work. They propose that design work encompasses:

- The structural arrangements that shape interactions among individuals and groups
- The dynamic processes through which information is gathered, processed, and applied as knowledge to tasks

Nonroutine work requires that the structure of the system be flexible and the knowledge explicit. Tacit knowledge is often hidden from examination and discussion.

Lloyd (1996), in discussing Compaq–Digital's organizational effectiveness practices, presents a new equation requiring us to interrelate values in order to learn how individuals, companies, and cultures can achieve a new level of knowledge that provides a sustainable and maintainable competitive advantage. He cites a set of assumptions for the knowledge era:

- Team and reteam capabilities
- Interest and reteam capabilities
- Interest in other selves reveals their capabilities and aspirations
- Transform both raw materials and raw ideas
- Leverage both capital and knowledge assets
- Cooperate and collaborate within and between companies

Linking previously untapped, unconnected expertise within an organization requires an environment conducive to sharing and learning. A knowledge business requires value changes in order to provide a sound basis for long-term success.

8.3 Towards an Alternative Framework: Integrating Communities of Practice and Sociotechnical System Perspectives

Sociotechnical system theory provides a broad conceptual foundation and insights into the nature of routine and nonroutine features of organizations. Knowledge-based organizations are viewed as non-routine organizations. At the most basic level, nonroutine organizations are composed of a social sub-system (the nature of the human assets — the people with knowledge, competencies, skills, attitudes), a technical subsystem (the inputs and the technology that converts inputs into outputs — or product-in-becoming) and an environment subsystem (including customers, competitors, and a host of other outside forces). Sociotechnical system design seeks to pull the three subsystems together toward optimal utilization of the firm's resources through knowledge management configurations and processes. The knowledge management configurations and processes are viewed as the engine that leads to knowledge creation, utilization of intellectual capital, and bottom-line business performance. These concepts are presented in Figure 2 below.

Recently, the framework of communities-of-practice (CP) was advanced to capture the dynamics of social knowledge (Brown and Duguid, 1991). Communities-of-practice are defined as people bounded by informal relations who share a common practice. According to Snyder (1997), "communities" refers to the informality and personal basis of many relationships in typical CPs; it also suggests that CP boundaries do not correspond to typical geographic or functional boundaries in organizations but rather to practice- and person-based networks. "Practice" indicates that CPs are centered on a shared practice, which may or may not correspond to an established function in the organization. Another interesting dimension of "practice" is the emphasis on "knowledge-in-action" (Schon, 1987) or "knowing" (Cook and Brown, 1996), and implies that practice is as much about learning as it is about doing (Snyder, 1997). As such, CPs emerge as people united in a common enterprise who develop a shared history as well as particular values, beliefs, ways of talking, ways of learning, and ways of doing things (Drath and Palus, 1994). They come together not so much on the basis of formal memberships or job descriptions, as by being involved with one another in action (Lave & Wegner; Raelin, 1997).

In the context of knowledge-intensive firms, the possible integration of sociotechnical system theory with communities-of-practice theory presents an alternative that might be a powerful way to investigate, manage, and guide theory and practice of the firm. As we have seen earlier, the sociotechnical system-based framework appears straightforward, even tautological, at first glance (Shani et al. 1992). Yet the

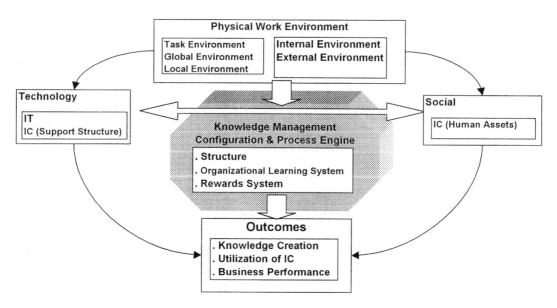

FIGURE 2.   The knowledge management engine.

theory, the design principles, and the change process focus our attention on complex processes within the firm, and between the firm and its environment (Pasmore, 1994). A fundamental axiom of STS is that whatever decisions are made about or within any one of the organizational subsystems, those decisions should meet the demands of the other subsystems. The scope of STS extends beyond work design to broader dimensions of organizational strategy, structure, and key managerial processes. STS provides a particular useful framework for the examination and analysis of the system-wide implications of new information technologies. Ensuring compatibility between the technical and environmental subsystems requires that new information technologies are effective in meeting the needs of customers and are capable of enhancing the competitive position of the firm. Hence, introducing new information technologies inevitably requires a redefinition of the relationship between the technical and environmental subsystems through adjustment to overall strategy. At the same time, compatibility between technical and social subsystems implies that a delicate balance must be struck between selecting the new information technologies that are most compatible with the existing social subsystem and changing key managerial processes, such as managerial accounting systems and human resources selection and training, to accommodate the requirements of the new information technology.

The business environment (the environmental subsystem or what we labeled earlier as the "mess") is made up of elements in the marketplace in which the organization competes. The key exogenous players in the environment are customers and competitors. As competition intensifies and customers become more sophisticated, the external environment becomes less stable and more complex. The business's strategic goals explain how the organization plans to compete in its industry. Different information technologies offer distinct benefits with regard to flexibility, productivity, quality improvement, efficiency, and integration of change. The primary requirements are that the information technology chosen is consistent with and supports the strategic goals of the firm and its human capacity to fully utilize the information technology (Shani and Sena, 1994).

The firm's social subsystem refers to human resources and human capital assets that work in the organization and the totality of their individual and social attributes. The social subsystem encompasses individuals' aptitudes, competencies and skills, knowledge-base, attitudes and beliefs, and relationships within and among groups. These include lateral and vertical relationships between supervisors and subordinates. They also include the relationships between the formal and informal systems and the components related to the culture and tradition of the organization, such as work habits and practices, assumptions, values, rites, rituals, and emergent role network. The social subsystem also establishes the foundation for knowledge creation. From a CP's perspective, knowledge creation is viewed as an integral part and output of the dynamics within the social subsystem.

The technical subsystem of an organization encompasses the technological resources, physical and financial assets, tools, techniques, devices, artifacts, methods, configurations, procedures, intellectual capital, and knowledge used by the organizational members to acquire inputs and transfer inputs into outputs (Pasmore, 1994). Noteworthy is the fact that important differences exist among different information technologies in terms of their impact on the firm's technical subsystem. The introduction of e-mail software has a limited and local impact; it leaves both the social and technical subsystems largely intact. Fully integrated local area networks involve transformations of both the technical and social subsystems.

The "design engine," or what was labeled as knowledge management configurations and process (and by STS theory as the optimization processes), provides the firm's tools and organization to achieve its strategy. This cluster includes multiple elements such as structural design of the firm, reward system, and learning systems. Top management's investment in new information technology requires the adjustment of organizational structure to accommodate needs of the information technology being adopted. Bureaucratic structures, with levels of functional specialization and numerous levels of hierarchy, are suited to efficient operation of highly mechanized operations under static conditions. Conversely, flexible organizations that are organized as communities-of-practice with high levels of interfunctional collaboration and decentralized decision making might be more suited for the knowledge intensive firms.

Finally, the business outcomes cluster refers to knowledge creation, utilization of intellectual capital, and business performance. A critical component of the knowledge management configurations and processes is the establishment of multiple methods and criteria to be used by the firm to measure its success, such as increase of the firm's capabilities and increase of its intellectual capital and knowledge creation.

8.4 Designing Knowledge Creation Systems

As we have argued earlier, knowledge management is seen as a response to a "mess," that is, environmental conditions that call for organization to develop methods for adaptation. It can be described as a set of *design requirements* and functions that an organization needs to fulfill for the adaptation to be successful. The organizational solutions to the requirements can be expressed in terms of *design mechanisms and dimensions*, alternative solutions, and line actions.

For knowledge creation to take place, certain organizational conditions and arrangements must be in place. The starting point is "the mess," a set of external and internal conditions that call for adaptation. In the case of knowledge creation and optimal utilization of intellectual capital, they usually are described as competitive pressures that require improved processes and organizational configurations that allow for improved operational efficiency and the creation of knowledge that is more relevant and potent to the success of the firm.

By design requirements we mean a minimum set of conditions that a manager aiming at knowledge creation and optimal utilization of intellectual capital would have to achieve. On the general level, the design requirements are related to the classic process criteria of work design: want to do, can do, know what to do. (Hackman and Oldham, 1980). As such they will incorporate some of the following managerial responsibilities (Marshall, Prusak, and Shpilberg, 1996):

- Determining the required levels of application-specific and managerial knowledge
- Enabling the centralized collection of that knowledge from sources internal and external
- Representation of current knowledge in documents, databases, and other clear and widely accessible formats

- Embedding of that knowledge in business rules, processes, policies, and control mechanisms
- Refinement and testing of that knowledge — for instance, stress testing the firm's existing models with worst-case scenarios
- Overseeing the transfer of knowledge to application-related decision making
- Overseeing the transfer of knowledge and information to senior management monitoring the current state of the firm
- Creation of an infrastructure to support all of these activities

We define design dimensions as the basic set of alternative solutions that the manager can choose from in order to meet the design requirements. The range of alternatives can be established from the literature or by benchmarking existing solutions. Our task here is to identify the choices that a manager wanting to achieve knowledge creation and optimal utilization of intellectual capital must make. As can be seen in Figure 2, we have identified for illustration purpose three design dimensions: structure, organizational learning system, and reward system. A wide range of choices is available for the manager on the type of structural configurations, the types of learning systems, and the nature of the reward system.

The following are some design dimension choices, based on sociotechnical-system and community-of-practice theories that managers need to study (and eventually make) around the firm's structural configurations and processes:

- Examination of alternative structural mechanisms for the process of creating knowledge
- Examination of the degree to which alternative structural mechanisms facilitate (or hinder) optimal utilization of intellectual capital
- Exploration of alternative deliberation forums (which can be found on a continuum from very formal to very informal structure)
- Exploration of alternative organizational learning mechanisms (which can be found on a continuum from parallel to integrated mechanisms) that can foster both knowledge creation and optimal utilization of intellectual capital
- Exploration of alternative reward mechanisms (which can be found on few continuums such as from individual to team-based rewards) that can foster both knowledge creation and optimal utilization of intellectual capital
- Exploration of alternative information technology configurations that can help integrate the three subsystems of the firm such that both knowledge creation and optimal utilization of intellectual capital can be achieved

8.5   Conclusion

The classical organization design orientations centered on the core production process. In the case of the knowledge intensive firm, the design must center on the design configurations and processes of knowledge creation and optimal utilization of intellectual capital.

We proposed an alternative perspective that integrates the theories of sociotechnical systems and communities-of-practice. As such, the framework centers on the need to identify the "mess," the design requirements and the design dimensions. The dynamics between the "mess," the social and the technical systems of the firm, sets the stage for the firm's ability to analyze its way of organizing and act on its findings. The engine of the firm is rooted in the knowledge management configurations and processes that are based on the design choices made by the managers along a variety of design dimensions. An optimally designed firm — a firm that utilized the proposed design framework — is likely to benefit for a better utilization of its intellectual capital, a prosperous knowledge creation process and outcomes, and a higher level of business performance.

References

Argyris, C. and Schon, D.A. (1978) *Organizational Learning: A Theory of Action Perspective*, Reading, MA: Addison-Wesley.

Broden, F. (1996) Hire great people fast. *Fast Company*, August–September, 132-136.

Brown, J.S. and Duguid, P. (1991) Organizational learning and communities-of-practice: Towards a unified view of working, learning and innovation, *Organization Science*, 2, 40-57.

Brown, J.S. and Duguid, P. (1997) *Organizing Knowledge*. Working paper, Xerox PARC, Palo Alto, CA.

Cook, S.D.N. and Brown, J.S. (1996) *Bridging Epistemologies: The Generative Dance between Organizational Knowledge and Organizational Knowing*. Working paper, Xerox PARC, Palo Alto, CA.

Dash, J. (1998) Turning technology into techknowledgey, *Software Magazine*, February, 64-73.

Davenport, T. and Prusak, L. (1998) *Working Knowledge*, Cambridge, MA: Harvard Business School Press.

Demarest, M. (1997) Understanding knowledge management, *Long Range Planning*, 30, 3, 374-384.

DiBella, A.J., Nevis, E.C., and Gould, J.M. (1996) Understanding organizational learning capability, *Journal of Management Studies*, 33, 3, 361-379.

Drath, W.H. and Palus, C.J. (1994) *Making Common Sense*, Greensboro, NC: Center for Creative Leadership.

Drucker, P. (1994) The age of social transformation, *The Atlantic Monthly*, 274(5), 54-80.

Edmondson, A.C. and Moingeon, B., (1996) Organizational learning as a source of competitive advantage, in Moingeon, B. & Edmondson, A.C. (Eds.), *Organizational Learning and Competitive Advantage*, Thousand Oaks, CA: Sage, pp. 7-15.

Edvinsson, L. (1997) Developing intellectual capital at Skandia, *Long Range Planning*, 30, 3, 366-373.

English, L. P. (1996) Turning information management into an effective business enabler. Information strategy, *The Executive Journal*, Winter, 16-27.

Friedman, G., Friedman, M., Chapman, C., and Baker, J. (1997) *The Intelligence Edge: How to Profit in the Information Age*, New York: Crown Publishing.

Furusten, S. (1995) *The Managerial Discourse: A Study of the Creation and Diffusion of Popular Management Knowledge*, Department of Business Studies, Uppsala University.

Griffith, V. (1997) Making information technology strategic, *Strategy and Business*, Booz-Allen & Hamilton, Fourth Quarter.

Inkpen, A. (1996) Creating knowledge through collaboration, *California Management Review*, 39, 1, 123-146.

Itami, H. (1987) *Mobilizing Invisible Assets*, Cambridge, MA: Harvard University Press.

Kaplan, R.S. and Norton, D. P. (1996) *The Balanced Scorecard*, Cambridge, MA: Harvard Business School Press.

Lillrank P., Shani, A.B. (Rami), Kolodny, H., Stymne B., Figuera, J.R., and Liu, M. Learning from the success of continuous improvement programs: An international comparative study, in Woodman, R. and Pasmore, W. (Eds.) *Research in Organizational Change and Development*, Volume 11, Greenwich, CT: JAI Press (in press).

Lloyd, B. (1996) Knowledge Management: The Key to Long-term Organizational Success. *Long Range Planning*, 29, 4, 576-580.

Marshall, C., Prusak, L., and Shpilberg, D. (1996) Financial risk and the need for superior knowledge management, *California Management Review*, 38, 3, 77-101.

Martin, J. (1994) *Enterprise Engineering*. Lancashire, U.K., Savant Institute.

Mitki, Y., Shani, A.B. (Rami), and Meiri, Z., (1997) Organizational learning mechanisms and continuous improvement, *Journal of Organizational Change Management*, 10, 5, 426-446.

Nonaka, I. (1994) A dynamic theory of organizational learning, *Organizational Science*, 5, February, 14-37.

Nonaka, I. and Takeuchi, H. (1995) *The Knowledge-Creating Company: How Japanese Companies Create the Dynamics of Innovation*. Oxford University Press.

Pasmore, W. A. (1993) Designing work systems for knowledge workers, *Journal for Quality and Participation*, July–August.

Pasmore, W.A. (1994) *Creating Strategic Change*, New York: John Wiley & Sons.

Purser, R.E., Pasmore, W.A., & Tenkasi, R.V. (1992) The influence of deliberations on learning in new product development teams, *Journal of Engineering and Technology Management*, 9, 1-28.

Roos, G, and Roos, J (1997) Measuring your company's intellectual performance, *Long Range Planning*, 30, 3, 413-426.

Schein, E.H., (1993) How can organizations learn faster: The problem of entering the green room, *Sloan Management Review*, 34, 2, 85-92.

Schon, D.A. (1987) *Educating the Reflective Practitioner*, San Francisco, CA: Jossey–Bass.

Senge, P.M. (1991) *The Fifth Discipline: The Art and Practice of the Learning Organization*, New York, Doubleday.

Shani, A.B. (Rami), Grant, G.M., Krishnan, R., and Thompson, E. (1992) Advanced manufacturing systems and organizational choice: Sociotechnical system approach. *California Management Review*, Summer, 91-111.

Shani, A.B. (Rami) and Mitki, Y., (1999) Creating the learning organization: Beyond mechanisms, in Golembiewski, R.T. (Ed.), *Handbook of Organizational Consultation*, New York, Marcel Dekker (in press).

Shani, A.B. (Rami) and Sena, J.A. (1994) Information technology and the integration of change: socio-technical system approach, *The Journal of Applied Behavioral Science*, 30, 247-270.

Snyder, W.M. (1996) *Organization Learning and Performance*, Unpublished doctoral dissertation, University of Southern California.

Stebbins, M. and Shani, A.B. (1995) Organizational design and the knowledge worker, *Leadership and Organizational Development Journal*, 16, 1, 23-30.

Tenkasi, R. (1995) The socio-cognitive dynamics of knowledge creation in scientific knowledge work environments, *Advances in Interdisciplinary Studies of Work Teams*, Vol. 2. JAI Press, pp. 163-204.

Tenkasi, R. and Boland, R. (1996) Exploring knowledge and diversity in knowledge intensive firms: A new role for information systems, *Journal of Organizational Change and Management*, 9, 1, 80-92.

Wiig, K. M. (1994) *Knowledge Management: The Central Management Focus for Intelligent-Acting Organizations*, Arlington, TX.

Wiig, K.M. (1997) Integrating intellectual capital and knowledge management, *Long Range Planning*, 30, 3, 399-405.

9
MetaKnowledge and MetaKnowledgebases

Ed Swanstrom
Knowledgment Management
Consortium

Using the term *agent* as it is used in economics, knowledge management is the intentional influence of an agent or group of agents on an organizational environment in which knowledge is produced, refined, and used by other agents. For example, if an organization's goal is to develop product ideas more rapidly, a knowledge management goal might be to provide tools, facilitators, and a method to increase the rate of product ideas of their product development group. If a pharmaceutical company needs to reduce the time to market for its drugs, a knowledge management activity might be to increase the rate of knowledge sharing between its scientists or provide better research resources. Software engineers could benefit from knowledge management if an environment was set up to promote code reuse and sharing of coding knowledge from other engineers. And so on.

Examining the last paragraph reveals two different roles of an agent in the knowledge management process. One role is that of a knowledge worker. A knowledge worker produces, refines, and uses knowledge. They are the product development group members, the scientists, and software engineers. Another role is that of a knowledge manager. They are the ones intentionally influencing the knowledge worker's environment to have an effect on the knowledge creation or usage process.

The relationship between these roles can be expressed as levels. Given that the world is level 0 with no consciousness of itself, the knowledge worker operates at level 1 on level 0 and the knowledge manager operates at level 2 on level 1. The knowledge worker uses level 1 knowledge drawing from a level 1 knowledgeabase, whereas the knowledge management uses level 2 knowledge drawing on a level 2 knowledgebase. Level 2 knowledge being used to operate on level 1 knowledge is called metaknowledge and its knowledgebase is called a metaknowledgebase. (*Meta* is used to mean a level above a target of discussion or study. Metaknowledge is knowledge about knowledge. This is the core of a knowledge management knowledgebase.) Understanding these levels and their relationship with each other is a primary key to successful knowledge management.

While the number of metalevels is unlimited, it is practical to discuss only the levels we can manage. Some organizations might be able to deal with two, others five. The limitation is related to how mature is the organization's knowledge management group and program. An example of how far these levels can go is found in the relationship between a professional society and a company member. The Knowledge Management Consortium (KMC) is a professional society of knowledge managers and engineers who are developing a knowledgebase about knowledge management knowledgebases and processes. The KMC knowledgebase is a level 3 knowledgebase. Within the KMC there exists a committee that examines and recommends improvements to the KMC processes. They are operating from level 4. And by talking about level 4 in this sentence, I am operating at level 5. Fortunately for a company, the KMC acts as the high-level metaknowledgebase.

9.1   Metaknowledge Management

There are many benefits to metaknowledge management:

- To reduce confusion when we communicate about knowledge
- To increase the power and leverage of knowledge management processes

Each will be discussed in turn.

Reducing Confusion

At a conference I attended recently, there was a discussion about knowledge management metrics. Minutes into the discussion, the group divided into two camps with different views on which KM metrics were important. Some said that measuring the effect on the business processes was important such as cycle-time and ROI, while others talked about the importance of measuring the quality of knowledge, knowledge production cycle-time, etc. A metaanalysis revealed that both were right, but were arguing from two different levels. For level 1 it is important to measure the processes that KM will affect: ROI, cycle-time, etc. At level 2, it is important to measure the actual KM processes and its products such as knowledge itself. Tying the two together we get a KM process that is being measured in relation to the business processes that the knowledge worker is engaged in. Once the group realized that they were coming at it from two different levels, the discussion became productively focused on one level at a time.

I have also witnessed people in other occasions deliberately or unknowingly switching levels to try to win an argument where an argument might be valid on one level, but not on another. This is a good way to confuse everyone to the point they don't know what they are arguing for or against.

Leverage Power

Operating at increasingly higher metalevels is the key to powerful knowledge management as well as organizational learning. We observe, analyze, and reflect on the effect of changes we make on level 2 on level 1. As we observe what works and doesn't work, we add that knowledge to our level 3 knowledgebase. Future knowledge management projects are driven from the upper-level knowledgebases. As those upper level knowledgebases change, the lower levels can change. The higher the level of change, the more powerful the change is for ripples down to the levels below.

9.2   Case of ABC Company

There is an actual case of a company trying to use knowledge management to achieve its goals faster and better. Due to competitive pressure, the ABC manufacturing company goal is to shift its IT organization from one technology to another within a year. Analysis of ABC's processes and structure revealed that the shift is not possible in such a short period of time. Current training and knowledge transfer techniques are not adequate.

The IT organization turned to the knowledge management group as consultants to provide assistance. The KM group accessed its own internal social and electronic knowledgebases, finding no knowledge capital related to this problem. They then accessed external knowledgebases. They accessed the KMC knowledgebase both by sending an inquiry to its members (the social knowledgebase) and by searching the Internet-accessible knowledgebase (electronic knowledgebase). With the help of the social knowledgebase, they found a process that was the accumulation of XYZ Company's lessons learned to deal with a similar problem. The KM group pulled this knowledge into their own metaknowledgebase and then set up a program based on that knowledge.

One technique that seemed to work for XYZ was increasing the rate of knowledge transfer from various software development teams by setting up lessons-learned meetings on a weekly basis for several months, then gradually moving the meetings to once a month. Another technique was the use of the company's Intranet to publish links to Internet knowledgebases that are related to the IT organization's goals and the lessons learned from the IT group itself. With this knowledge, the KM group put a KM program into place with a one-hour meeting each week for team lessons-learned meeting. A scribe was assigned to record the results of the meetings and publish them on the Intranet.

At the start of the program, the KM group set up a process that monitored the program's effectiveness in helping the IT organization achieve its goals. The KM group held weekly and then monthly meetings to access and, if necessary, make corrections to the program. The strategies and changes that worked were recorded as lessons learned in the KM group knowledgebase. They also have sessions where they review the effectiveness of their knowledgebase search and their process of making improvements to any KM group process.

9.3 Metaanalysis of the ABC KM Group

Metalevels

Looking over the activities of the KM group, several levels are identified. Level 1 is obvious, the knowledge needs of the IT organization. Level 2 is the knowledgebase of the KM group. Level 3 is the knowledgebase that is used to guide the strategy the KM group uses to set up programs and search other level 2 knowledgebases. Level 4 contains the knowledge for guiding level 3 knowledge.

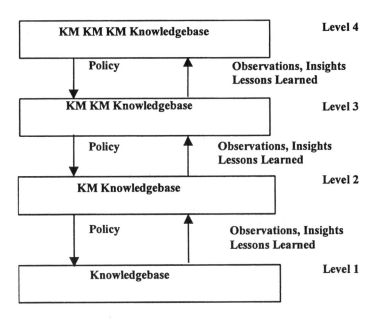

Metametrics

There are two types of metametrics. The first type is the one that is targeted to metalevel itself, such as measuring the quality of knowledge transfer. The other type is measuring the effectiveness of a lower-level metric, such as a metric for measuring the effectiveness of the metric to measure knowledge transfer. The first type is called an intrametric, while the other is a true metametric.

Metaanalysis of the KM group activity reveals mostly intrametrics. Breaking the metalevels down with possible metrics:

Level 1:   Software developers and managers
           Metric: Time and cost of migrating to a new technology
Level 2:   The KM group knowledgebase
           Metric: Effectiveness of KM in helping level 1 achieve their goals faster, better, with less cost
           Knowledge sharing
Level 3:   The KM change and learning process and knowledgebase
           Metric: Effectiveness of the philosophy on helping the KM group be effective.
Level 4:   The KMC knowledgebase
           Metric: Effectiveness of the KMC knowledgebase on helping level 2 activities

9.4   Building Multilevel MetaKnowledgebases

To build metaknowledgebases using a KM approach, a knowledge management group plays three roles: knowledge gatherers, creators, and users.

As knowledge gatherers, they need to "stand on the shoulders of giants" by gathering data that is relevant to organizational knowledge. First they need a knowledge-gathering method and process with knowledgebase that guides them in knowledge gathering. Observations, insights, and lessons learned are used to improve the gathering process. They then gather data from primary sources such as the sociology of knowledge and related disciplines as well as enterprise specific journals and reports.

The history of science and engineering is full of stories of successful knowledge management in one form or another. How did Socrates transfer knowledge to his students? How was knowledge managed during the building of the Egyptian pyramids? How did Robert Oppenheimer and General Grove manage to develop the atomic bomb in such a short period of time with so many gifted scientists with such stubborn personalities? How did NASA and their contractors manage the knowledge of its workers during the Apollo mission to produce a successful moon landing? How do scientists create and manage scientific knowledge in general? The answer to these questions helps answer questions about knowledge management for a business.

As a creator, the knowledge manager observes and records the natural knowledge management behavior of the organization as well as the effect of changes caused by the knowledge managers. Like scientists, they must study how organizations create, validate, accept, and use knowledge with and without interference. They are focused on how groups and teams manage the knowledge of their peers, managers, and subordinates. But first they need to develop a process and method with a knowledgebase for knowledge creation. Observations, insights, and lessons learned are collected and then used to improve the knowledge creation process.

As practitioners, the KM management group makes use of the their own knowledgebase to guide them in developing KM processes that have direct effect on the business environment. Each KM process requires a direct link to business process with clear business case of perceived value to the business. Each KM process requires a clear link to stakeholder needs along with metrics to monitor the fulfillment of those needs. If any technology can be used to help the process, then the technology needs to be driven by the process rather than the other way around.

Building the MetaKnowledgebase

When knowledge is intentionally managed, a team leader will provide conditions that are intended to increase the creation or use of knowledge — providing information related to the problem and increasing communication among the team members.

The problem is there are not enough studies to provide all the evidence that can help managers begin. So all teams and managers must reflect on how to improve the following processes:

- Knowledge transfer from person to person
- Knowledge creation
- Knowledge sharing
- Knowledge extracting from people
- Knowledge extracting from artifacts
- Knowledge usage
- Knowledge gathering

By providing a facilitator to promote communication, or a scribe to record lessons learned, a manager is attempting to influence the knowledge process. By recording the results of his actions to see if knowledge creation or sharing has been increased, he or she starts accumulating metaknowledge about knowledge management.

Jump-starting with a prebuilt metaknowledgeabase is a good beginning step. Organizations such as the KMC are working on this type of resource. Commercial programs will be available within the next year.

General MetaMetrics

The following are examples of metrics that can be applied at three different levels:

Metrics at level 1
   ROI
   Productivity
   Quality
   Cycle-time
Metrics at level 2
   Rate of knowledge transfer
   Ratio of thought leaders to general population of employees
   Range of influence
   Meme (Idea) propagation rate
   Meme authority
   Meme effectiveness
Metrics as level 3
   Rate of knowledge management process improvement
   Ratio of knowledge managers to knowledge workers
   Effectiveness of KM group activities

In the future, the use of metametrics and metaknowledgebases will become an important part of knowledge management systems as the KM field further develops over the years.

10

Investigating a Theoretical Framework for Knowledge Management: A Gaming Approach

Robert de Hoog[1]
University of Amsterdam

Gertjan van Heijst
CIBIT

Rob van der Spek
CIBIT

John S. Edwards[2]
Aston University

Ron Mallis[2]
AED Inc.

Bart van der Meij[2]
Reekx

Robert M. Taylor[2]
KPMG Management Consulting

10.1 Introduction

Over the last two years interest in knowledge management has suddenly exploded worldwide. Living an obscure existence in the decade before, one can now easily spend the better part of the year visiting knowledge management conferences, symposia, etc. This soaring attention is also reflected in the literature: special issues of journals (*International Journal of Strategic Management*, vol. 30(3), June 1997; *Expert Systems With Applications*, vol. 13(1), July 1997; *Knowledge Management*, vol. 1(1), Aug.–Sept. 1997) as well as books published (Wiig, 1993, 1994, 1995; Stewart (1997b), Edvinsson (1997), Sveiby (1997)). This development has its benefits and drawbacks. The main benefit is the reinstating of knowledge as one of the most important production factors in advanced economies. After a period in which business process reengineering was seen as the major road to improved competitiveness, it is now realized that there is a significant danger of throwing out the baby with the bathwater, the baby being the knowledge in an organization. The drawback is that many people from very diverse backgrounds will join the bandwagon, all claiming to do knowledge management, thus watering down what could be subsumed under this term. The clear danger is that knowledge management becomes "everything" that

---

[1]Written while employed temporarily by CIBIT.

[2]Participants of the game that contributed to the production of this article.

goes on in an organization, and as a corollary the dangerous idea that after we have mastered knowledge management paradise is just around the corner. Knowledge management should avoid the trap of trying to manage all knowledge save its own.

A cursory glance at the current state of the art shows that three main approaches are followed to put more flesh on the knowledge management concept:

- Top-down approaches: developing general theoretical frameworks of varying complexity
- Bottom-up approaches: case studies in specific business environments
- Tool-centered approaches: application of a particular tool (e.g., Lotus Notes™)

These three approaches all suffer from some major shortcomings:

- Top-down: lack of empirical evidence for proposed theories
- Bottom-up approaches: localized, difficult to generalize
- Tool-centered approaches: too strongly linked to one particular solution

In order to make knowledge management more than just another management and/or IT fad, it should be developed into a *discipline* with its own methodology (a way of working). Thus the goal should be to develop a methodology for knowledge management that acts as the main backbone for the discipline (just like software engineering is the methodological backbone for the programming discipline). This methodology can be developed only by a combination of the three approaches mentioned above. We will take in this chapter an existing theoretical methodology framework as the starting point.

One of the substantial problems, however, is the empirical confirmation of ideas, theories, tools, etc. It seems evident that conventional "scientific" experimentation is out of the question, because it will be almost impossible to find cases that are realistic in their complexity and still fit into the experimental paradigm of controlled conditions and naive subjects. The accumulation of case studies is possible, but will always be surrounded by the flavor of selecting nonrepresentative cases (e.g., only the successful ones) performed in a unique setting by nonreplaceable individuals.

The goal of the investigation reported here is to use gaming as a way to sidestep some of the disadvantages mentioned above:

- *Realistic case*: Because more time is available the case can be made more complex and more realistic than in time-constrained experiments.
- *Different people*: A game has many different players, thus providing for variety in possible actions, which is absent in case studies.
- *Sequence of actions*: A game allows for a sequence of actions and events, which is almost impossible to achieve in one-shot experiments.
- *Theoretical framework*: The design of the game can be driven by theoretical considerations that almost never play a role in more or less "accidental" case studies.

In our opinion a game fills a niche between purely experimental approaches and case study research, that can be fruitfully used for knowledge management research. In short, the objectives of the game described in this paper were:

- *Case building*: to use the theoretical framework to design a credible case and a series of events that can act as triggers for knowledge management actions
- *Theory testing*: to investigate whether the actions of the players reflect the "predictions" of the theory/framework
- *Investigation of gaming as a technique*: to determine whether gaming, perhaps in a different form, can be used as a major investigation technique for knowledge management in the future

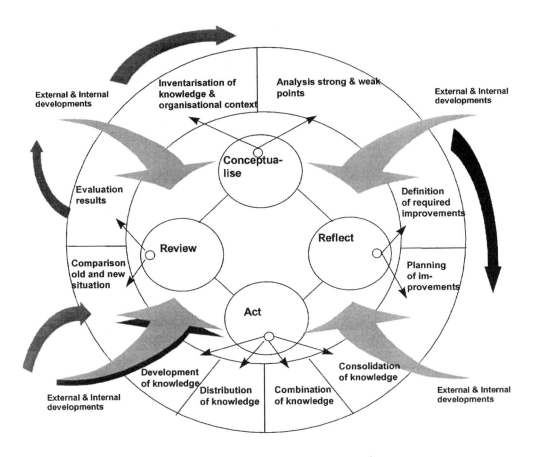

FIGURE 1 The knowledge management cycle.

This paper tries to find an answer to the main issues listed above. First we will elaborate the theory behind the game. Next the set-up of the game; the business case and the events are described in more detail. We will show how parts of the business case and the events can be related to the theory. The main body of the text is the analysis from the angle of the theory of the actions the players proposed. This is followed by a more general analysis of the games' results. The paper ends with a summary of the major conclusions.

10.2   Theoretical Framework

Essentially the knowledge management game is a simulation game. Like any simulation, it is based on a model or a reference system. The reference system used in this game is the knowledge management cycle, a process model for implementing knowledge management (see Figure 1, taken from van der Spek & Spijkervet (1995) and Wiig et al. (1997)).

In the KM cycle, knowledge management is perceived as a cyclic process comprised of four activities: review, conceptualize, reflect, and act. Review means checking what has been achieved in the past, what the current state of affairs is. Conceptualize is sitting back and trying to get a view on the state of the knowledge in the organization, and analyzing the strong and weak points of the knowledge household. Reflect is directed toward improvements: selecting the optimal plans for correcting bottlenecks and analyzing them for risks that accompany their implementation. Act is the actual effectuation of the plans chosen previously.

The analysis, plans and actions are usually formulated in terms of the four *basic operations on knowledge* that can be found in organizations: development, distribution, consolidation, and combination.

- *Developing knowledge:* Companies survive by the continuous development of new knowledge based on creative ideas, the analysis of failures, daily experiences, and work in R&D departments. If, for some reason, the development of new knowledge halts or breaks down, special action should be undertaken (e.g., lessons-learned programs, machine learning on databases).
- *Consolidating knowledge:* Knowledge must be safeguarded against loss due to different causes (e.g., people retiring, documents that cannot be accessed anymore, etc.). Consolidation could be supported by, for instance, corporate memories, KBSs, knowledge transfer programs, etc.
- *Distributing knowledge:* Knowledge must be actively distributed to those who can make use of it. The turnaround speed of knowledge is becoming increasingly crucial for the competitiveness of companies. Actions to improve knowledge distribution include the installation of help desks and the use of intranets.
- *Combining available knowledge:* A company can perform at its best only if available knowledge areas are combined in new products. If an organization is unable to combine the knowledge available, it will miss opportunities and eventually lose market share. Knowledge combination could be improved by a variety of organizational measures (e.g., job rotation).

The relationship between the setup of the KM game and the theoretical model is straightforward. The latter model is the "norm", the way a "sensible" knowledge responsible person should carry out the task. The game players are members of a special KM task force, responsible for knowledge management in the organization. Thus, they should:

1. Undertake all the activities distinguished in the KM cycle (Figure 1). Each time an event is created, the teams should review the state of affairs by looking at the event description and the business case. During the conceptualize activity they should try to determine the knowledge-related problems and opportunities in the current situation. Then, in the reflect phase, an action plan should be developed to deal with the knowledge-related problems and opportunities.
2. Propose actions that reflect the basic knowledge operations that underlie the events as defined by the leaders of the game (see also "The Events").

The actual behavior of the players of the game is used to validate the norm model. If they all behave according to the model, one can say with some evidence that the model seems to be a valuable guide for directing knowledge management work. If not, the model needs adjustment or one could investigate the players' actions in depth in order to check whether they behaved in a suboptimal way. To put it differently: either the model or the players are wrong. Investigating these discrepancies constitutes the main goal of the game, providing more insight into the do's and don'ts of knowledge management.

A weak point of the current game set-up is that the actions taken do not really change the state of the world as an input for the next cycle of the game. To do this, a strong (computational) model is required of the way in which KM actions affect the knowledge household of an organization. Currently we don't have such a model at our disposal.[3] We have tried to cater for this by asking the teams to specify the expected effects of the actions proposed as well as the likelihood of the action's success. In this way we hoped they provided themselves with the needed feedback concerning the state of the world.

---

[3]A simulation tool called Tango! is marketed by the Celemi company in Sweden (see http://www.celemi.se/). It claims to be based on the work of Sveiby. It is currently used by Price–Waterhouse for training purposes. Stewart (1997a) describes a session with Tango! Teams playing the game. It is not clear from the information at our disposal whether this tool is just another business/management game or does have knowledge management specific features. It would be interesting to compile the lessons learned during sessions and investigate how they do or don't reflect the theoretical framework used in this chapter.

10.3   Game Set-up

In this section we will elaborate the main components of the game: the players, the communication channel, the business case and the events.

The Players and the Communication Channel: Using the Internet

In order to minimize the cost of playing the game, it was decided to make use of the Internet for distributing information and communication between participants. Apart from efficiency considerations, this arrangement could also provide some ideas about the feasibility of playing a game when participants are located all over the globe.

Participants were invited by a call posted on the Web site of the International Knowledge Management Network in March 1997. This call was immediately successful; more than 20 people expressed interest. In order to keep the number of participants within reasonable limits without disappointing too many people, it was decided to form international teams of two or three people. In composing the teams we tried to achieve a mixture in each team of people working in industry and academia. By this we hoped to install different perspectives in each team. One person was appointed as the team leader. Team members communicated with each other also through the Internet. Finally, there were six teams comprising 17 people.

It is stressed that playing the game with teams was not originally part of the design. We intended to play the game with individuals, but, as has already been said, we did not want to disappoint too many people. The introduction of teams raises interesting new questions about team dynamics in distributed and electronically communicating teams, but these are outside the scope of this chapter.

Below are reproduced the rules that were set by the organizers of the game.

Teams participating in the Knowledge Management Game agreed to the following rules and conditions:

1. Teams don't know who the members of the other teams are.
2. All teams will receive an identical description of the organization they are working in, including its products, customers, and markets.
3. Their goal is to optimize the knowledge household by taking measures to deal with internal and external events.
4. Events will be generated by the leaders of the game at certain time intervals.
5. The play will run in "simulated time": the time scale for the developments and actions will be something like five years, but the actual play time will not be more than two months at the most.
6. After the generation of an event all teams must decide on what actions to take, the actions taken must be sent to the leaders of the game accompanied by a description of what other actions have been considered, why the actions were chosen, what the costs of the action(s) are, and what the expected effects of the actions are. Before deciding on actions the teams must provide an estimate of the success of the identified actions.
7. Actions undertaken by teams are known only to the teams themselves and the leaders of the game.
8. When the game is over, all teams submit a description of what they think the final state of their organization is, including the knowledge household.
9. Teams/participants agree not to make public the results of the game before the common paper has been published and refrain from writing papers about the game before the common paper has been written and submitted for publication.

Rule 1 was set because one of the goals of the game was to investigate whether diversity would occur in the measures taken by the teams. If the teams knew who were in the other teams they might contact each other to find out what other people were doing, thereby reducing the potential diversity. Rule 6 had as its goal to enforce some kind of uniformity in the aspects the teams had to consider. Without this rule the diversity might become too large, leading to actions that cannot be compared. Rule 7 is a corollary of rule 1 and rule 9 is meant to give all participants and the organizers a common stake in the game.

As has been said above, all communication went through the Internet, utilizing the Web site of the International Knowledge Management Network. The part of this Web site devoted to the KM Game was only accessible to the participants and the game leaders. Every team had its own discussion pages on the site. These pages were only accessible to the members of that particular team. The game leaders had access to all pages.

The task of the players was as follows:

The board of directors has recognized that knowledge is a key asset for Coltec. To develop a better understanding of the roles of knowledge in the organization and the ways it should be managed, a special KM task force has been put together. Your team is this special task force. Your task is to initiate specific activities that improve the efficacy of the knowledge household of Coltec. You are expected to propose both proactive and reactive actions.

In the next subsection we will describe the Coltec Company referred to above.

Business Case: Coltec[4]
History

Coltec is a manufacturer of adhesives and coatings, headquartered in Utrecht, The Netherlands. Coltec was established in 1968. Initially, Coltec operated in the market of custom-formulated adhesives and coatings. During this period the company developed a unique competence in the development and manufacturing of coatings and adhesives for extreme temperatures. Based on this competence, Coltec developed in the seventies a series of standardized products for the industrial market. In 1981 the Namco Group, a leading U.S.-based consortium in the chemical industry, acquired Coltec. In the 1980s, Coltec extended its activities to include consumer products (do-it-yourself glues, etc.). Within the Namco Group Coltec operates as an independent company. It develops, manufactures, and sells its own products. Since the acquisition by Namco, Coltec has steadily extended its range of products. In 1997, Coltec offers over 250 products, ranging from high-performance adhesives used in space engineering to D.I.Y. products. Coltec currently operates in 23 countries in Europe and the Middle East. It has production plants in 12 European countries and employs over 5000 people.

Products

Coltec produces more than 250 products, divided over 7 product divisions. The product divisions are

- Custom-made construction adhesives
- High-performance adhesives
- Waterproof membranes
- Tiling adhesives and additives
- Vinyl adhesives
- Coatings
- Abrasives

Market

Coltec operates in a high-tech, steadily growing market. The market is characterized by short product life cycles. Coltec has an average market share of about 27% on the European market, but almost no market share elsewhere. There are few new entrants in the market, but the number of competitors is growing, mainly because more and more American adhesives manufacturers are beginning to operate in Europe. Coltec is planning to extend its activities to other parts of the world, but has not yet decided on a strategy to do so.

---

[4]For reasons of space we have shortened the business case description. The complete one contained more information on market share, profits, and organizational structure. Coltec does not exist as a real company and all supplied information about products is fictional.

Organization

In Coltec there are three main functions: marketing and sales, research and development, and manufacturing. Further, there are four staff functions: human research management, strategic planning, finance, and information technology. They are described cursorily below.

Marketing and Sales
Marketing and Sales is organized on a country/regional basis. Within each country it is subdivided following the seven main product divisions. There are marketing and sales organizations in 23 countries in Europe and the Middle East.

Research & Development
Research & Development is located in Delft, the Netherlands. Coltec spends about 20% of its turnover on research and development.

Manufacturing
The manufacturing department is divided into seven product divisions. Most of the product divisions have plants in multiple countries:

- *High-performance adhesives:* one plant in Sweden, one plant in Spain, and one plant in Austria
- *Waterproof membranes:* one plant in Ireland
- *Tiling adhesives and additives:* one plant in Portugal and one plant in Poland
- *Vinyl adhesives:* two plants in France (one in Lille, one in Clermont Ferrand)
- *Coatings:* one plant in the Netherlands, one in England, and one in Italy
- *Abrasives:* one plant in Belgium, one plant in Germany
- *Custom-made construction adhesives:* one plant in the Netherlands.

Human Resource Management
The goal of this staff function is to develop company-wide strategic directions for the acquisition, deployment, and use of human resources in Coltec.

Strategic Planning
This staff function investigates future opportunities and markets for Coltec, and develops business strategies.

Finance
Finance develops and enforces company wide accounting standards and reports quarterly to the board.

Central Computing Department
The role of the central computing department is to keep track of the various computer systems that are used in Coltec. At this moment, Coltec mainly uses mainframe computer systems with character-based terminals. The R&D department uses a PC network.

The Events

The game was played in five cycles, all of which were triggered by an event (actually, the last cycle was triggered by two events). The designers of the game created these events. This was not a random selection. They were devised based on the four basic knowledge operations described in the section titled "Theoretical Framework." Thus, for every operation we created an event and as a consequence we expected the actions proposed by the players to reflect this background. In other words: for events created based on, for example, "Developing knowledge," we hypothesized that most actions would be instances of this type. This can be seen as the second "theory" behind the game, which can be checked against the players' behavior.

We will now briefly describe the events and show how they are related to the four basic knowledge operations. This "rationale" for each event was unknown to the players.

Event 1   Decline in the Number of Chemistry Students

You receive from the Human Resources Department a report that summarizes the results of an investigation carried out by the European Union in Brussels. It analyzes the predicted fluctuations in students graduating from universities and polytechnics in the coming five years. The main conclusions are:

- The number of students graduating in chemistry will decline markedly in Sweden, Spain, The Netherlands, Italy, and Germany.
- The preference of high school students for studying science has steadily declined over the last four years all over the European Union.

Predictions beyond those five years are not possible, due to lack of reliable data.

Generating basic knowledge operation: Consolidate knowledge. Clearly, there is a risk of losing knowledge if no appropriate action is taken. How can you prevent this?

Event 2   Differences in Productivity

An internal study shows that there are substantial differences between the productivity of the different plants. This holds in particular for the plants producing tiling adhesives and the plants that produce coatings. In tiling adhesives, the plant in Portugal is approximately twice as productive as the one in Poland. With respect to coatings, the plant in the Netherlands is about 20% less productive than the other two (in England and Italy). The study also shows that these differences cannot be attributed to external circumstances, such as government policies, labor unrest etc. A likely culprit is the machine failure rate.

Generating basic knowledge operation: Distributing knowledge. Apparently, Coltec has the knowledge to keep the machines up and running. However, the knowledge is not available at all places. There is insufficient dissemination of best practices in Coltec.

Event 3   New Legislation

Because of new environmental legislation, from 1 January 1999 on it is no longer allowed to use polyethydimydyl[5] in adhesives manufactured and sold in countries of the European Union. This is a major threat for Coltec, because polyethydimydyl is used in all its adhesives to prevent premature coagulation. Coltec has no immediate replacement for polyethydimydyl. Shortly after the announcement of the new legislation, Gluco, a smaller competitor has put on the market a tiling adhesive that does not contain polyethydimydyl, which is rapidly gaining market share.

Generating basic knowledge operation: Developing knowledge. This is a problem of inadequate business intelligence. Coltec was not informed quickly enough about changing business conditions, compared to its rival Gluco.

---

[5] This is again entirely fictional. We have no idea whether such a material really exists, and if so whether it has anything to do with the coagulation of adhesives.

Event 4 Former Employee

You receive a memo from the CEO of Coltec that he has heard that the technology Gluco uses to manufacture adhesives without polyethydimydyl was developed by a group led by a former employee of Coltec. The rumor goes that this employee, when he was still working for Coltec, already was developing ideas about adhesives without polyethydimydyl, but that nobody was interested at the time. Therefore, he moved to Gluco, where he got a job as project manager at the R&D department.

Generating basic knowledge operation: Consolidating knowledge. This is a case of an ill-functioning "lessons learned" program. Coltec is unable to detect and value new ideas developed at the shop floor.

Event 5 Reorganization[6]

The outcome of the strategic repositioning process is that we have decided to move from a product–oriented organization to a market segments–oriented organization. This means that we will no longer have a structure and strategy that is primarily based on (groups of) products but one directed to serving specific segments of the market. As a consequence the current product–based way of operating will be changed into a market segment–based way of operating. The seven existing product divisions will be dissolved and replaced by market segment–directed divisions. The following divisions will come into existence in the coming year:

- Consumer market (do-it-yourself)
- Construction market
- Shipbuilding market
- Aerospace-engineering market
- Car manufacturing market

Personnel currently working in the existing product divisions will be allocated to one of the new divisions. The manufacturing of our products will take place in the existing factories for the foreseeable future, but production will be based primarily on demand from the marketing and sales organizations. These will still be organized on a country/regional basis, but will follow locally the new market segment divisional structure. The country/regional marketing and sales organizations will be managed as separate business units with a great deal of autonomy. They must buy products from the plants and sell these in the markets they are operating in.

The KM taskforce should submit one month from now a concise report covering the following questions:

- What are the major effects on the way knowledge is (not) managed of the reorganization?
- What measures should be taken to deal with these effects?
- What are the risks we are running when we do not take those measures?
- Does the new organization offer new opportunities for managing the knowledge household?
- Is there a need also for a repositioning of the way our R&D is organized, and if so what kind of repositioning is needed?

Generating basic knowledge operation: All four. With a reorganization a company should rethink its whole knowledge infrastructure.

---

[6] The original event consisted of two letters of Coltec's CEO. For reasons of space we combined both in a short description of the event.

10.4   Results

In reporting the results of the game, we will follow a systematic approach. After each event the teams/players submitted their reactions covering the aspects indicated. We will analyze these reactions from the angle of the normative theory from Section 2: is it in line with the "norm" model (Figure 1) and do the proposed actions reflect the assumptions we made concerning the basic knowledge operations generating the event. We classified all relevant information into these categories (if possible). To make sure the analysis can be followed, we will employ in each of the following subsections the two tables shown below, but filled with results of the analysis.

	Review		Conceptualize		Reflect	
	R1	*R2*	*C1*	*C2*	*RF1*	*RF2*
Team 1						
Team 2						
Team 3						
Team 4						
Team 5						

In the table template above, the columns are the basic steps in the KM cycle minus the Act step (Figure 1), and the rows represent teams. The three main steps are subdivided in substeps, just as in Figure 1:

R1:   Comparison of old and new situation
R2:   Evaluation results
C1:   Inventarization of knowledge & organizational context
C2:   Analysis of strong and weak points
RF1: Definition of required improvements
RF2: Planning of improvements

In the cells it will be indicated, by means of an **X**, whether we could detect that the teams performed these steps when reacting to the event. It was not always easy to identify these steps, because the players are not familiar with them (rightly so!). By means of the rules of the game, in particular rule 6, we have hinted somewhat in this direction. The way in which these slots are filled by the teams is indicative for entries we have decided on in the cells.

	Develop	*Consolidate*	*Distribute*	*Combine*	*Miscellaneous*
Team 1					
Team 2					
Team 3					
Team 4					
Team 5					

The second table template has in the columns the four basic knowledge operations (and the catchall miscellaneous) and in the rows the teams. In each cell the frequency will be indicated in which class of operation each proposed action was placed. The "theoretical" operation class will be highlighted (like Develop in the template). This reflects our hypothesis about the event-generating background. This template is a more detailed specification of the Act step in Figure 1, which was omitted from the first template.

As a consequence we can find actions of the teams in either the first or the second template. If actions are indicative of actions in Figure 1, they will appear in the first, otherwise in the second.

Decline in the Number of Chemistry Students

First we analyze the three steps of the KM cycle in Table 1 below. One has to keep in mind that filling in Table 1 is a highly subjective matter (this holds for all the upcoming tables of this type), because the players did not know (rightly so!) the "theory" behind the game and could not respond in the terms mentioned in Figure 1. However, some of the rules of the game, in particular rule 6, hinted at aspects that could be relevant for the three steps. By paying attention to the topics mentioned in that rule there was a certain guarantee that some of the (sub)steps would be covered.

TABLE 1   Analysis of KM Cycle Steps, Event 1

	Review		Conceptualize		Reflect	
	R1	*R2*	*C1*	*C2*	*RF1*	*RF2*
Team 2			X		X	
Team 3					X	
Team 4					X	
Team 5			X	X		

From Table 1 it becomes clear that the Review step is covered by none of the teams. This is not surprising, because event 1 is the first one and there does not seem much to review from the previous step. Two teams propose Conceptualize activities, but two others don't. They seem to "jump" to the actions in Table 2 without trying to analyze how the event will impact on the future knowledge of Coltec, in which areas Coltec is knowledge strong or knowledge weak, what Coltec's current standing is in the student acquisition race, etc. This can be caused by the fact that the business case does not describe this in detail. However, the players were free to initiate actions that could furnish additional information not given in the case description. Most teams covered the Reflect step, but none paid attention to the risks that could endanger their plans (RF2).

In Table 2 the results for the Act step can be found.

TABLE 2   Analysis of Actions, Event 1

	Develop	*Consolidate*	*Distribute*	*Combine*	*Miscellaneous*
Team 2	0	2	0	0	1
Team 3	1	2	1	0	2
Team 4	0	2	1	0	0
Team 5	0	2	0	0	0
Total	1	8	2	0	3

Of the 14 actions proposed, 8 (57%) belong to the Consolidate category. As this was the basic knowledge operation that generated the event, this can be seen as a corroboration of the notion that this underlying generating operation invokes parallel behavior in the players. If we omit the three Miscellaneous actions, the percentage is 72.

Differences in Productivity

Again we will begin with analyzing the first three steps in KM cycle from Figure 1 (see Table 3).

TABLE 3 Analysis of KM Cycle Steps, Event 2

	Review		Conceptualize		Reflect	
	R1	*R2*	*C1*	*C2*	*RF1*	*RF2*
Team 1					X	
Team 2			XX		X	
Team 3	X		X			
Team 4					X	
Team 5					X	

Compared to the previous event there is at least one instance of Review. C2 (Conceptualize, strong and weak points) is absent. This could be caused by the fact that in the description of the event some notion of "weak" and "strong" was already present through the identified differences in productivity. Precisely detecting these weak and strong points (which is what was proposed quite a few times), can be seen as a kind of C2 step. Reflect has been covered by most of the teams, but RF2 (Planning of improvements, risks) is still lacking.

TABLE 4 Analysis of Actions, Event 2

	Develop	*Consolidate*	*Distribute*	*Combine*	*Miscellaneous*
Team 1	1	1	0	0	0
Team 2	3	0	0	0	0
Team 3	1	0	3	0	2
Team 4	1	0	1	0	0
Team 5	2	0	3	0	0
Total	8	1	7	0	2

Of the 18 actions in Table 4, 8 are in the Develop category (44%). The predicted Distribute category accounts for 7 (39%). This co-occurrence is not surprising. Careful reading of the event shows that the knowledge about the discrepancies is not yet available, calling for an investigation of possible causes (i.e., a Development activity concerning knowledge about breakdowns). However, it does not seem to make sense to keep this knowledge hidden as soon as it is created, calling for a Distribute activity. So most teams lumped them together. This resulted in a kind of "new" event consisting of two "part events":

- *Event 2a*: there are discrepancies between factories (what shall we do to find the cause?)
- *Event 2b*: the nature of the discrepancies has been found, it is known what the cause is for the differences (what shall we do with this knowledge?)

This "new" event leads to Develop and Distribute actions, thus in a sense supporting the idea of convergence between the underlying basic knowledge operation and parallel player behavior. If we take the "new" instead of the original event, 15 out of 18 actions (83%) are of the "right" type.

New Legislation

The KM cycle steps are analyzed in Table 5.

TABLE 5 Analysis of KM Cycle Steps, Event 3

	Review		Conceptualize		Reflect	
	R1	*R2*	*C1*	*C2*	*RF1*	*RF2*
Team 1					X	
Team 2		X	XX			
Team 3		X				
Team 4			X			
Team 5					X	

Table 5 shows again a slight increase in Review activities, while Conceptualize is stable. Surprising is the small number of Reflect activities; only two teams consider alternative actions. Our hypothesis is that this is caused by the very specific "emergency"-like nature of the event, which seems to shift the focus of the actions from *knowledge management* actions to *general management* actions. Stated differently: there is shift from preventive actions to curing actions. We will return to this when discussing Table 6.

TABLE 6 Analysis of Actions, Event 3

	Develop	*Consolidate*	*Distribute*	*Combine*	*Miscellaneous*
Team 1	1	0	0	0	0
Team 2	4	0	0	0	2
Team 3	2	0	0	1	2
Team 4	0	0	1	0	1
Team 5	1	0	0	0	1
Total	**8**	**0**	**1**	**1**	**6**

Compared to Tables 2 and 4, Table 6 shows a significant increase in actions classified as Miscellaneous (37% in Table 6, 19% in Table 2, 11% in Table 4). This seems to corroborate the hypothesis, put forward above, that the nature of the event leads to "curing" actions that are mainly of a general management type. We think that one of the proposed actions "Buy Gluco" is a general management action, which belongs to the discretion of the CEO or other high-level management people. Though it might "solve" the current problem, there is no guarantee that this will not happen again after buying Gluco. In our view the task of the knowledge manager is mainly to propose measures that will *prevent* the reoccurrence of this event. The same holds, though in a lesser degree, for proposed Develop activities concerning alternatives for polyethydimydyl. Of course, it is important to find replacements for polyethydimydyl, but does this prevent the same thing happening to any other material Coltec is currently using? Thus Develop can also be directed toward obtaining knowledge about what can happen in the future in the outside world, i.e., business intelligence. Given this interpretation of the Develop actions proposed, we can conclude that 50% coincide with the "theoretical" background.

Former Employee

The analysis for the KM cycle steps can be found in Table 7.

TABLE 7 Analysis of KM Cycle Steps, Event 4

	Review		Conceptualize		Reflect	
	R1	*R2*	*C1*	*C2*	*RF1*	*RF2*
Team 1			X		X	
Team 3	X	X				
Team 4	X	X	X	X		
Team 5			X			

Table 7 shows a sudden increase in the Review activities. This could be so because this event is a direct sequel to event 3; prompting looking back to what was done and achieved for the previous related event. Just as in the other tables Reflect is not very well covered, probably again due to the specific nature of the event, eliciting rather focused measures.

TABLE 8   Analysis of Actions, Event 4

	Develop	Consolidate	Distribute	Combine	Miscellaneous
Team 1	1	2	0	0	1
Team 3	0	1	0	1	1
Team 5	1	0	0	0	1
Total	2	3	0	1	3

From Table 8 it can be seen that the number of actions has decreased compared to previous events. There is substantial variety. The "theoretical" action category Consolidate covers 30% of all actions, the same as Miscellaneous. Closer inspections of the proposed actions show that the tendency to move into general management actions instead of specific knowledge management actions observed before persists. For example, the action "Sue Gluco for the property of polyethydymydyl" can in our opinion hardly be seen as a knowledge management action, it has more to do with legal actions. The core problem seems to be that Coltec is not able to detect and/or value new inventions made in the organization. For knowledge management this seems to imply that measures must be taken to prevent this from happening again. Suing Gluco is a measure to cut losses, but will in no way contribute to better internal knowledge management in Coltec.

This raises interesting questions about the validity of the normative framework, to which we will return in the final section.

Reorganization

For unknown reasons most teams did not respond to this event. This can be due to the holiday period (July, August 1997), but also to the complex nature of the event, which required too much time and effort from the participants.

Interpreting the tables below has to be done carefully, because they cover only one team. In order to be as complete as possible we have included them nonetheless.

TABLE 9   Analysis of KM Cycle Steps, Event 5

	Review		Conceptualize		Reflect	
	R1	R2	C1	C2	RF1	RF2
Team 3	X	X	X		X	X

In Table 9, five of the six steps are taken, reflecting the complex nature of the event. Only C2 (Analysis of strong and weak points) is lacking.

TABLE 10   Analysis of Actions, Event 5

	Develop	Consolidate	Distribute	Combine	Miscellaneous
Team 3	2	1	0	1	3

In Table 10, three of the four generating basic knowledge operations are present, save distribute. The absence of the latter is not entirely surprising given the encompassing nature of the event, but attention could have been paid to knowledge distribution problems in a market segment oriented organization.

10.5 Overall Analysis and Conclusions

Case Building

One of our objectives was to use the framework to design a credible case and a series of events that can trigger knowledge management-related actions. We believe we have partly succeeded. The case and the events did generate a lot of discussion about knowledge management-related issues and, as described in the previous section, often the issues that were expected from our theoretical framework. However, from the feedback of the players we also learned that the case description was at some moments not sufficiently complete to decide on the suitability of particular actions. So one lesson learned is that for a gaming experiment like this one, a very detailed case description is required.

Another drawback of the current version of the game is its static nature. Actions taken by the teams do not really affect the state of the organization. As explained before, this is the result of our lack of understanding of the dynamics of knowledge management. We have tried to circumvent this problem by asking the teams to report the expected effects of the actions that they undertake. However, apparently this was not sufficient to make the game truly realistic.

Theory Testing

In this chapter we have employed the theory as described in Section 2. It is clear that apart from results from the game, the theory is still far from complete. One should keep in mind that we aim at a normative theory: a set of empirically validated prescriptions (or less ambitious: advice) about how to carry out a specific task. As has been detailed in van der Spek & de Hoog (1995) such a theory or methodology should cover all layers of the so-called Methodological Pyramid shown in Figure 2.

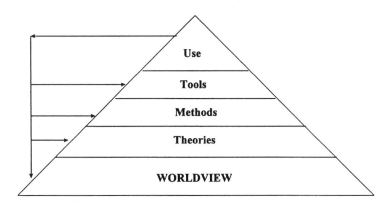

FIGURE 2   The methodological pyramid.

The meaning of the different layers in the pyramid can be summarized as follows:

Worldview   The principles and assumptions underlying the methodology.
Theories   The knowledge available in the domain of the methodology.
Methods   Methods and techniques that support the user of the methodology in carrying out the work as prescribed by the methodology.
Tools   The devices that operationalize the methods: documents, sheets, lists, computer programs etc.
Use   The touchstone. Using the methodology will reveal shortcomings and the need to revise (parts of) the layer(s).

The layers of the pyramid are far from "fleshed out" for knowledge management, though many bits and pieces do exist (see for example the work by Wiig and others mentioned in the references), albeit without a coherent underlying framework. In this chapter we have employed a part of the theory layer: the knowledge management cycle of Figure 1 and the basic knowledge operations. These components were not new, as they were taken from existing sources. Thus, we cannot claim to have contributed to the initial theory formation, but by investigating the "links" with the Use layer we will be able to make statements concerning the plausibility of the theory.

In the analyses for each of the events we did already comment on the actions of the teams, including some reflections on the theory. The theory consisted of two aspects:

- Do players actually execute the knowledge management tasks as prescribed by Figure 1?
- Do players propose actions that reflect the theoretical basis of the events (i.e., the basic event generating actions)?

We will address each aspect in turn.

Executing the knowledge management tasks

Table 11 below summarizes the steps taken for each of the events as well as the overall distribution of steps.

TABLE 11   Frequency of Occurrence of KM Cycle Steps over Events

	R1	R2	C1	C2	RF1	RF2
Event 1	0	0	2	1	3	0
Event 2	1	0	3	0	4	0
Event 3	0	2	3	0	2	0
Event 4	2	2	3	1	1	0
Event 5	1	1	1	0	1	1
Total	5	5	12	2	11	1

It should be stressed that the numbers in Table 11 (as they are in Section 4) are based on the reports submitted by the teams. We were not able to monitor all considerations the teams passed over internally. The discussion forum on the Web site was not obligatory, and even if we wished to make it so, we had no means to enforce it. This made interpretation beyond what was explicitly supplied by the teams too hazardous. However, the conclusion should be read with this *caveat* in the back of the mind.

In Table 11 three groups can be discerned:

- *Frequently occurring steps*: C1 (Inventarization of knowledge & organizational context), RF1 (Definition of required improvements)
- *Moderately frequently occurring steps*: R1 (Comparison of old and new situation), R2 (Evaluation results)
- *Steps that are almost absent*: C2 (Analysis of strong and weak points), RF2 (Planning of improvements)

From a theoretical point of view, the steps from Figure 2 are obligatory; one would expect a more or less equal frequency for all six steps. Table 11 clearly contradicts this expectation and the empirical data are not in line with the prescriptive theory. We will return to this after dealing with the second aspect. Here it suffices to say that the absence of C2 is surprising. From our point of view it is very difficult to imagine that one can move from C1 to RF1 without first analyzing the strong and weak points of the current situation in the light of the new event.

Players' actions and theoretical actions

Table 12 below summarizes the types of actions proposed by the players for each of the events.

TABLE 12   Frequency of Occurrence of Basic Knowledge Operation Actions over Events

	Develop	Consolidate	Distribute	Combine	Miscellaneous
Event 1	1	8	2	0	3
Event 2	8	1	7	0	2
Event 3	8	0	1	1	6
Event 4	2	3	0	1	3
Event 5	2	1	3	1	3
Total	21	13	10	3	17

In Table 12 the "hit" between the theoretical underlying generating measure and the actual actions proposed by the teams are highlighted. Total number of "hits" is 30 out of 64 (47%). The main discrepancies can be found in Miscellaneous, especially for events 3 and 4. The low frequency for the Combine action is due to the fact that this was only relevant in the 5th event, which was handled by one team. Taken together there is some weak evidence supporting the theory from which the events were generated, because 47% is about half way between a 100% hit score and a 0% hit score.

Combining the analysis of Tables 11 and 12 we can say that the empirical support for the theory as a prescriptive framework is not encouraging.

As has been said in Section 2 this can be because:

- The theory is wrong.
- The players behave in a suboptimal way.
- The design of the game is such that it biases the players actions.

The last option was investigated at the beginning of this section. Below we will investigate the other two options. It seems that the issue between "players wrong" and "theory wrong" revolves mainly around events 3 and 4. Both events have a strong emergency aspect, much more than events 1 and 2. The main discrepancies between theory and behavior can be found for these two. As was already said in the section "The Results," we don't think that buying or suing Gluco are "bad" actions, we only questioned whether they can be seen as *knowledge management* actions. This brings us to the question also raised in the introduction: if we want to prevent knowledge management from becoming a catchall, we need to have some clear ideas about what constitutes "proper" knowledge management and what does not. The only way to establish this is to construct a prescriptive methodology that defines boundaries. As we don't want to start a fundamental philosophical discussion about what knowledge management "really" is, we are prepared to live with almost any definition as long as it can be used to demarcate knowledge management from other management areas.

Returning to events 3 and 4, we find it very hard to believe that any "demarcating" knowledge management definition will consider actions like "buying" and "suing" to be knowledge management actions. If it does, then one might consider general management and knowledge management as the same, which is something people promoting knowledge management want to prevent. As a consequence we tend to conclude that for events 3 and 4 the players were wrong by forgetting the role they were supposed to play (see the role definition in the section titled "Game Set-up"). Of course, keeping a sharp eye on your role, especially when you have to play many at the same time, is hard, but this is a key issue if one wants to be clear about the discretion your current role brings with it. So if we "skip" events 3 and

4, practice and theory are much closer, showing that what people do is not always too different from what they "ought" to do. At the same time the presence of events 3 and 4 has highlighted the need to be clear about the proposed scope of knowledge management as a managerial (sub)discipline.

As for the absence of analysis of strong and weak points we tend to accept the explanation that the limited nature of the game, especially the static aspect, prevented the players from going deeper into this area. The ability to "play" with a dynamic simulation could have fostered a more in depth approach of weak and strong points. The absence of detailed information in the case description might also have contributed.

Gaming as a Technique

In this investigation, gaming has been used as a research technique. We have devised an environment in which we could validate our theoretical framework against the behavior of specialists in the field of knowledge management. At some moments in the game there were evident mismatches between the behavior predicted by the model and the behavior of the game players. As mentioned before, this could mean that the normative model is wrong or otherwise that the game players did not act in an optimal way. However, there is also a third possibility. It might be the case that the game players did not have the same perception of the case and the events as the game developers, because some of the information about the situation of Coltec was not sufficiently visible to the players. If this is the case, the obvious question is: what information was not perceived? Answering this question could help identifying crucial indicators for knowledge management.

Besides, as a technique for research, Gaming can also be used as a training tool. For this application, it is even more important to have feedback mechanisms. In the current version of the game, the only form of feedback is the feedback from team members during the discussions about which actions to take. Although not very elaborate, game players have reported that they found this a very fruitful aspect of the game.

References

Edvinsson, L. & M.S. Malone (1997). *Intellectual Capital.* Harper Business, New York.

Expert Systems With Applications, vol. 13(1), Elsevier/Pergamon Press, Oxford, UK, July 1997.

Journal of Intenational Strategic Management, Vol. 30(3), June 1997.

Knowledge Management, vol. 1(1), Aug./Sept. 1997.

Spek, R. van der & R. de Hoog (1995). "A Framework for a Knowledge Management Methodology." In: K.M. Wiig: *Knowledge Management Methods: Practical Approaches to Managing Knowledge.* Schema Press, Arlington, TX, pp.379-393.

Spek, R. van der & A. Spijkervet (1995). *Knowledge Management: Dealing Intelligently with Knowledge.* Kenniscentrum CIBIT, Utrecht, The Netherlands.

Stewart, T.A. (1997a). *The Dance Steps Get Trickier All the Time.* Fortune, May 26.

Stewart, T.A. (1997b). *Intellectual Capital: The New Wealth of Organizations.* Doubleday/Currency, New York.

Sveiby, K.E. (1997). *The New Organizational Wealth: Managing and Measuring Knowledge Based Assets.* Berrett-Koehler, San Francisco.

Wiig, K.M. (1993). *Knowledge Management Foundations: Thinking about Thinking — How People and Organizations Create, Represent, and Use Knowledge.* Schema Press, Arlington, TX.

Wiig, K.M. (1994). *Knowledge Management: The Central Management Focus for Intelligent-Acting Organizations.* Schema Press, Arlington, TX.

Wiig, K.M. (1995). *Knowledge Management Methods: Practical Approaches to Managing Knowledge.* Schema Press, Arlington, TX.

Wiig, K.M., R. De Hoog & R. van der Spek (1997). "Supporting Knowledge Management: A Selection of Methods and Techniques." *Expert Systems With Applications,* vol. 13 (1), pp.15-27.

Section IV
Knowledge Management:
Knowledge Technologies

11

Intelligent Agents for Knowledge Management — Toward Intelligent Web-Based Collaboration within Virtual Teams

Seung Baek
Saint Joseph's University

Jay Liebowitz
University of Maryland–Baltimore County

Srinivas Y. Prasad
The George Washington University

Mary Granger
The George Washington University

11.1 Introduction

Today's organizations are experiencing an extremely competitive and turbulent business environment. In this environment, they are under pressure to react more rapidly and accurately to changes in customers' needs and competitors' actions. Many organizations have coped with this pressure through radical decentralization of their hierarchical structures (Halal, 1996; Mohrman et al., 1995). More recently, the proliferation of personal computers and communication networks has enabled organizations to acquire and retain distributed organizational structures (Ahuja, 1996; Tapscott & Caston, 1993; Tapscott, 1995). By using a computer network, geographically distributed people with common goals can communicate, coordinate, and collaborate their work efforts across time and space barriers. These groups have been called "virtual teams" (Geber, 1995). Jessup et al. (1996) define virtual teams as "turbo task forces, with teams forming and disbanding as needed, with team size fluctuating as necessary, and with team members coming and going as they are needed." Since members in a virtual team may, at times, participate in multiple teams and the life of their team may be very short, they must have easy, flexible access to other members, meeting contexts, and information (Jessup et al., 1996; Snizek, 1995). The virtual teams are used to support various kinds of collaborative efforts ranging from routine, mundane works to complex, creative works (Geber, 1995; Snell, 1994; Snizek, 1995). Because the virtual teams can bring together the right mix of people who have the appropriate set of knowledge, skills, information, and authority to solve difficult problems quickly and easily, they are receiving considerable attention from knowledge workers (Boldyreff et al., 1996; Jessup et al., 1996; McGuire, 1996). These knowledge workers are characterized as highly qualified individuals who need to make decisions under nonroutine, unstructured, and uncertain environments (Knight et al., 1993). Therefore, the close collaboration among them is more important than the collaborations among other kinds of workers. As the numerous benefits and advantages of the virtual teams in increasing effectiveness and efficiency of knowledge workers becomes widely recognized, organizations face a new challenge in coping with their new organizational structure (Davidow & Malone, 1993). Since many knowledge workers think of themselves as professionals, and expect a certain amount of autonomy based on their expertise, it may be hard to integrate their different specialties toward achieving a common goal (Mohrman et al., 1995). The challenge that organizations face is to turn the scattered, diverse knowledge of their knowledge workers who are working in a virtual team into a well-structured knowledge repository (Spek & Spijkervet, 1996; Wiig, 1993). A well-structured knowledge repository enhances the probability of seamless, flexible knowledge acquisition, sharing, and integration among the knowledge workers by making their goals, assumptions, and beliefs to a problem-solving approach apparent to all of them. Although a well-structured knowledge repository is called different names such as "team memory" (Walz et al., 1993), "group memory" (Nunamaker et al., 1991), and "organizational memory" (Conklin, 1992; Walsh & Ungson, 1991), all these definitions agree that it provides an instrument that brings people together so that they can share and refine their expertise.

The availability of individual team members' knowledge in a well-structured knowledge repository is expected to enhance communication and integration across the different kinds of knowledge (Wiig, 1993). Many empirical studies have shown that the sharing of knowledge among team members increases the group's performance across various domains (Cooprider & Victor, 1994; Salaway, 1987; Walz et al., 1993). Many organizations, to support knowledge sharing inside their virtual teams, have implemented a knowledge repository using Lotus Notes or Internet technology (Halal, 1996; Heijst et al., 1996; Stewart, 1991, 1995; Quinn et al., 1996). As the knowledge repository begins to emerge as a viable solution for managing the collaboration of geographically distributed teams, virtual teams, there is an increasing need to ensure that the developed knowledge repositories are robust, reliable, and fit for their purpose (Heijst et al., 1996). At present, the majority of existing knowledge repositories are designed and developed in an ad hoc fashion — following little or no formal methodology for organizing and structuring knowledge in such a way that they contribute to the effectiveness of teams (Heijst et al., 1996; Spek & Spijkervet, 1996; Wiig, 1993). Although some methodologies for organizing and structuring knowledge, repositories, such as concept maps (Gaines & Shaw, 1996), knowledge maps (Howard, 1989), spider-based method (Boland et al., 1994), and issue-based information systems (IBIS) method (Conklin, 1992, 1996), are available, they are still not practical. Because they support the processes of representing knowledge possessed by individuals or groups, they are less suitable for supporting the processes of knowledge communication and sharing. Methodologies for representing and organizing only individual arguments and thoughts are insufficient to cope with the dynamic environment of virtual teams in which new knowledge is constantly added and existing knowledge is constantly updated. In terms of indexing new knowledge and updating old knowledge, the methodologies for representing and organizing knowledge depend heavily on users. As the size and complexity of a knowledge repository increase, they are no longer effective tools for knowledge sharing and communication.

Knowledge management (KM) is suggested as a methodology for creating, maintaining, and exploiting a well-structured knowledge repository (Stewart, 1991, 1995; Wiig, 1993). KM is defined as the collection of processes that support the creation, dissemination, and utilization of knowledge between appropriate individuals, groups within an organization, and independent organizations (Spek & Spijkervet, 1996; Wiig, 1993). Whatever information technology is used for implementing a knowledge repository, it should support the processes of knowledge creation, dissemination, and renewal, as well as the structures of retention facility of a knowledge repository (Hosseini, 1995).

This chapter focuses on developing a conceptual model for KM and a framework for the roles of intelligent agents in the conceptual KM model. Furthermore, it presents the intelligent agent-based knowledge management system that helps multimedia designers share design knowledge on the Web.

11.2 Knowledge Management (KM)

To stay ahead in today's highly unstable and competitive business environment, organizations try to develop new products with better quality, faster time to market, and higher customer satisfaction. It has become increasingly apparent that potential bottlenecks in achieving these goals lie not just in labor or capital management, but in the ability to manage effectively their employees' knowledge (Quinn et al., 1996). Especially as more organizations are defined by working relationships governed by functional interdependencies rather than organizational boundaries, knowledge management is a major challenge facing modern organizations.

The rapidly growing importance of knowledge is highlighted by the fact that many organizations now attempt to organize and to make available the relevant collective knowledge of their employees by building an organizational knowledge repository (Halal, 1996). The organizational knowledge repository is managed by executives with titles like Chief Knowledge Officer, Intangible Asset Manager, Director of Lessons Learned, or Director of Corporate Learning (Halal, 1996; Quinn et al., 1996; Spek & Spijkervet, 1996). The reason for the development of knowledge repositories is the recent realization that knowledge comprises the key strategic asset of modern business (Quinn, 1992; Wiig, 1993). However, because organizations have very ambiguous ideas about how to discern what their organizational knowledge is worth, and how to convert that knowledge into useful products and services in order to maximize its earning potential, they still feel unable to manage knowledge (Davenport, 1996).

What Is Organizational Knowledge?

One of the basic questions of epistemology concerns the notion of knowledge (Popper, 1967). Many philosophers have tried to define knowledge for centuries, but they are still far from fully understanding human knowledge. Likewise, although many organization theorists and psychologists attempt to define organizational knowledge by establishing a bridge between individual knowledge and collective knowledge, organizational knowledge is still not well understood.

Spek and Spijkervet (1996) state that organizational knowledge consists of five different types of knowledge: "knowing which information is needed (know what); knowing how information must be processed (know how); knowing why information is needed (know why); and knowing where information can be found to achieve a specific result (know where); knowing when which information is needed (know when)." According to Spek and Spijkervet (1996), individual employees construct organizational knowledge by sharing these types of knowledge with other employees.

Nonaka (1991, 1994) distinguishes two types of knowledge: tacit and explicit knowledge. Explicit knowledge is specified either verbally or in writing. For this reason, it can be easily communicated and shared. Tacit knowledge is difficult to articulate, communicate, formalize, and encode, because it is an organic and intangible entity. Nonaka argues that organizational knowledge can be created through continuous dialogues between peoples' explicit and tacit knowledge.

Quinn et al. (1996) propose that individual knowledge in an organization consists of four different types of knowledge: "know-what" is the basic knowledge that individuals can acquire through extensive training; "know-how" is the ability to apply "know-what" knowledge to complex real-world problems; "know-why" is deep knowledge of cause-and-effect relationships; and "self-motivated creativity" is the highest level of knowledge, and it consists of will, motivation, and adaptability. Quinn et al. (1996) argue that the value of organizational knowledge can increase markedly as an organization helps its employees develop self-motivated creativity, and leverage this type of knowledge throughout the organization.

Although these definitions are somewhat arbitrary, we can draw two key characteristics of organizational knowledge from these definitions. First, organizational knowledge is knowledge that is shared among organizational members. All definitions emphasize that although organizational knowledge is created via individual knowledge, it is more than the sum of individual knowledge. Complete organizational knowledge is achieved only when individuals keep modifying their knowledge through interactions with other organizational members. Second, organizational knowledge is distributed. Organizational knowledge is created and managed by individuals who act autonomously within a decision domain. These conflicting characteristics make organizational knowledge management more difficult. Therefore, a powerful, intelligent information infrastructure for supporting KM should be constructed (Halal, 1996). To build the information infrastructure for KM, first we need to understand a transformation process between individual knowledge and organizational knowledge. The following section will review three different theoretical perspectives for the transformation process.

How Is Individual Knowledge Transformed into Organizational Knowledge?

The transfer mechanism between individual and organizational knowledge has been studied widely. We can find three organization theories that present different transformation strategies: organizational knowledge creation (Nonaka, 1991, 1994), organizational learning (Argyris & Schon, 1978; Garvin, 1993; Senge, 1990), and absorptive capability (Cohen & Levinthal, 1990). These theories provide theoretical foundations for developing a conceptual model about the process through which individual knowledge is transferred to organizational knowledge.

Knowledge-Creating Organization

Nonaka's (1991, 1994) descriptions of the "knowledge-creating" organization provide a useful starting point for theorizing about how an individual's personal knowledge can be transformed into organizational knowledge that is valuable to the company. Nonaka (1991, 1994) argues that organizational knowledge can be created through interactions between tacit knowledge (i.e., knowledge not easily expressed and communicated) and explicit knowledge (i.e., knowledge codified and expressed in formal language). He identifies four distinct interaction modes:

1. Tacit knowledge to tacit knowledge (e.g., an employee amplifies individual knowledge through sharing of tacit knowledge)
2. Tacit knowledge to explicit knowledge (e.g., an employee articulates the tacit knowledge into explicit knowledge, thus allowing it to be communicated to others)
3. Explicit knowledge to explicit knowledge (e.g., all team members combine their explicit knowledge by putting it into a manual or work book)
4. Explicit Knowledge to Tacit Knowledge (e.g., team members begin to internalize explicit knowledge into an organization; that is, they intuitively incorporate the explicit knowledge in their products and services)

While tacit knowledge held by individuals is the core of the knowledge-creating process, according to Nonaka (1991) practical benefits can only be realized by the dynamic interaction of all the aforementioned modes of interaction. In the knowledge-creating company, employees are continually improvising, inventing new methods to deal with unexpected difficulties and to solve immediate problems, and sharing these innovations with other employees through effective communication channels. Nonaka's knowledge-creating company focuses on the interaction between individual and organizational knowledge, rather than only using the former as an analogy of the latter.

Organizational Learning

Companies that built competitive advantages through effective information and knowledge management must continually refresh and update their intellectual capital (Garvin, 1993). This is the process of organizational learning. Recently, the concept of organizational learning has gained much attention among large organizations as they attempt to develop structures and systems that are more adaptable and responsive to internal and external changes (Garvin, 1993; Senge, 1990). Over 30 years ago Cyert and March (1963) addressed organizational learning as a process by which organizations as collectives learn through interaction with their environments. In this process, members of the organization act as learning agents, responding to changes in the internal and external environments of the organization, and sharing their experiences throughout the organization for other members (Kim, 1993). Garvin (1993) formally defines organizational learning as a process "for creating, acquiring, and transferring knowledge, and modifying individual behaviors to reflect new knowledge and insights." Organizational learning is viewed as a metaphor derived from cognition of individual members of the organization (Senge, 1990). Organizational learning requires organizations to have "a shared memory" where learning agents' (individual employees') discoveries, inventions, and evaluations are embedded. The revisions to the memory are not accomplished by the simple exchange of new knowledge for old knowledge, but by single-loop and double-loop learning within the organization (Argyris & Schon, 1978). Single-loop learning occurs when the organization adds new knowledge to its knowledge base without altering the fundamental nature of the organization's activities, whereas double-loop learning occurs when the organization questions and modifies existing norms, procedures, policies, and objectives (Dodgson, 1993). This theory explains the process through which individual learning advances organizational learning by using human memory as an analogy (Senge, 1990).

Absorptive Capability

Cohen and Levinthal (1990) argue that the ability to recognize the value of new ideas, to assimilate them, and to apply them to commercial ends depends in part on the base of prior related knowledge, both tacit and explicit. This ability constitutes what they call a firm's "absorptive capability." Using human memory as an analogy, they suggest that accumulated prior knowledge increases both the ability to put new knowledge into a shared memory and the ability to recall and to use it. They stress the role of centralized facilitators to transfer processes of knowledge across and within organizations. Since the group process for creating and maintaining knowledge is hardly self-managed, the role of the facilitator who understands group processes, and therefore can assist a group to understand its problems and finds solutions for them, is a valuable asset to any group. The facilitator augments the individual's capability for making novel linkages and associations to prior knowledge. The theory explains the roles of a shared memory and the facilitator in the knowledge management process.

Spek and Spijkervet (1996) identify four basic KM activities: knowledge creating, knowledge securing, knowledge distributing, and knowledge retrieving. Figure 1 shows an overview of KM process.

Table 1 summarizes the different perspectives of three theories regarding the transformation process that links individual and organizational knowledge.

11.3 Intelligent Agents

The recent popularity of the World Wide Web has provided a tremendous opportunity to expedite the dispersement of various knowledge creation/diffusion infrastructures (Chen & Gaines, 1996, 1997; Ives & Jarvenpaa, 1996). Because the Web enables organizations to create a knowledge repository and to extend the scope of collaboration in an easy and cost-effective manner, it creates the possibility of developing global collaborative KM platforms (Barua et al., 1995; Davenport, 1996). However, the unstructured nature of the Web creates an information overload problem. While the Web allows various kinds of knowledge to be created and disseminated across time and space barriers, it does not support the processes of using and updating the knowledge in a timely manner. Rasmus (1996) and Silverman et al. (1995) suggest the use of intelligent agents as a promising solution for assisting and facilitating these processes.

Intelligent agent technology is a rapidly growing area of research and new application development. However, because many people use the term "intelligent agent" without precisely defining it, it is poorly understood (Nissen, 1995). Even among scientists and engineers, there is, as yet, no commonly agreed on definition of an intelligent agent (Wooldridge & Jennings, 1996). As long as many people are successfully developing interesting and useful agent applications, it hardly matters that they do not have a single

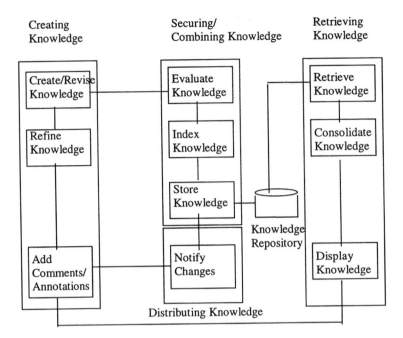

FIGURE 1   An Overview of the KM process.

universal definition of an intelligent agent. However, there is the danger that unless the issue is discussed, the term "intelligent agent" might create much confusion in the marketplace and in the research community (Wooldridge & Jennings, 1996).

What Is an Intelligent Agent?

Researchers and practitioners involved in intelligent agent technology have offered a variety of definitions, each hoping to explicate their use of the term "intelligent agent." Russell and Norving (1995) define an agent as "anything that can be viewed as perceiving its environment through sensors and effects." Selker (1994) defines an agent as "a computer program that simulates a human relationship by doing something that another person could otherwise do for you." Maes (1994) states that "agents assist users in a range of different ways: they hide the complexity of difficult tasks, they perform tasks on the user's behalf, they can train or teach the user, they help different users collaborate, and they monitor events and procedures." Maes (1995) defines intelligent agents as "computational systems that inhabit some complex dynamic environment, sense and act autonomously in this environment, and by doing so realize a set of goals or tasks for which they are designed." Hayes-Roth (1995) states that "intelligent agents continuously perform three functions: perception of dynamic conditions in the environment; action to affect conditions in the environment; and reasoning to interpret perceptions, solve problems, draw inferences, and determine actions." Foner (1994) defines intelligent agents as systems that collaborate with the users to improve the accomplishment of the user's tasks. FTP (1995) states that an intelligent agent is "an autonomous computer program that is self-contained and performs tasks on the behalf of a user or user-initiated process." IBM (1995) defines intelligent agents as "software entities that carry out some set of operations on behalf of a user or another program with some degree of independence or autonomy, and in so doing, employ some knowledge or representation of the user's goals or desires." Although there is no firm consensus on what constitutes an intelligent agent, the following capabilities are often associated with the notion of an intelligent agent: reactive, autonomous, adaptive, goal-oriented, cooperative, flexible, self-starting, mobile, temporal continuous, and personality (Etzioni & Weld, 1995; Foner, 1994; Roesler & Hawkins, 1994; Wooldridge & Jennings, 1996).

TABLE 1   A Summary of Three Organization Theories

Theory/Activity	Creating Knowledge	Securing Knowledge	Distributing Knowledge	Retrieving Knowledge
Knowledge-Creating Company (Nonaka, 1991, 1994)	Knowledge is created through interactions between tacit knowledge and explicit knowledge at two different levels: the individual and the group levels.	Once the task of a team is completed, team members incorporate tacit knowledge acquired and created in the project with explicit knowledge in the forms of documents and reports.	Knowledge is disseminated by building a cross-functional, self-organizing team.	The knowledge created is tested by various departments within the organization. This is called a "crystallization process."
Learning Organization (Senge, 1990)	Knowledge is created through communication of individual learning among co-workers.	Besides formal knowledge, informal knowledge in the form of tacit know-how, letters, memos, informal conversations should be captured, shared, and reused.	A variety of mechanisms can be used for spreading knowledge quickly and efficiently throughout the organization; this includes written, oral, and visual reports; site visits and tours; personal rotation programs; education programs and seminars.	Through double-loop learning, individuals continuously update the existing norms, procedures, and policies in the organization based on their experiences.
Absorptive Capability (Cohen & Levinthal, 1990)	Knowledge is created based on prior knowledge. In other words, the process of creating knowledge can be characterized as the process of assimilating new knowledge into preexisting knowledge.	When new knowledge is added, the existing linkages and associations among different knowledge sources need to be modified.	Individuals who stand between subunits within the organization, capture, translate, and disseminate external information in order to allow other co-workers to share it.	All available knowledge can be combined by establishing new linkages with preexisting knowledge. Diverse knowledge sources are closely linked in a shared memory.

An intelligent agent, when a new task is delegated by the user, should determine precisely what its goal is, evaluate how the goal can be reached in an effective manner, and perform the necessary actions (goal-oriented, flexible, and personality). An intelligent agent should also be capable of learning from past experience and responding to unforeseen situations with its reasoning strategies (adaptive, self-starting, and temporal continuous). It should be reactive/autonomous so that it can sense the current state of its environment and act independently to make progress toward its goal (reactive and autonomous). Finally, it often needs to interact with other agents to perform its tasks (cooperative and mobile). Table 2 shows descriptions of the general characteristics of intelligent agents.

Table 3 summarizes the characteristics of intelligent agents developed by different researchers.

Different Types of Intelligent Agents

Intelligent agents are a popular research topic in such fields as psychology, sociology, and computer science. Although an exact definition and set of characteristics are yet to be stated and chosen, many applications are developing under the term "intelligent agents." Two major applications of an intelligent agent technology can be found: personal assistants and communicating agents (Woelk et al., 1995).

TABLE 2 General Characteristics of Intelligent Agents

Property	Meaning
Reactive	Agents perceive their environment and respond in a timely fashion to changes that occur in it.
Autonomous	Agents perform their tasks without the direct human intervention, and they have control over their own actions and their own internal state.
Adaptive	Agents automatically customize themselves to the preferences of their user on the basis of previous experience and adapt to changes in their environments.
Goal-Oriented/Reasoning	Agents do not simply act in response to their environment, they exhibit opportunistic, goal-directed behavior, and take the initiative where appropriate.
Cooperative/Social	Agents interact, when they deem appropriate, with other agents and humans in order to complete their problem solving and to help others with their activities.
Flexible	Actions are not scripted; agents can dynamically choose their actions based on their perceptions to environment.
Self-Starting	Agents can decide when to act by sensing environmental changes.
Mobile	Agents can transport themselves from one machine to another and across different system architectures and platforms.
Temporal Continuous	Agents are continuously running to monitor environmental changes.
Personality	Agents have well-defined, believable personality that facilitates interaction with human users.

TABLE 3 Different Definitions of Intelligent Agent

	(1)	(2)	(3)	(4)	(5)	(6)	(7)	(8)	(9)	(10)
Hayes-Roth (1995)	X	X		X		X	X		X	X
IBM (1995)		X		X	X		X		X	X
Maes (1994)		X		X		X	X		X	X
Maes (1995)	X	X		X		X	X		X	X
FTP (1995)	X	X		X			X	X		X
Russell & Norving (1995)		X		X			X		X	X
Fonor (1994)		X	X	X	X		X		X	X
Selker (1994)	X	X		X			X		X	X

(1) Reactive (2) Autonomous
(3) Adaptive (4) Goal-Oriented
(5) Cooperative (6) Flexible
(7) Self-Starting (8) Mobil
(9) Temporal Continuous (10) Personality

Personal Assistants

The first type of intelligent agent focuses on the interaction between a user and a computer (Woelk et al., 1995). The intelligent agents are computer programs that perform repetitive, burdensome tasks on behalf of their human users (Sycara & Zeng, 1995). They have been developed for a wide range of applications, such as automatic disk back-up, information filtering, information retrieval, mail management, and meeting scheduling. Maes (1994) refers to these kinds of intelligent agents as "personal assistants." These agents differ from regular software mainly by (1) their use of machine learning techniques so that they can adapt to user habits and preferences (Maes, 1994), and (2) their use of automated reasoning so they can decide when to help the user, what to help the user with, and how to help the user (Riecken, 1994). Table 4 summarizes the most popular intelligent agent applications in this category.

While most intelligent agents provide their users with significant value when used in isolation, there is an increasing demand for programs that can collaborate — to exchange information and services with other agents, and thereby solve problems that cannot be solved alone (Indermaur, 1995).

TABLE 4 Sample of Personal Assistants

Agent	Technique	Domain
Coach (Selker, 1994)	Machine Learning	Coach records user experience to create personalized user help.
AppleSearch[1]	Rule-Based	AppleSearch is an information search and retrieval application that provides an easy-to-use, cost-effective way to search and retrieve information on local-area networks and on the Internet.
Maxim (Maes, 1994)	Machine Learning	Maxim prioritizes, deletes, forwards, sorts, and archives e-mail messages on behalf of the user.
NewT (Maes, 1994)	Machine Learning	NewT helps the user filter Usenet Netnews.
Open Sesame![2]	Machine Learning	Open Sesame! is a learning agent that observes user actions in the background, finds repetitive patterns, and automates them upon approval.
Object Lens (Crowston & Malone, 1988)	Rule-Based	Object Lens filters incoming e-mail messages based on the rules the user specifies.
Bargain Finder[3]	Rule-Based	Bargain Finder is an agent that shops for the best price for CDs and cassettes from several on-line music stores.
Fire Fly[4]	Machine Learning	Fire Fly attempts to know the user's preferences in music styles and artists and recommends a list of CDs that conform to these preferences.
Topic Agent[5]	Rule-Based	Topic Agent searches, filters, or categorizes information and delivers it according to the user's preferences.
Letizia (Lieberman, 1995)	Machine Learning	Letizia assists the user browsing the WWW by tracking user behavior and attempting to anticipate items of interest.

[1] [http://kamaaina.apple.com:80/], December 1, 1996
[2] [http://www.opensesame.com/], December 1, 1996
[3] [http://bf.cstar.ac.com/], December 1, 1996
[4] [http://www.agents-inc.com/], December 1, 1996
[5] [http://www.verity.com/products/tas_data.html], December 1, 1996

Communicating/Collaborative Agents

The second type of intelligent agent focuses more on the interaction among computing agents (Woelk et al., 1995). The basic issues addressed are those concerned with interactions among geographically distributed agents executing on heterogeneous hardware platforms. In this context intelligent agents are programs that communicate together in a universal language. Such interactions might greatly improve an agent's efficacy — one intelligent agent simply asks another agent for information instead of attempting to find information on its own. To operate with other agents, intelligent agents need to communicate more intelligently and exhibit sufficient flexibility to build up their dialogues dynamically and interactively. The ability to perform intelligent dialogues is particularly crucial when programs, written by different people, at different times, in different languages, need to collaborate with each other with little or no information about each other's capabilities.

Intelligent agents can collaborate with other agents — even if they have different underlying implementations — since they use the same communication language. There are currently two different approaches regarding a communication language for intelligent agents: procedural approach and declarative approach (Woelk et al., 1995).

The procedural scripting approach causes execution of a remote task by sending a procedural script for interpreted execution at the remote site. In other words, the recipient executes the procedure in order to perform some tasks on behalf of the sender. General Magic's Telescript is an example (White, 1995). In contrast with the procedural approach, the declarative approach is based on the idea that communication can be modeled as the exchange of declarative statements. The recipient performs an inference process in order to derive results from the sender's declarative statements. An example of this approach is knowledge query manipulation language (KQML) (Finin et al., 1993). It is unclear whether the procedural or declarative approach is better. The procedural approach is suggested when the sender is

TABLE 5   Sample of Communicating/Collaborating Agent

Agent	Communication	Domain
Visitor Hoster (Sycara & Zeng, 1995)	Declarative Approach	Visitor Hoster helps a visitor to arrange an appointment with faculty whose research interests match the interests that the visitor has expressed in his or her visit request.
Warren Inc[1] (Sycara & Zeng, 1995)	Declarative Approach	Warren Inc manages portfolio over time by finding, filtering, and evaluating relevant information over the Internet.
ADEPT[2] (Jennings, 1994)	Declarative Approach	ADEPT supports a distributed decision making environment.
SHADE[3] (Kuokka et al., 1993); PACT[4] (Cutkosky et al., 1993)	Declarative Approach	SHADE and PACT projects are primarily concerned with the information sharing aspect of the concurrent engineering problem.
MAGMA[5] (Tsvetovatyy & Gini, 1995)	Declarative Approach	The system is a prototype of the virtual free market. It allows agents to sell and buy physical and electronic goods.
Mobile Agents (TCL, Telescript) (Indermaur, 1995)	Procedural Approach	Telescript agents are mainly used for electronic commerce, and TCL agents are used for distributed information retrieval.

[1]Web Agents for Retrieval of Reliable Economic News for INvestment Counseling (Warren Inc)
[2]Advanced Decision Environment for Process Tasks (ADEPT)
[3]SHAred Dependency Engineering (SHADE)
[4]Palo Alto Collaborative Testbed (PACT)
[5]Minnesota Agent Marketplace Architecture (MAGMA)

requesting a task that the recipient does not know how to entirely perform (Woelk et al., 1995). The declarative approach is suggested for knowledge-intensive applications in which assertions need to be exchanged, or in which planning and inference are required (Woelk et al., 1995). Table 5 summarizes existing communicating/collaborative intelligent agents.

11.4   Intelligent Agents for Knowledge Management

The importance of the World Wide Web (WWW or Web) on the Internet has increased dramatically in recent years. Specifically, the potential of the Web as a commercial medium is widely recognized (Kalakota & Whinston, 1996). The Web allows companies to create a virtual marketplace where their customers can order products and services without going to a physical marketplace (Jones & Navin-Chandra, 1995), and where companies can transact business operations with their suppliers (Erkes et al., 1996).

Groups and organizations use the Web as a way to share business knowledge within a group or an organization. The result is a significant gain in efficiency over other ways of sharing knowledge (Chen & Gaines, 1996). Traditionally, team members have shared knowledge through file servers, e-mail, and groupware. However, none of these tools are fundamentally designed for facilitating an exchange of knowledge, particularly not among team members who are geographically distributed. File servers are designed to provide shared access to files and applications by a team. They focus on providing fast and reliable access to shared files rather than on helping users actually to find knowledge from them. E-mail is designed to send messages to one or more people. It is not intended to distribute shared knowledge or to enable group communication. Groupware, like "Lotus Notes" and "Key File," allows users to share knowledge, but it requires them to have the same groupware system in order to share knowledge. The emergence of the Web creates open channels for human communication and collaboration within an office, across geography, and across time (Barua et al., 1995; Ives & Jarvenpaa, 1996). Chen and Gaines (1996) state that "the Web can be considered as a large scale groupware for facilitating knowledge creation/dissemination in special interest communities." Barua et al. (1995, p. 417) also argue that "the Web has created the possibility of developing global collaborative platforms for supporting interactions between professionals in various disciplines." Recently, many Web-based groupware systems, such as

Domino of Lotus Notes[1] and Collabra of Netscape,[2] are used for collaboration and knowledge sharing among teams of people inside and outside an organization.

Many companies have used the Web as a technology for enabling knowledge sharing at both workgroup and company levels (Davenport, 1996; Halal, 1996). As the concern about security of the Web grows, companies are setting up internal Web sites called "intranets." These "intranets" facilitate collaboration and knowledge sharing over the Internet that is restricted to teams of people inside an organization (Netscape Communications Corporation, 1996). Although intranets protect private, company-specific knowledge from unauthorized users, the unstructured organization of the Web makes information overload problems unavoidable, even with state-of-the-art natural language processing techniques, index mechanisms, or the assistance of powerful search agents (Euzenat, 1996). While intranets facilitate the creation of a tremendous amount of knowledge, it is very difficult to glean knowledge efficiently from knowledge repositories in intranets. Information overload problems undermine the Web's efficiency as a knowledge management tool. Although many users have powerful search tools to access massive amounts of knowledge on the intranet, they frequently experience difficulty in finding the right knowledge (King & O'Leary, 1996). Presently, most existing search tools depend heavily on one or more keywords given by a user, and they are domain-independent searching tools. They assume that the user is capable of formulating the right set of keywords in order to retrieve the desired knowledge. Querying with the wrong, too many, or too few keywords might cause irrelevant knowledge to be retrieved. Or a keyword search might not retrieve closely relevant knowledge because it does not contain synonymous keywords. Even if the user formulates a single query with the correct keywords, it is possible that the user can still retrieve irrelevant knowledge, because the keywords can have different meanings in different circumstances or contexts.

Further, searching tools are presently unable to provide the user with time-sensitive knowledge. To perform knowledge-searching tasks quickly and efficiently, the user needs to be aware of changes in an organization's knowledge. There problems interfere with the effective use of the Web as a knowledge management tool. These are tremendous opportunities for intelligent agents in solving these problems (Crowston & Malone, 1988; Etzioni & Weld, 1995; O'Leary, 1996; Rasmus, 1996). Table 6 summarizes how intelligent agents can solve these problems to facilitate KM processes. The intelligent agents perform repetitive and mundane tasks to enhance KM on behalf of the user.

The next section will present the intelligent agent-based knowledge management system for multimedia systems design.

11.5 A Case Study: The KMS-Based Collaborative Environment for Multimedia Systems Design

An example of knowledge workers who coordinate, cooperate, and collaborate in a virtual team for solving an unstructured and uncertain problem can be observed in an information systems design team making requirement determinations and design decisions. As the complexity and the size of today's information systems are increasing rapidly, successful information systems design requires managers and designers with varied backgrounds, abilities, and cultures to coordinate, cooperate, and collaborate intensely (Granger & Schroeder, 1994). Many studies have been conducted in order to examine how individual team members acquire, share, and integrate project-relevant knowledge inside an information systems design team in the early phases of the design process. There has been a growing recognition that achieving successful information systems design requires effective coordination of ideas (Curtis et al., 1988; Kraut & Streeter, 1995; Malhotra et al., 1980; Walz et al., 1993). Because information systems design is generally performed by a heterogeneous group of people under an uncertain environment, communication bottlenecks and breakdowns for knowledge exchange are still a big problem (Curtis et al., 1988). Curtis et al. (1988), Sonnenwald (1995) and Walz

---

[1]See: [http://domino.lotus.com], November 29, 1996
[2]See: [http://home.netscape.com/collabra], November 29, 1996

TABLE 6   Roles of Intelligent Agents in KM

KM Activity	Related Theory	Findings	Sample IAs	System Features
K Creating	Gestalt theory (Mayer, 1992)	In Gestalt theory, problem representation rests at the heart of K creating activity. The theory suggests that tools and techniques should be invented for helping people represent problems in useful ways.	ObjectLens	Intelligent K creating templates Big picture view Remind/cue Automatic agenda management A structured K creating mechanism
K Securing	ACT (Anderson, 1993)	ACT theory presents a comprehensive cognitive architecture. It states that human memory consists of three major components: declarative memory, productive memory, and working memory.	IBIS CBR GroupLens	Intelligent indexing and structuring mechanisms Social indexing Issue-based indexing
K Distributing	Information theory (Schramm, 1969)	Information theory argues that the ability of individuals to generate and transmit knowledge has the potential to promote interdependency among individuals. It states that mutual awareness is an important issue for supporting communication.	CHRONO	Intelligent K Distributing Various views of changes Filtering changes Dynamic and automatic K Distribution
K Retrieving	Cognitive flexibility theory (Spiro et al., 1988)	Cognitive flexibility theory states that if users access various perspectives for solving a problem, they might get a deeper, clearer understanding about the problem. Because of the limited capability of human memory, too much knowledge makes users experience cognitive overload problems.	Internet Searching Agents	Intelligent K retrieving engine Content-based retrieving Context-based retrieving Structured presentation Automatic Query Generation

et al. (1993) argue that to reduce the coordination problems, KM processes in information systems design should be supported. Although they suggest various KM strategies that help design participants interact more efficiently and ultimately improve the quality of design outcomes and the design communication process, they do not explore the possibilities and opportunities of current information technology for the KM strategies inside an information systems design team.

Creating an environment to enhance KM among designers is particularly important when the details and implications of an information system are not understood because of its complexity (Kedzierski, 1988). Most studies addressing the coordination problems assume that the problems in small or mid-sized projects are not serious (Curtis et al., 1988; Kraut & Streeter, 1995; Walz et al., 1993). However, as the use of interactive multimedia gains popularity in education and training software development, even small-sized software development projects will involve complex communication and coordination processes. Multimedia systems, compared with text-based systems, require deeper and broader application domain and technical knowledge. In order to unify the team members with different skills and specialties, the KM processes in designing multimedia systems need to be modeled and supported with collaboration tools.

Multimedia Systems Design and KM

Multimedia systems design consists of three basic cognitive activities, which are usually performed by a group of designers: selecting information, structuring contents, and selecting media (England & Finney, 1996). By analyzing application domains and target audiences, designers decide what information should be contained in a multimedia system. Then they identify a hierarchical relationship (content structure) among the selected information sources. Finally, they decide how the selected information should be presented in a multimedia system. These activities are very closely related, rather than separated. The storyboard allows designers to streamline these activities (Kiddo, 1992). The storyboard shows how one frame relates to adjacent ones as well as to the whole concept, and what media are needed to achieve a particular effect (Kiddo, 1992). By using storyboards, design team members can explicitly communicate and negotiate how narration, images, texts, special effects, and background music/images are brought together and linked together to form a final presentation. By helping designers create, exchange, and share storyboards, we can enhance their knowledge communication and sharing activities. In order to design a knowledge management system that can support these activities, we examine the knowledge communication and sharing processes of multimedia designers.

Knowledge management (KM) is the collection of processes that support the creation, dissemination, and utilization of knowledge between individuals, groups within an organization, or independent organizations (Liebowitz & Wilcox, 1997). KM enhances the probability of seamless, flexible knowledge sharing among a group of people by making their knowledge apparent to all of them. In the context of multimedia systems design, KM can be defined as a method for systematically and actively managing and leveraging design ideas among team members while developing storyboards.

11.6 The KMS-Based Design Environment

An intelligent Web-based collaboration environment for multimedia systems design consists of three intelligent agents: user agent, knowledge manager, and knowledge agent. The intelligent agents are designed to support the KM activities on the Web. Each agent consists of a knowledge base and a set of production rules that manage its knowledge base and generate adequate interfaces for designers. In order to perform their tasks, the agents communicate with other agents through HTTP communication protocols. An agent might request information from other agents. The knowledge bases are implemented using Microsoft Access. The rules are implemented using Java Script. For the interface between Java Script and Access, we use the CFML (cold fusion markup language). The overall system architecture is shown in Figure 2. By communicating with each other, agents dynamically generate the interfaces.

By implementing these three intelligent agents on the Web, we built the KMS-based design environment for multimedia systems design. By using the knowledge management system (KMS) that consists of three intelligent agents, designers can share and communicate their storyboards freely on the Web.

Figure 3 shows a sample code of an intelligent agent. When agents are loaded, they read their databases to update their knowledge (Part 1). Whenever changes are detected in any database, all three agents are reloaded. Based on their knowledge, agents generate proper interfaces for designers (Part 2). If agents need more information from other agents, they formulate queries (Part 3).

Table 7 summarizes system requirements of intelligent agents.

Implementation of Intelligent Agents

This section presents a step-by-step description of how a designer works in the KMS-based collaboration environment. Figure 4 is the initial screen of the knowledge management system (KMS). When a designer enters his or her user name and password in the initial screen, the user agent records his or her login

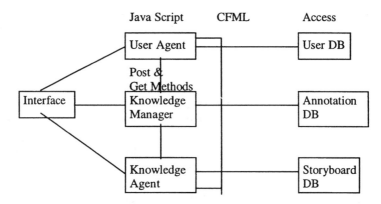

FIGURE 2   Overall architecture of KMS.

```
<CFQUERY NAME="GET_SB"   DATASOURCE="SB">          PART 1
       SELECT * FROM STORYBOARD
       WHERE ID = #URL_NO#
</CFQUERY>
```

```
<HTML><BODY>
```

```
<SCRIPT LANGUAGE="JAVASCRIPT">

Function DisplaySB(no)
{
       <CFOUTPUT QUERY="GET_SB">
       if (GET_SB.ID==no)
       {       doc.document.write('<b>......</b>');.......        PART 2
       }
       </CFOUTPUT>
}

if (UserRequest=="ShowSB") {DisplaySB(id);}

</SCRIPT>
```

```
<FORM NAME="REQUESTANNOTATION" ACTION="KMGR.CFM?NO=#ID#&|
      USER=#NAME#" METHOD="POST">

</FORM>
```

```
</BODY></HTML>                                      PART 3
```

FIGURE 3   Sample code.

date and time. Then the user agent sends a query to the knowledge manager and the knowledge agent to retrieve the storyboards and the annotations that have been modified since the last login date and time. Once it receives all information, it generates and prioritizes the agendas. Figure 5 shows the mailbox of a designer. The mailbox consists of the public mailbox and the personal mailbox. The public mailbox contains public announcements. Additionally, the personal mailbox contains the agendas that designers need to perform. If other designers comment on a storyboard or change a storyboard, its originator gets an urgent message in his or her personal mailbox. The user agent manages the mailboxes.

TABLE 7   System Requirements of Intelligent Agents

Intelligent Agents	System Requirements
User Agent	• Remember all design activities while designers use the KMS.
	• Dynamically organize designers' agendas.
	• Provide templates for knowledge creation.
Knowledge Manager	• Monitor all changes that occur in a knowledge repository and forward them to the User Agent.
	• Reformulate the designer's queries based on the designer's need.
Knowledge Agent	• Index design knowledge.
	• Detect inconsistency and generate recommendation.

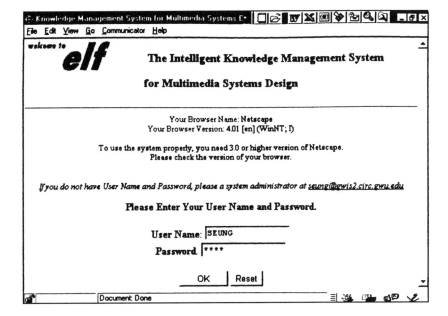

FIGURE 4   Login screen.

After checking the agendas, a designer goes into the main screen for viewing the storyboard library (Figure 6). The knowledge manager and the knowledge agent create the screen. From the screen, a designer can actually start his or her design activities. By entering a keyword, he/she can retrieve the storyboards and the annotations related to the keyword. The knowledge manager generates a query to retrieve all information that is related to the keyword. By clicking one of the storyboard numbers, he/she can view the detailed information about the selected storyboard (Figure 7). The storyboard consists of a description and hot spots. The description provides general information about what information the storyboard presents, and how the information is presented in the storyboard. The hot spots are the objects that users can interact. In Figure 7, the lower frame keeps tracking the storyboard changes. When a designer changes a specific storyboard, the knowledge manager accumulates the changes and shows a list of changes when the storyboard is retrieved. In a storyboard, a designer can also view its annotations and add new annotations (Figure 8).

The user agent has three different templates for creating storyboards, hot spots, and annotations. Figure 9 shows a template for creating storyboards. A designer can create either a superstoryboard or a substoryboard under a superstoryboard. By filling out the title and the description of a storyboard in the template, he or she can create a new storyboard. Figure 10 shows a template for creating hot spots. By clicking one of the hot spots in the library, a designer can reuse the hot spots. He or she also can create new hot spots. All serial numbers of storyboards and hot spots are generated by the KMS based on the created contents. Figure 11

FIGURE 5   Agenda screen.

FIGURE 6   Main screen.

shows a template for creating annotations. For the purpose of indexing, three different types of annotations are identified: programming note, content note, frequently-asked question.

Before a new storyboard is stored in the knowledge repository, the knowledge agent checks its consistency with the existing storyboards. If he or she detects any duplication and inconsistency, he or she provides recommendations to a designer (Figure 12). The designer might modify the storyboard until the knowledge agent does not find any duplication and inconsistency. To show the relationships among storyboards, the knowledge agent generates a matrix (Figure 13). From this matrix, a designer can navigate various storyboards.

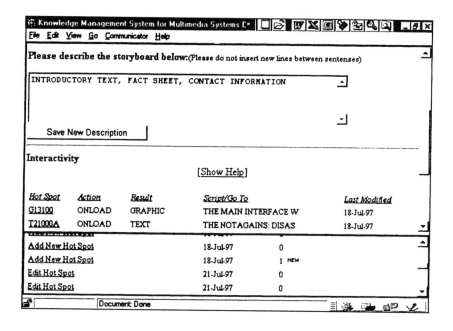

FIGURE 7   Storyboard screen & change tracking screen.

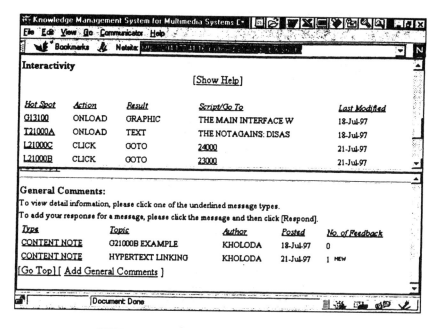

FIGURE 8   Storyboard screen & annotation screen.

11.7   Summary

By implementing three intelligent agents on the Web, we construct a collaborative design environment in which designers can share their design knowledge freely among themselves on the Web. Under this environment, designers create, revise, and share their knowledge in very structured ways. The structured design process might enhance the KM activities of designers while designing a multimedia system within

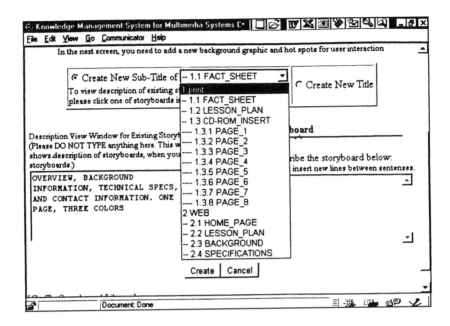

FIGURE 9   A template for creating storyboards.

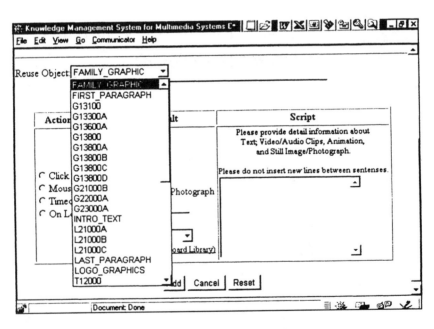

FIGURE 10   A template for creating hot spots.

a virtual working space. Without having the structured process, they might face difficulty in coordinating their design efforts.

Many intelligent agents have been developed for assisting users to retrieve knowledge from the Web. However, few intelligent agents have been developed for supporting other activities (i.e., knowledge creating, knowledge distributing, and knowledge securing activities). The paper proposes the roles of intelligent agents for supporting and streamlining KM activities.

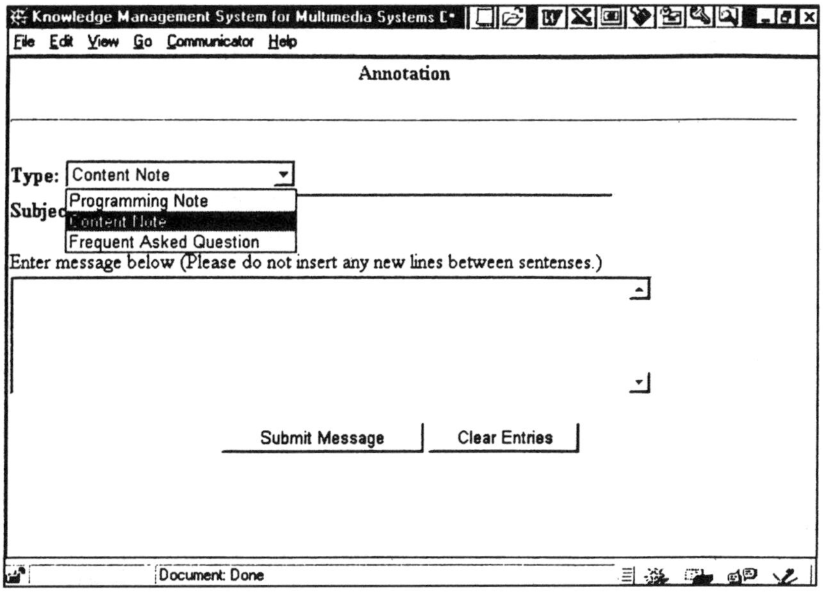

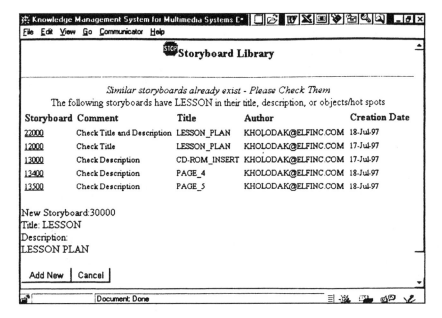

FIGURE 11 A template for creating annotations.

FIGURE 12 Consistency checking screen.

Acknowledgments

The authors thank the Electronic Learning Facilitators (ELF), Inc. (Bethesda, MD), who supported the research. Especially, the authors would like to express sincere appreciation to Marshall Lewis at ELF, Inc., whose assistance played an essential role throughout the research.

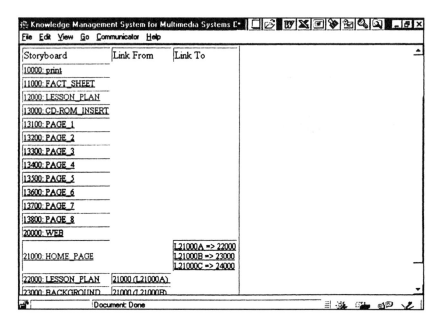

FIGURE 13 A matrix screen for viewing relationships among storyboards.

References

Ahuja, M.K. (1996), "An Empirical Study of Network and Individual Performance in Distributed Design Groups," Working Paper, University of Pittsburgh.

Anderson, J.P. (1993), *The Architecture of Cognition*, Harvard University Press, Canbridge, MA.

Argyris, C. and Schon, D.A. (1978), *Organizational Learning: A Theory of Action Perspective*, Addison-Wesley, Reading, MA.

Barua, A., Chellappa, R. and Whinston, A.B. (1995), "Creating a Collaboratory in Cyberspace: Theoretical Foundation and an Implementation," *Journal of Organizational Computing*, 5(4), 417-442.

Boland, R.J., Tenkasi, R.V. and Te'eni, D. (1994), "Designing Information Technology to Support Distributed Cognition," *Organization Studies*, 5(3), 456-475.

Boldyreff, C., Newman, J. and Taramaa, J. (1996), "Managing Process Improvement in Virtual Software Corporations," *[http://wwwagr.infomatik.uni-kl.de/~maurer/WETICE96_Papers/Julien_Newman.html]*.

Chen, L.L. and Gaines B.R. (1996), "Knowledge Acquisition Processes in Internet Communities," Forthcoming in *Tenth Knowledge Acquisition for Knowledge-Based Systems Workshop*.

Chen, L.L. and Gaines, B.R. (1997), "A Cyber Organization Model for Awareness in Collaborative Communities on the Internet," Forthcoming in *International Journal of Intelligent Systems*.

Cohen, W. M. and Levinthal, D.A. (1990), "Absorptive Capability: A New Perspective on Learning and Innovation," *Administrative Science Quarterly*, 35, 128-152.

Conklin, E.F. (1992), "Capturing Organizational Memory," *Proceedings of Groupware*, pp. 133-137.

Conklin, E.J. (1996), "Designing Organizational Memory: Preserving Intellectual Assets in a Knowledge Economy," *[http://www.cmsi.com/business/info/pubs/desom/body.htm]*, September 19, 1996.

Cooper, J.G. and Victor, K.M. (1994), "The Contribution of Shared Knowledge to IS Group Performance," *ICIS Proceedings*, pp. 285-299.

Crowston, K. and Malone, T.W. (1988), "Intelligent Software Agents," *BYTE*, December, pp. 267-271.

Curtis, B., Krasner, H. and Iscoe, N. (1988), "A Field Study of The Software Design Process for Large Systems," *Communications of the ACM*, 31(11), 1268-1287.

Cyert, R.M. and March, J.G. (1963), *A Behavioral Theory of the Firm*, Prentice-Hall, Englewood Cliffs, NJ.

Davenport, T.H. (1996), "Some Principles of Knowledge Management," *Strategy, Management, Competition,* Winter, pp. 34-40.

Davidow, W.H. and Malone, M.S. (1993), *The Virtual Corporation,* Harper Business, New York.

Dodgson, M. (1993), "Organizational Learning: A Review of Some Literature," *Organization Studies,* 14(3), 375-394.

England, E. and Finney, A. (1996), *Managing Multimedia,* Addison-Wesley, New York.

Erkes, J.W., Kenny, K.B., Lewis, J.W., Sarachan, B.D., Sobolewski, M.W. and Sum, R.N. (1996), "Implementing Shared Manufacturing Services on the World-Wide Web," *Communications of ACM,* 39(2), 34-45.

Etzioni, O. and Weld, D.S. (1995), "Intelligent Agents on the Internet: Fact, Fiction, and Forecast," *IEEE Expert,* August, pp. 44-49.

Euzenat, J. (1996), "Corporate Memory through Cooperative Creation of Knowledge Bases and Hyper-Documents," Forthcoming in *Tenth Knowledge Acquisition for Knowledge-Based Systems Workshops.*

Finin, T., Weber, J., McGuire, J., Shapiro, S. and Beck, C. (1993), "Draft Specification of the KQML Agent Communication Language," *[http://www.cs.umbc.edu/kqml/kqmlspec/spec.html].*

Foner, L.N. (1994), "What's An Agent, Anyway?: A Sociological Case Study," *[http://www.media.mit.edu/people/foner/Julia/Julia.html].*

FTP Software, Inc. (1995), "Introduction to Intelligent Agents," *White Paper, [http://www.ftp.com/Cyber-Agents],* Andover, MA.

Gaines, B.R. and Shaw, M.L. (1996), "Porting Interactive Learning Tools to the Web," *Ed-Media '96 Tutorial Notes.*

Garvin, D.A. (1993), "Building a Learning Organization," *Harvard Business Review,* July–August, pp. 78-91.

Geber, B. (1995), "Virtual Team," *Training,* April, pp. 36-40.

Granger, M.J. and Schroeder, D. (1994), "Integrating the Internet into the Business Environment," *Industrial MGT and Data Systems,* 94(8), 37-40.

Halal, W.E. (1996), The New Management, Berrett-Koehler Publishers, San Francisco, CA.

Hayes-Roth, B. (1995), "An Architecture for Adaptive Intelligent Systems," *Artificial Intelligence,* 72, pp. 329-365.

Heijst, G., Spek, R. and Kruizinga, E. (1996), "Organizing Corporate Memories," Forthcoming in *Tenth Knowledge Acquisition for Knowledge-Based Systems Workshop.*

Howard, R.A. (1989), "Knowledge Maps," *Management Science,* 35(8), 903-922.

IBM (1995), "Intelligent Agent Strategy," *White Paper, [http://activist.gpl.ibm.com:81/WhitePaper/ptc2.htm].*

Indermaur, K. (1995), "Baby Step," *BYTE,* March, pp. 97-104.

Jennings, N. and Wooldridge, M. (1996), "Software Agent," *IEEE Review,* January, pp. 17-20.

Jessup, L., Valacich, J., Dennis, A. and Wheeler, B. (1996), "Real-Time Collaboration on the World Wide Web: Supporting Virtual Teams Across Space and Time," *Proceedings of American Information Systems (AIS) Conference, Available as [http://hsb.baylor.edu/ramsower/ais.ac.96/papers/LENAIS96.htm].*

Jones, D.H. and Navin-Chandra, D. (1995), "IndustryNet: A Model for Commerce on the World Wide Web," *IEEE Expert,* October, pp. 54-59.

Kedzierski, B.I. (1988), "Communication and Management Support in System Development Environment," In I. Greif (ed.), *Computer-Supported Cooperative Work: A Book of Reading,* Morgan Kaufmann, San Mateo, CA., pp. 253-268.

Kiddoo, K. (1992), "Discovering the Rules: Initiative and Teamwork in Multimedia Project Design," *Multimedia Review,* 3(4), pp. 46-51.

Kim, D. H. (1993), "The Link between Individual and Organizational Learning," *Sloan Management Review,* Fall, pp. 37-50.

King, D. and O'Leary, D. (1996), "Intelligent Executive Information Systems," *IEEE Expert,* December, pp. 30-35.

Knight, D., Murray, F. and Willmott, H. (1993), "Networking as Knowledge Work: A Study of Strategic Interorganizational Development in the Financial Services Industry," *Journal of Management Studies,* 30(6), 975-995.

Kraut, R.E. and Streeter, L. (1995), "Coordination in Software Development," *Communications of the ACM,* 38(3), 69-81.

Lieberman, H. (1995), "Letizia: An Agent That Assists Web Browing," *Proceedings of the International Joint Conference on AI.*

Liebowitz, J. and Wilcox, L. eds. (1997), *Knowledge Management and Its Integrative Elements,* CRC Press, Boca Raton, FL.

Maes, P. (1994), "Agents that Reduce Work and Information Overload," *Communications of the ACM,* 37(7), 31-40.

Maes, P. (1995), "Intelligent Software," *Scientific American,* 273(3), 84-86.

Mayer, R.E. (1992), *Thinking, Problem Solving, Cognition,* W.H. Freeman and Company, New York.

McGuire, E.G. (1996), "Coordination of Distributed Teams in Process-Driven Development," *Proceedings of American Information Systems (AIS) Conference, Available as [http://hsb.baylor.edu/ram-sower/ais.ac.96/papers/mcguire.htm].*

Mohrman, S.A., Cohen, S.G. and Mohrman, A.M. (1995), *Designing Team-Based Organizations,* Jossey-Bass Publishers, San Francisco, CA.

Netscape Communications Corporation (1996), "The Netscape Intranet Vision and Product Roadmap," *[http://home.netscape.com/comprod/at_work/white_paper/intranet/vision. html].*

Nissen, M. (1995), "Intelligent Agents: A Technology and Business Application Analysis," *[http://haas.berkeley.edu/~heilmann/agents.html].*

Nonaka, I. (1991), "The Knowledge Creating Company," *Harvard Business Review,* Nov–Dec, pp. 96-104.

Nonaka, I. (1994), "A Dynamic Theory of Organizational Knowledge Creation," *Organization Science,* 5, pp. 14-37.

Nunamaker, J.F., Dennis, A.R., Valacich, J.S., Vodel, D.R. and George, J.F. (1991), "Electronic Meeting Systems to Support Group Work," *Communications of the ACM,* 34(7), 40-61.

O'Leary, D.E. (1996), "AI and Navigation on The Internet and Intranet," *IEEE Expert,* April, pp. 8-10.

Popper, K.R. (1967), "Knowledge: Subjective vs. Objective," in D. Miller (ed.), *Popper Selections,* Princeton University Press, pp. 58-77.

Quinn, J.B. (1992), *Intelligent Enterprise,* The Free Press, New York.

Quinn, J.B., Anderson, P. and Finkelstein, S. (1996), "Managing Professional Intellect: Making the Most of the Best," *Harvard Business Review,* March–April, pp. 71-80.

Rasmus, D.W. (1996), "Mind Tools: Connecting to Groupware," *PC AI,* September/October, pp.32-36.

Roesler, M and Hawkins, D.T. (1994), "Intelligent Agent," *ONLINE,* July, pp. 18-32.

Russell, S. and Norving, P. (1995), *Artificial Intelligence: A Modern Approach,* Prentice-Hall, Englewood Cliffs, NJ.

Salaway, G. (1987), "An Organizational Learning Approach to Information Systems Development," *MIS Quarterly,* June, pp. 245-264.

Selker, T. (1994), "Coach: A Teaching Agent that Learns," *Communication of the ACM,* 37(7), 92-99.

Senge, P.M. (1990), *The Fifth Discipline,* Doubleday, New York.

Silverman, B.G., Bedewi, N. and Morales, A. (1995), "Intelligent Agents in Software Reuse Repositories," *CIKM Workshop on Intelligent Information Agents.*

Snell, N.W. (1994), "Virtual HR: Meeting New World Realities," *Compensation and Benefit Review,* November–December, pp. 35-43.

Snizek, W.E. (1995), "Virtual Offices: Some Neglected Considerations," *Communications of the ACM,* 38(9), 15-17.

Sonnenwald, D.H. (1995), "Contested Collaboration: A Descriptive Model of Intergroup Communication in Information System Design," *Information Processing and Management,* 31(6), 859-877.

Spek, R. and Spijkervet, A.L. (1996), "A Methodology for Knowledge Management," *Tutorial Notes of The 3rd World Congress on Expert Systems,* Seoul, Korea.

Spiro, R., Coulson, R., Feltovich, P. and Anderson, D. (1988), "Cognitive Flexibility Theory: Advanced Knowledge Acquisition in Ill-Structured Domains," in D. Nix and J. Jehng (eds.), *Cognition, Education and Multimedia*, Erlbaum, Hillsdale, NJ.

Stewart, T.A. (1991), "Brainpower," *FORTUNE*, June 3, pp. 44-60.

Stewart, T.A. (1995), "Mapping Corporate Brainpower," *FORTUNE*, October 30, pp. 209-212.

Sycara, K. and Zeng, D. (1995), "Coordination of Multiple Intelligent Software Agents," *Forthcoming in International Journal of Cooperative Information Systems*.

Tapscott, D. and Caston, A. (1993), *Paradigm Shift*, McGraw-Hill, New York, NY

Tapscott, D. (1995), *The Digital Economy*, McGraw-Hill, New York, NY.

Thomas, J.C. and Carroll, J.M. (1979), "The Psychological Study of Design," *Design Studies*, July, 1(1), pp. 5-11.

Walsh, J.P. and Ungson, G.R. (1991), "Organizational Memory," *Academy of Management Review*, 16(1), pp. 57-91.

Walz, D.B., Elam, J.J. and Curtis, B. (1993), "Inside a Software Design Team: Knowledge Acquisition, Sharing, and Integration," *Communications of the ACM*, 36(10), pp. 63-77.

Walz, D.B., Elam, J.J., Krasner, H. and Curtis, B. (1987), "A Methodology for Studying Software Design Teams: An Investigation of Conflict Behaviors in the Requirements Definition Phases," in G. Olson, E. Soloway and S. Sheppard (eds.), *Empirical Studies of Programmers*, Vol. 2, Albex, Norwood, NJ, pp. 83-99.

White, J.E. (1995), "Telescript Technology: An Introduction to the Language," *White Paper*, [http://www.genmagic.com/TeleScript], General Magic, Inc., Sunnyvale, CA.

Wiig, K. (1993), *Knowledge Management Foundation*, Schema Press, Arlington, TX.

Woelk, D., Huhns, M., and Tomlinson, C. (1995), "Uncovering the next generation of active objects," *Object Magazine*, July/August, pp. 33-40.

Wooldridge, M.J. and Jennings, N.R. (1996), *Intelligent Agents*, Springer-Verlag, Berlin.

12

Groupware: Collaboration and Knowledge Sharing

David Coleman
Collaborative Strategies

Groupware, software that supports the ability for two or more people to communicate and collaborate, is the cornerstone for most electronic knowledge sharing today. Our studies show that many organizations that have implemented groupware (mostly Lotus Notes) do not have a high degree of knowledge sharing. Much of this is due not to the software, but to our competitive, rather than collaborative, culture. The diagram below shows the coevolution of collaborative culture and technology.

12.1   The Evolution of Collaboration and Knowledge Management

At the bottom of Figure 1 we see corporate cultures evolving from networked cultures to e-mail cultures (where most companies are today). The leading companies today are migrating from e-mail to collaborative (groupware) cultures. This is the area where most of the companies we benchmarked were. We believe once these technologies are in place and adopted by an organization, the collaboration that ensues will cause a huge increase in the generation of knowledge (actionable information). This increase in knowledge generation will require corporate cultures to evolve to the next level of managing this knowledge. A few companies we benchmarked are moving into this arena, especially Arthur Andersen, whose

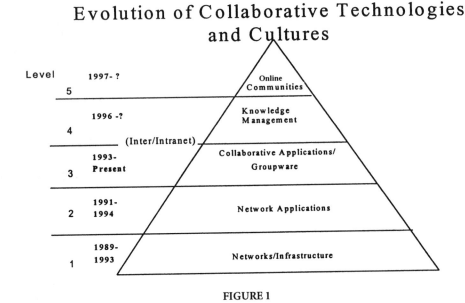

FIGURE 1

product is knowledge. We see the following trend. Companies start to organize, access, index, manage, and apply knowledge. Once this happens knowledge tends to gather around two organizing principles, relationships and people.

When organizations realize the importance of these organizing principles (relationships and people) to managing knowledge and intellectual capital, they will start to build community. Most organizations now are a network of small communities. What we mean here is that they build the electronic and cultural infrastructure to support these communities. This includes not being too heavy-handed in managing the process. Those that have built purposeful communities have seen great benefits from them.

The common thread through all stages is "collaboration," the ability to work together and exchange information and knowledge. We believe collaborative practices need to identify or show clear economic benefits in order to be successful as well as sustained. But what exactly do we mean by "electronic collaboration" or groupware?

12.2   What Is Groupware or Electronic Collaboration?

Groupware is an umbrella term describing the electronic technologies that support person-to-person collaboration. Groupware includes e-mail, electronic meeting systems (EMS), desktop video conferencing (DVC) as well as systems for workflow and business process reengineering (BPR). Technologies that support collaboration are in greater demand today than ever before, and in recognition of that fact vendors are integrating collaboration technologies into their products. Distributed workforces, information overload, and getting products to market as quickly as possible are just a few of the motivations pushing collaboration technology development. In this chapter, we will discuss many of the issues fundamental to groupware strategy and success.

12.3   What Does Groupware Really Do?

First and foremost, groupware supports the efforts of teams and other paradigms, which require people to work together, even though they may not actually be together, in either time or space. Groupware maximizes human interaction while minimizing technology interference.

12.4   Definitions of Groupware

Groupware is a relatively new term, first coined in 1978. The following most commonly used definitions are presented by industry leaders:

> *Intentional group processes plus software to support them.* Peter and Trudy Johnson-Lenz, 1978
> *A co-evolving human-tool system.* Doug Englebart, 1988
> *Computer-mediated collaboration that increases the productivity or functionality of person-to-person processes.* David Coleman, 1992

12.5   Groupware Taxonomy

The twelve functional categories listed below form a logical taxonomy, which includes a separate category for groupware services, a new category for groupware applications, and a special category for the emerging Internet-based collaborative applications and products.

1. Electronic mail and messaging
2. Group calendaring and scheduling
3. Electronic meeting systems
4. Desktop video and real-time data conferencing (synchronous)
5. Non-real-time data conferencing (asynchronous)
6. Group document handling
7. Workflow
8. Workgroup utilities and development tools
9. Groupware services
10. Groupware and KM frameworks
11. Groupware applications
12. Collaborative–Internet-based applications and products

Following is a partial list of products associated with each category and a sampling of outstanding issues for each category. These issues direct attention to some of the technical, organizational, and cultural challenges associated with each category and often present questions one might ask if considering the use of this product category in a specific organization. It is important to realize that many products fit into more than one category. For example, Lotus Notes fits into many categories because of its broad range of functionality.

1. Electronic Mail and Messaging:

Includes messaging infrastructures and e-mail systems

Sample products:

cc:Mail/Notes mail — Lotus/IBM	Eudora — QUALCOMM Inc.
Microsoft Mail/Exchange	Banyan Intelligent Messaging — Banyan Systems Inc.

Issues:

- Standards, XAPI, MAPI, X.400, X.500 (directory services)
- How to integrate multiple mail systems in one enterprise
- Security and who owns my e-mail?
- Etiquette and the efficient use of e-mail
- Filters, agents, and the ability to deal with 100s of messages a day

2. Group Calendaring and Scheduling

Products for calendar, meeting and resource coordination

Sample products:

Lotus Organizer — IBM/Lotus	OnTime — Open Text Corporation
Synchronize — CrossWind Technology	Meeting Maker — ON Technology Corporation
Microsoft Schedule +	

Issues:

- Proliferation of meetings because they are now easier to schedule
- Privacy for personal calendars (big brother is watching!)
- Enough users in the company to make it worthwhile
- Scheduling across multiple time zones

3. Electronic Meeting Systems (EMS)

Real-time conferencing systems (local and remote) as well as collaborative presentation systems

Sample products:

Group Systems — Ventana	MeetingWorks — Enterprise Solutions
Council Services — CoVision	Option Finder — Option Technologies
Facilitate.com — Facilitate.com, Inc.	TeamTalk — Trax SoftWorks

Issues:

- Integration with calendaring/scheduling systems
- Post-meeting follow-through; action items, goals, commitments
- Affordability of desktop videoconferencing
- Availability of multipoint conferencing
- Lack of standards limiting the application of the technology
- Acceptance within the corporate culture

4. Desktop and Real-Time Data Conferencing:

The focus is on real-time, rather than BBS or Notes. All products in this category store documents, and/or allow others to see and work on documents simultaneously, or on each other's screen or on a whiteboard.

Sample products:

PlaceWare Conference Center — PlaceWare	PictureTalk — Picture Talk
NetMeeting — Microsoft	FarSight/Net.120 — DataBeam/Lotus

Issues:

- Control of the cursor on the screen?
- Number of people who can conference efficiently?
- Role of the facilitator. Is a facilitator needed?
- Interaction/baud rates, equipment compatibility
- Internet and intranet availability
- Post-meeting follow through; action items, goals, commitments

5. Non-Real-Time Conferencing

Synchronous conferencing is most like a bulletin board, where you carry on a conversation over time, leave a message for someone and they answer it, and you can respond to them later. These messages can be public (as in a BBS) or private (as in a Notes discussion database).

Sample products:

Collabra — Netscape
eRoom — Instinctive Technologies
Lotus Notes/Domino — IBM/Lotus

WebBoard — O'Reilly
Groupwise — Novell
FirstClass Intranet Server — SoftArc Inc.

Issues:

- Number of people who can conference efficiently?
- Role of the facilitator. Is a facilitator needed?
- Maximizing the benefits of conferences/discussion databases; ROI
- Replication, network topologies, scalability
- Transaction-based vs. store and forward databases
- Support for worldwide locations
- Integration with legacy systems
- Integrating with electronic calendaring and scheduling systems
- Post-meeting follow through; action items, goals, commitments

6. Group Document Handling

Group editing, shared screen editing work, group document/image management and document databases

Sample products:

Domino.Doc — Lotus
MarkUp — Mainstay Software
Livelink — Open Text

Enterprise Document Management System — Documentum, Inc.
OnGo Document Management — Uniplex

Issues:

- Page mark-up standards such as SGML, HTML, and CALs.
- Support for word processors and page layout programs
- Version control and document security
- Integration with enterprise document/image databases or repositories
- Where does group document management stop and multimedia begin?
- Data integrity and integration with other documents and repositories
- Compression issues

7. Workflow and Process Tools

Workflow process diagramming and analysis tools, workflow enactment engines, electronic forms routing products

Sample products:

Action Works Metro — Action Technology
Staffware — Staffware
JetForm — JetForm Corp.

LiveLink — OpenText
MQ Series Workflow — IBM

Issues:

- Workflow coalition standards
- Passing documents and information between products
- Automating poor processes
- Integration with EDI and other customer services

8. Workgroup Utilities and Groupware Development Tools

Utilities to support group working, remote access to someone else's computer and specific tools for workgroup applications development.

Sample products:

Windows for Workgroups — Microsoft	ReplicAction — Casahl Technologies
Lotus Notes/Domino — Lotus Notes Pump	

Issues:

- What functionality should be part of the OS and what functionality should be part of the application?
- What are the decision-making issues when deciding whether to develop for the OS, GUI, or network?
- How to ensure issuer compatibility; standards; object-oriented (reusable) code; licensing (network, multimedia, intellectual property rights).

9. Groupware Services

Services to support collaboration.

Sample:

Planning and implementation	Business process reengineering
Application development	Knowledge management
Training and maintenance	Electronic meeting facilitation
Collaborative assessment	On-line community building
Change management	Consulting

Issues:

- Expertise is a most valuable commodity in the groupware market. It is highly unusual to find all the necessary expertise in house. Additionally, no single vendor offers a complete groupware solution and reengineering often requires a multiple of products and service vendors in order to collaborate. How do you identify and pull together the resources best suited to your organization?
- How are meetings facilitated successfully?
- What tools are best suited for reengineering?
- How to identify the problems with the greatest potential for turnaround from groupware?
- How are consultants best used? What do they know that people in your organization don't?
- It is imperative that top management and all stakeholders support any process change. How do you enlist and sustain their support.
- How to evaluate the return on investment of your groupware?

10. Groupware and KM Frameworks

This meta-category focuses on products that help integrate "islands of collaboration" to make seamless across computer platforms, operating systems, e-mail systems and network architectures.

Sample products:

GroupWise — Novell   Lotus Notes/Domino — Lotus/IBM
TeamOffice — ICL/Fujitsu   Netscape SuiteSpot Servers — Netscape
GoldMetal — GoldMetal, Inc.

Issues:

- Integrating the desktop while supporting collaborative efforts
- Security
- Can frameworks products help collaboration outside of the organization?
- Will establishing groupware standards make frameworks less attractive

11. Groupware Applications

Vertical applications that use collaborative technologies to either enhance processes or support collaboration in a specific work environment

Sample products:

BAI-5000 Distribution Management System —
    Business Automation   MedTrak — MedTrak Systems
CustomerFirst — RTI (Repository Technologies, Inc.)   CenterPoint — Bank of Montreal

Issues:

- Customizing applications; infrastructure and cost issues
- Vertical market competition
- Does application solve specific collaborative business need?
- Integration with existing legacy systems

12. Collaborative Internet-Based Applications:

Many collaborative functions are moving to the WWW and use the Internet as the input and output while still using traditional groupware on the LAN.

Sample products:

Domino/e-suite — IBM/Lotus   Knowledge Manager — Aeneid/Gale Group
LiveLink — Open Text   Wisebot — Tetranet Software
Learning Space — Lotus   CommonQuest — Imana

Issues:

- Application customization for seamless collaboration on the WWW
- Costs of publishing to/from the Web
- Data/information storage
- Balance between security and collaboration
- Limitations of traditional groupware relative to Web applications
- Limitations of Web applications relative to traditional groupware
- Integration with existing legacy systems

Although a taxonomy is useful for classification, do not get trapped into believing your product or the product you have selected will fit neatly into one category. The Web has changed everything for collaboration. Vendors of traditional groupware are moving their products onto the Web and adding functionality that lets them support multiple categories. The ability of the Web to offer a common platform through a browser has made collaboration much more popular. The large learning curve that Notes had is now gone, and collaboration is no longer so expensive that only large companies can investigate and apply it. Now anyone with a server and software can support collaboration. There are even services on-line like e-room or instinctive or Changepoint that all offer their products as services for small companies. Even ISP's like Concentric are starting to offer this as a service. Lotus, once offering Notes for $64,000 a license, is even going after the smaller business with their Teamroom and e-suite products.

However, you choose to focus on collaboration, through the tools or the people, it is important to realize that collaboration is a change in communications behaviors. If you change communications behaviors you change the culture and also the structure of the organization. How will you the collaborative champion deal with these changes? Will you plan for them? Will you react to them? Will you deal with the people and reward issues proactively, or will you focus on the technology?

12.6 Managing the Changes Collaboration Causes

Collaborative Strategies' research, respected experts, and empirical evidence all support the fact that organizational change is disruptive and requires commitment from everyone involved if the change is going to result in improved corporate performance. Insofar as introducing collaborative working into mature companies does represent significant change, enlisting support and commitment from key stakeholders who are able to identify economic incentives that justify instituting new practices is essential. Although all the companies we benchmarked were able to "intuit" the value of collaboration, very few actually developed a business case for it.

This chapter is focused on knowledge sharing, which we see as the intersection of KM and groupware/electronic collaboration. Now that we have established our conceptual framework, we will work through some of the issues around knowledge sharing in the rest of the chapter.

12.7 What We Do (and Don't) Mean By Knowledge Management

In reality KM is a misnomer and an oxymoron. No one wants his or her knowledge managed. Rather, they want their environment or the products of the knowledge managed. After all, how can you manage something so intangible as knowledge? Compounding the difficulty of understanding and using this term is growing confusion about what services fit under the definition of "knowledge management." There are now over 400 products identified in the KM market. For example, many of the document and image management vendors, with wonderful products to help manage explicit information, have jumped on the KM bandwagon. This has muddied the waters a bit in the KM market and has made a new and emerging market even unclear.

KM is really an umbrella term for a wide variety of interdependent and interlocking functions:

- Knowledge creation
- Knowledge valuation and metrics
- Knowledge mapping and indexing
- Knowledge transport, storage, and distribution
- Knowledge sharing

Not all of these functions happen all the time, and some must follow others in a process in order to produce something of value. You can't retrieve knowledge until you create and store it in a way that makes it retrievable. You can't distribute and transport knowledge until you can retrieve it. And just because you can retrieve information or knowledge does not mean it has value as intellectual capital. Intellectual capital is defined as "intellectual material — knowledge, information, intellectual property, experience — that can be put to use to create wealth" (Thomas Stewart, *Intellectual Capital,* 1997 Doubleday, New York). What this definition does not say is that knowledge has value only if it meets time and target criteria. For example, if I told you, an OD professional, all the secret tricks for DOS 2.0 you probably would not care. Why? Because this hard-won knowledge is (1) out of date, since we are currently on DOS 6.0 (wrong time) and (2) you are the wrong target for this knowledge since you do not have the context or interest to apply it. Therefore, it has no value to you as intellectual capital.

Another issue regarding intellectual capital is measurement. How do you measure and value intellectual capital. Our old accounting system deals mostly with the costs of tangible assets of a manufacturing firm, yet Microsoft, which manufactures nothing, has a stock value many times its tangible assets. Accountants are baffled, and it often falls to HR/OD professionals or business unit managers to determine how to calculate the value of a firm's intellectual capital. Marc Auckland, an OD professional in the Global Training Department at British Telecom, was given the task of justifying a several billion pound valuation of the BT brand to the telecommunications regulators.

HR/OD professionals also can get involved in KM in many other ways. In our work with British Telecom, these professionals were trained to assess and evaluate a group's ability to collaborate. They also provided workshops to business units about KM created executive briefings on KM and set up a network of top executives at the company to share knowledge. They also become the stewards of KM and are a clearinghouse for information on all the different KM projects taking place at BT. As OD/HR professionals they also have been involved in the initial implementations of many of the KM projects that were started at BT in the last 18 months, many of which focused on knowledge sharing.

I should note that about 80% of the collaboration and KM projects we do work on are initiated and involve IT/MIS professionals. Because they always focus on the tools, the first question we are asked is "what tool should we use for…?" The standard answer to this question is "what do you want to do in your collaboration and KM program?" Asking this question is similar to saying, "I have a hammer, now I can build a house." In reality, we look at tool selection at the end of Phase 2 of our methodology, and insist that our clients do an assessment and then develop clearly defined goals for their KM or collaboration projects. Finally, we have them examine their current opportunities for KM and collaboration to ensure that the right tools are being picked for the right job. We do all this before looking at selecting tools.

12.8 Focus on Knowledge Sharing

Many organizations try to build KM programs and architectures before building the cultural/collaborative or business foundations for these programs. KM programs, like programs in electronic collaboration, are of little value unless tied directly to easily seen business benefits. No matter how good the technology systems that support knowledge sharing are, without these foundations negative attitudes, poor return-on-investment (ROI), and resistant behaviors are generally the result. When not dealt with at a cultural level, those asked to contribute often see such requests as:

- Extra work
- Not of great benefit to them (what's their ROI?)
- Not of great benefit to the organization, because they are not sure who will see and use it
- Possibly of benefit to others, but of little value to them

With such attitudes and beliefs, it is no wonder that people have no motivation to share and therefore are unwilling to give away what they have learned (sometimes at great expense) to others for free.

In reality, sharing knowledge can be hard work. It requires you to reflect and think about what you know before you transfer it to paper or computer and share it with others. It is in this process of reflection that knowledge is developed.

Reward structures often need to be developed to encourage knowledge sharing. Such structures must answer the question "what's in it for me?" This is a radical change for technical organizations that think they are putting in a new information system, and end up dealing with compensation issues. It comes as a shock to them, and is generally outside their area of expertise. Their expertise is not in changing behavior, but in programming and infrastructure. However, this is one area where having OD professionals on the team can really make a difference. Especially if these OD professionals have expertise in change/transformation issues. Their ability to deal with the "human" side of this issue makes them just as valuable as the technology implementers in dealing with KM and collaboration technologies. Without their help, two thirds of these collaborative information systems projects are doomed.

A recent example can help illustrate the importance of the change management component to the success of a KM project: One group we know of worked on a knowledge management project with an oil company. The company spent $15 million on the project, which included five groups. Four of the groups received extensive coaching and mentoring on communications and on ways to support collaborative and knowledge-sharing behaviors. Over $9 million of the project budget was spent on this, rather than on technology. The only group that failed was the one that focused on technology and did not receive coaching.

Sharing knowledge is different from sharing information. Knowledge is not as easily duplicated as information. Knowledge is "sticky" and has a highly context-sensitive nature. What I am writing in this chapter is information; the way I write and think is knowledge. Obviously, it is much harder to duplicate or transfer knowledge. Those that truly understand this difference between knowledge and information are usually more willing to share their knowledge because they know that they have lost nothing: when you share knowledge you still retain what you know. But in reality it really comes down to how much you believe in and value yourself. Many people in organizations believe that their value to the organization is just what they know. However, we know it is so much more, people are more than their knowledge, experiences, and behaviors.

The question for an organization is how does it value each individual and their contribution. Today most value is again accounted for in dollars, but even that is shifting. Today, here in Silicon Valley, where intellectual capital is king, and competition for good people is fierce, the only way to keep someone in your organization is stock. By giving employees stock or equity, you are essentially granting them a piece of all the intellectual capital that is part of the value of the company that the market sees. Why then do employees, even those with stock, hoard knowledge and information with the mistaken belief that if they share it, they will be "worth less" to their organization?

Today, to be successful in most endeavors, the sharing of both information and knowledge is critical. However, there is a core belief in many people, which is reinforced by most organizational cultures, that an individual's only value to an organization is what and who they know and how much work they produce. Given the need to share knowledge in today's turbulent environment, hoarding does not seem like a rational act. This is especially true in competitive organizations and environments where either you win or I win. People asked to share information can have a legitimate fear that others will copy or use that information without crediting their contribution. Often in organizations, people are not credited for the knowledge and information they contribute, nor is value attached to their knowledge. Without these things it is very hard for even a team player to trust that they will be acknowledged or compensated for their contribution. When trust is not a component of an organization's culture, knowledge sharing is unlikely to occur. Trust is a critical prerequisite to knowledge sharing.

But how do you establish trust with someone you have never met, from another company, another culture, who lives on the other side of the world? In a world filled with "virtual relationships" where we may not be able to see or hear that other person, how do we develop trust networks to aid us in our increasingly complex business endeavors? One of the challenges of the OD professional is to find ways to help create a shared context in virtual relationships, which will in turn promote trust and the ability to share knowledge. One of the best references on this topic is a book called *Global Work, Bridging Distance, Culture and Time* (Jossey-Bass, 1994), by Mary O'Hara Devereaux and Robert Johansen, which describes teams working together in different parts of the world. O'Hara Devereaux and Johansen found that when team members from the U.S. stayed in the homes of hosting team members in Sweden, there was a much higher degree of collaboration than existed among those where the U.S. team members had stayed in a hotel. They attributed this better teamwork to increased context for those with "home stay" as opposed to hotel stay.

One of the best ways to create trust for knowledge sharing is to create "context." Although context generally refers to an understanding of the external world, it is also a mental framework in which you place someone. It gives you an idea of what their internal world is, their values, ideas, beliefs, as well as their character and where there are commonalties and differences. Context has value in enabling people to discern the meaning, rationale, and justification for observed behaviors and actions. With that knowledge, the individual can then begin to develop an idea of how the other might treat them and assess trustworthiness

In the above research, they hypothesized that "home stay" helped not only in building context, but also in "bonding" between workers. The ability to trust and ultimately to share knowledge works both ways. Once you have bonded and gained the trust of Employee A and he says Employee B is "OK" and trustworthy, generally you are able to extend your trust to Employee B, even though you do not know him, because you trust your filter (Employee A in this case). This process has even greater validity in "virtual relationships."

12.9   Tools and Technologies for Knowledge Sharing

Today, the formalized process of Web-based knowledge sharing is very immature. The good news is that you're getting in on the ground floor. The bad news is the train has left the station and it is picking up speed. As a matter of fact, we have just all realized that it is a "bullet" train. Five years ago if you had asked anyone if they were using Web-based KM tools they would have looked at you as if you were a Martian. Today there are hundreds of vendors extolling the virtues of such tools.

What is important here for OD/HR professionals is to realize that these tools are enablers. They support the inexpensive transportation, creation, distribution and sharing of knowledge. Since the economics of KM have changed so radically, it has become a function that more and more organizations are realizing they have to deal with. Since most organizations, like the accountants mentioned above (who only deal with the costs of labor and tangible assets), like to grapple with something tangible, they first look at the infrastructure and tools available for KM and collaboration. In focusing their efforts on the easier tangible side of KM, they often neglect the more difficult, softer social issues of knowledge sharing. The rule here for the OD/HR professional is to understand the KM process in most organizations and help to guide resources to the right place at the right time, and to not get bogged down in tool selection or infrastructure building. To accomplish this you must first educate those in the KM project as to what some of the intangibles for knowledge sharing are, and then determine if the infrastructure they are building will support them.

12.10   So What Is Required for Knowledge Sharing?

Here are some of the intangibles that we see as critical:

- Trust, trust, and trust!
- The ability to communicate clearly and with enough bandwidth to transfer meaning
- A common context or language
- A reason or goal for sharing
- The space to think and reflect
- The ability to interact with others in a nonpurposeful way
- The autonomy to share
- Awareness that knowledge is local and sticky and often does not transfer easily
- A flexible organizational structure that supports knowledge sharing (a rigid command and control structure does not support knowledge sharing)
- The infrastructure to support knowledge and information sharing

Notice that infrastructure is last. It is last because it is probably the least important thing on the list, or the easiest thing to correct, although it is the first thing that most people focus on because it is the most tangible. If you are looking at collaboration and knowledge sharing, try to deal with the people/cultural issues, the hard stuff, first.

12.11   Building a Foundation for Knowledge Management

The best way to learn to do something difficult is to see what others have done in this area. We know of no one who has done a lot of basic research on how people collaborate and how collaboration is an integral part of the culture and technology for knowledge sharing in business organizations. As a result we built a metric to look at the factors needed for successful collaboration and knowledge sharing. This benchmark was the focus of a recent study to look at some of the top collaborative companies in the U.S. Not surprisingly, many of them were using their lead in collaboration to initiate knowledge management (KM) projects. This recent benchmark examined collaboration on intranets and extranets in eight of the top U.S. companies. Our benchmark consisted of some metrics and an extensive questionnaire. Although you can collaborate without managing knowledge, and you can generate new knowledge without collaboration, generally, with human interactions this is not the case. In many of the companies we benchmarked, the collaborative infrastructure they created was for a purpose, and about 50% of the time it was used for knowledge management. Here are some of the findings from this study:

12.12   Why Were These Companies Successful?

- They were early adopters of collaborative technologies.
- They believed intuitively that collaboration was critical to their future success.
- They all had support from the management team for these collaborative projects.
- They believed there was significant economic benefit from collaboration.
- They were able to quickly show/prove this economic benefit from collaboration.
- They were innovative in their approach, trying new technologies and processes to solve difficult problems.
- They all believed in the value of intellectual capital and tacit knowledge in their organizations and wanted to leverage it.
- Most of the benchmarked companies were leaders in their industry and wanted to stay that way.
- Some of the companies had a culture or mission statement focused on supporting collaboration, others supported technical excellence and small team collaboration allowed them to achieve excellence.
- The more top management supported and used collaborative technologies, the more successful the project or company was.
- Many of the companies started collaboration on Notes and were migrating to intranets. Once they got internal collaboration going, they focused on extranets.
- Lower level people collaborated more frequently or more effectively than upper level or top management people.
- Organizations unsuccessful with collaboration had a "do as I say" culture, not a "do as I do" culture.
- Collaboration between the company, customers, and/or suppliers was encouraged.
- Most compensation plans did not support collaboration or teaming directly; most were supported indirectly through stock options.
- Many of the benchmarked firms were experimenting with compensation and reward structures to support collaboration.

- Most companies realized that they had to compete globally and that collaboration was necessary in order to achieve this.
- Collaboration evolved from e-mail to groupware, from groupware to knowledge management and from knowledge management to community.
- Problems occurred because the technology was revolutionary while the culture was evolutionary.
- An impedance mismatch between technology and culture was the big problem.

Most hi-tech firms we benchmarked had less of a mismatch, because they were cultures that were used in adopting new technologies, especially ones they had created or made. We believe most organizations are made up of a group of communities. The question is how to initiate, support, and sustain these communities electronically. We believe collaborative technology is a way to jump the evolutionary curve and collapse the groupware, knowledge management, and community steps.

One of the most successful areas to apply collaboration tools for the leveraging of expertise is in the consulting firms or consulting and service arms of various companies. We have seen this fail when the culture was not collaborative, i.e., when they are not rewarded for sharing information and knowledge and the software requires a large behavioral change.

Of the companies we benchmarked we believe that Cisco has the most collaborative culture and Microsoft the least; although Microsoft does support some types of collaboration well, it is through the availability of technology not because of its culture. Even though both companies are technology oriented, and both offer for sale and use their own networking product (routers and network OS respectively) these companies are only similar in their fanatical focus on marketing. Almost all of the companies we benchmarked are the top in their field or market niche; however, we were not able to determine if these companies' success is due to collaboration. We also did not hear that these companies built extensive business cases for collaboration, but rather the CEO, CIO, or someone else with vision intuitively knew that a collaborative infrastructure supporting the transfer and sharing of knowledge was critical to their organization continuing as the leader in the field.

Cisco has become a best-in-class company for collaboration largely through their very collaborative, supportive company culture. Upper management has stood behind Cisco's effectiveness in electronic collaboration as well through doing it themselves. They have also communicated the message that collaboration will help showcase their own products by utilizing their world-class network. They have an excellent and heavily leveraged intranet that is a very important part of how they collaborate, but they have not yet used groupware to collaborate electronically. However, Cisco realizes the potential power of groupware and implemented a number of applications on its internal network in the summer of 1997. As this chapter is being written Cisco has WebTV and online video training on their intranet, as well as a variety of collaborative applications based on an exchange backbone. Though the reward and incentive systems have not explicitly rewarded collaboration, and the dollar benefits of collaborative technology have not been measured or tracked to date, Cisco's compensation and promotion systems do indirectly reward collaboration. Effective teamwork is rewarded through peer performance evaluations.

Of the companies we interviewed, only Hewlett Packard has an explicit mission statement for collaboration. All the interviewed companies recognize that collaborative practices enhance their bottom line, but only a few (HP, Arthur Andersen, and Eli Lilly) have explicit policies or strategies to support collaboration. Where those policies exist, they are centered around competency centers (AA), and partnerships and research teams (EL). These findings are consistent with other collaborative strategies research where we found that collaboration is not usually articulated as a formal policy or practice. In many cases, we believe that companies have supported collaborative practices without realizing that they were doing something unusual.

We found that the level of collaboration often correlates well with your level in the organization, i.e., the lower you are, the more you collaborate. All participants reported that collaboration is encouraged between individuals, and only Shell reported that collaboration is not encouraged between workgroups/teams. As we rise in the organizational structures of Microsoft, SeaLand, and Shell, collaboration is less obvious between departments at Microsoft, SeaLand, and Eli Lilly; collaboration was encouraged, but ran into more barriers between divisions. At the enterprise level, all companies reported encouragement for collaboration with the exception of Microsoft, who reported only "minimal" collaboration. Collaboration between the enterprise and customers/suppliers is encouraged at all companies, and many firms saw this as the "next frontier."

Compensation programs in all companies interviewed are primarily based on individual performance. In the case of Cisco, salary is based 100% on individual performance, but salary is considered only 50% of overall compensation. The remaining 50% is awarded in stock options and bonuses, which are determined by organizational performance. This is consistent with Cisco's assumption that collaborating is an essential characteristic of how they do business.

In contrast, Microsoft's culture includes a tension between collaboration and competition. Microsoft is very project oriented, as is Cisco, but at Microsoft innovation is expected from individuals, not necessarily the project team. So, individuals are rewarded for their accomplishments and project teams are how the work gets done. When compensation is based 100% on individual performance, the incentive to share and collaborate is significantly reduced. Still, Microsoft reported that their corporate culture strongly supports sharing, and everyone at Microsoft has stock. To add to the contradiction in Microsoft's reporting, interviewees characterized their culture as supporting both collaborative and cowboy-like work styles.

Overall we see that compensation and reward structures are an area where many of these companies are beginning to experiment. SeaLand is starting to do this in some groups or on some projects. We believe that collaborative cultures are in the act of evolving, while the technologies are moving at a revolutionary pace. This causes an impedance mismatch, which is what trips up most implementations. The organizations we benchmarked were more successful than most because their cultures were more accustomed to adopting new technologies (hi-tech firms). As a result of this the mismatch was less, or there was clear support for the technology and change in behavior by top management (Arthur Andersen, Microsoft, Cisco).

12.13 Summary

This chapter has covered a lot of ground, from introducing the concepts and definitions of collaboration to the evolution of collaboration in the sociotechnical matrix of today's organization. We also examined best practices and how collaboration supports knowledge management. We looked at a variety of different types of collaborative tools and a classification/taxonomy for these tools, with examples and issues for each category. We examined the issues around cultural change that groupware catalyzes as well as looking at how these tools and methodologies support and form a base for knowledge sharing.

We looked at some of the challenges of knowledge sharing and KM projects, as well as some of the intricacies in determining the value of the intellectual capital in your organization. We looked at trust, and how critical creating trust is in knowledge sharing, as well as in building value-added communities. We looked at tools and criteria that can be used in knowledge sharing. We examined some research results of corporations that have done well with collaboration, and tried to learn from what they did (both right and wrong). We gave you the tools to start to begin to assess your own organization for its current level of collaboration.

We looked at early practices for KM implementation, such as yellow pages and best practices databases, but also made some suggestions as to how these could be made more dynamic and valuable to the organization. Finally, we tried to encompass suggestions for OD professionals, IT professionals, and general business professionals, as we believe all three of these groups have skills necessary for successful implementation of collaboration and KM methodologies.

References

Global Work, Bridging Distance, Culture and Time, Mary O'Hara Devereaux and Robert Johansen, Jossey-Bass, 1997.

Groupware: Collaborative Strategies for Corporate LANs and Intranets, David Coleman, Editor, Prentice Hall, 1997.

Groupware: Technology and Applications, David Coleman and Ramman Khanna, Editors, Prentice Hall, 1995.

Information Ecology: Mastering the Information and Knowledge Environment, Thomas H. Davenport, Oxford University Press, 1997.

Intellectual Capital: The New Wealth of Organizations, Thomas A Stewart, Doubleday, 1997.

Intellectual Capital: Realizing Your Company's True Value by Finding Its Hidden Roots, Leif Edvensson and Michael S. Malone, Harper Business, 1997.

Knowledge Management, A Real Business Guide, Stuart Rock, Editor, Caspian Publishing Ltd, 1998.

Shared Minds, The New Technologies of Collaboration, Michael Schrage, Random House, 1990.

The Faster Learning Organization: Gain and Sustain The Competitive Edge, Bob Guns and Kristin Anundsen, Pfeiffer and Company, 1996.

The Knowledge Creating Company: How Japanese Companies Create the Dynamics of Innovation, Ikujiro Nonaka, et al., Oxford University Press, New York, 1995.

Working Knowledge: How Organizations Manage What They Know, Thomas H. Davenport and Laurence Prusak, Harvard Business School Press, 1998.

13

Knowledge Discovery

Glenn Becker

Thomson Labs

We are drowning in information, but are starved for knowledge.

– John Naisbitt

... pattern-finding is close to the core, if not the core, of intelligence.

– Douglas Hofstadter

An increasing percentage of people in the workforce are knowledge workers, as predicted by Peter Drucker in the 1960s. Knowledge workers collect and analyze information to generate reports, designs, or research. To do this work efficiently, knowledge workers require ready access to information through search and data visualization tools. The technologies that support knowledge work are constantly changing, just as the variety of information they need to access continues to increase. Knowledge is not specific to humans. Anything that can learn and organize what is learned can acquire and manipulate knowledge. In this context, *knowledge* can be defined as the right collection of information at the right time. Other, more philosophical, definitions of knowledge abound, but this definition seems appropriate for most practical business applications. Later, in the introduction, this definition will be analyzed more closely.

Knowledge discovery is a process that attempts to identify and interpret patterns in information that are important to performing some task. What makes patterns interesting is defined by the task at hand. Depending on the task, knowledge discovery can be either partially or entirely automated. The need for more automated knowledge discovery has recently become more acute, because the Internet has made vast amounts of database information and real-time data available to a worldwide audience. Coincident with this, affordable computers have become powerful enough to support deeper analysis in practical time frames, and more scalable algorithms have been found to handle huge data sets.

This chapter is a basic introduction to knowledge discovery and some of its algorithms and applications. In an attempt to cover the wide variety of topics generally called knowledge discovery, most of these topics will be covered at a high level. A few topics that are not discussed widely in other texts will be presented here in greater detail.

13.1 Introduction

The term *knowledge discovery* helps to focus the discussion. *Discovery* means to find something that already exists. *Knowledge creation* forms novel associations among items and information by trying to identify coincident timing, location, or participants. Just because discovery deals with existing things doesn't mean, however, that it is necessarily a quick or easy process. Consider this analogy about the process of discovering knowledge from ancient Egyptian writings.

The civilization of ancient Egypt accumulated vast amounts of knowledge during its almost 3000 years of existence. Egyptian knowledge about architecture, sculpture, agriculture, anatomy, medicine, and government was the envy of the ancient world. The Egyptians recorded much of this knowledge and history by carving or painting stories and records on building walls and monuments. These stories and records were written in hieroglyphs and a cursive demotic script. This information allowed the Egyptians to retain their history and identity over several thousand years, despite being conquered several times.

For more than a thousand years, from the fall of the Roman Empire until the late 1800s, no one knew how to read Egyptian writing. Although many Egyptian artifacts were found during this time, the inscriptions could not be deciphered. Hieroglyphs were generally assumed to be pictograms, with each symbol representing a word or concept, but many new discoveries contradicted these prevailing theories. Then in the summer of 1799, the Rosetta Stone was discovered. This slab of basalt was engraved with three versions of the same story, written in hieroglyphs on top, Egyptian cursive demotic script in the middle, and the more familiar Greek at the bottom. Researchers spent three years deciphering the Greek text and then most of the remaining century deciphering the demotic script and hieroglyphs. It turned out that the demotic script was a less formal version of the hieroglyphic characters. Both hieroglyphic and demotic scripts are phonetic, with each character representing a sound or a short series of sounds. With the ability to understand hieroglyphs, researchers have translated volumes of information about the politics, religion, and everyday life in ancient Egypt. Access to this information has allowed researchers to discover some of the knowledge that the Egyptians had accumulated and recorded throughout their history. The story carved on the Rosetta Stone was actually ancient propaganda supporting the then current Pharaoh, Ptolemy V.

Organizing information, including business rules and statistical summaries, is an important part of collecting and analyzing it for knowledge. The definition of knowledge offered earlier explicitly mentioned context for information in two senses. One context is grouping information together with other related information. The second context is finding the right information at the right time in a workflow process. Both of these contexts require information about the environment where this information will be used. Anyone who has ever had to work on source code written by someone else can probably sympathize with the archeologists trying to read hieroglyphs prior to the Rosetta Stone. They had the information but not the context. In the case of Egyptian writing, some basic assumptions could be made about how it might be organized. All languages have nouns, verbs, and some sentence structure for grouping these words together. These knowledge structures, called ontologies, are an important part of knowledge discovery. Ontologies define the relationships among various bits of information in a collection. These relationships add context to the information for a particular task.

Nonaka (1995) divides knowledge into two types: *explicit* and *tacit*. Explicit knowledge includes textual information, numeric data, spreadsheets, and images. Tacit knowledge is information about processes or policies. Tacit knowledge can be discovered by monitoring processes in the workplace, both internally and externally. Ontologies can be used to organize information for the extraction of explicit knowledge, tacit knowledge, or both.

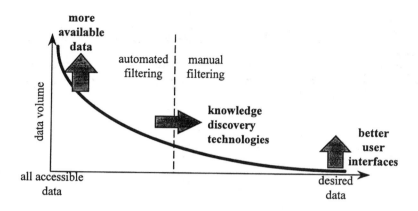

FIGURE 1 Information filtering..

The power of ontologies lies in the ability to apply more than one of them to the same information. This means that the knowledge discovered in a set of information is not always the same as the knowledge that went into it. Someone once said that real genius lies in looking at old facts in a new way. Ontologies will be discussed later in more detail.

Knowledge discovery systems add value to information by making it more accessible, tractable, and usable. The information is more accessible because improved search engines allow knowledge discovery systems to collect pertinent information from rapidly growing databases. The information is more tractable, because the information can be displayed at the right time in a workflow or in higher density using data visualization techniques. The information is more usable because discovered rules or constraints can be more tightly integrated into workflow and collaborative network systems.

Figure 1 shows a graphical representation of the large and ever-growing volume of information on the left and the demand for more customized, precise information on the right. A great deal of unwanted data or inappropriate rules can be automatically filtered, but some manual filtering is still required. The vertical dashed line represents the boundary between automated and manual filtering. Suppose the left side of the graph represents all the information available on the World Wide Web. The Web can be searched by entering a keyword or simple query into any of the common Web-based search tools. These search tools automate the filtering of the billions of available Web pages down to a few thousand. The process of manually looking at the remaining pages can still be very time-consuming. The goal of knowledge discovery tools is to use knowledge about the user and the user's goals to move the automated filtering boundary toward the right. At the same time, better data visualization techniques are being developed that allow a user to see and work with hundreds or thousands of data points simultaneously, instead of the list of 10 or 20 sites currently returned by most Web-searching tools. Data visualization can be considered human-in-the-loop knowledge discovery.

An example of one such data visualization tool, shown in Figure 2, is the WebCite prototype built by Mark Turner, Bill Majoros, and Ed Quackenbush in the Natural Language Processing Lab of Thomson Labs. WebCite was built to support researchers looking for relevant documents in large archives. First the user queries the database to retrieve a set of documents that appear to be relevant based on their contents. This set of documents can then be plotted as a scatter diagram or citation network (shown). In the citation network, each document is represented by a small square. These squares are arranged horizontally by their publication date. Their vertical arrangement is meaningless other than to avoid line crossings. Documents that cite other documents in the set are connected to their predecessors by a line to show their relationship. This interface facilitates bibliographic research by allowing users to navigate through the citation Web, add cited documents to the view, and identify important documents that are cited often.

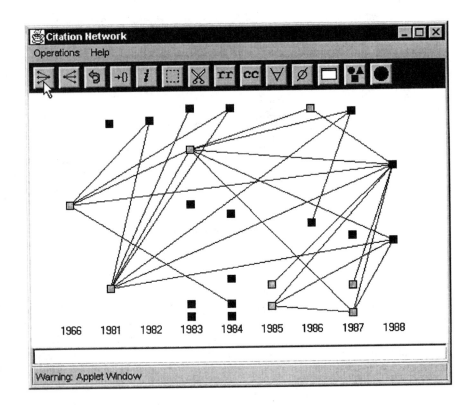

FIGURE 2 WebCite example.

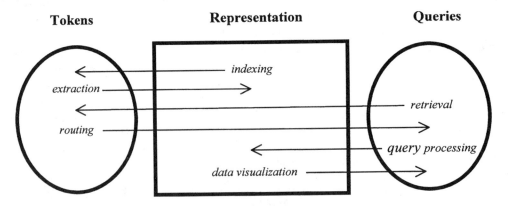

FIGURE 3 Terminology model.

Before the applications and technologies of knowledge discovery can be discussed in detail, some basic definitions are needed. To reiterate the definition of knowledge: knowledge is the right collection of information at the right time. Knowledge discovery was defined earlier as a process that attempts to identify and interpret patterns in information. The following terms, which will be used throughout this chapter, describe many of the fundamental actions generally considered part of knowledge discovery. Steven Finch, of the Thomson Labs, deserves the credit for developing the following list of definitions and the model shown in Figure 3 that relates them.

A *token* is a logical unit of data, like a document, spreadsheet, or record, generally stored in a database.

A *representation* is an abstraction of a token or group of tokens in a form that a user can manipulate in a user interface. The abstraction could be a set of rules, explicit data examples, or a summary of the data.

A *query* is an expressed information need for some task. Queries can be in a natural language, like English, or a contrived query language like SQL.

Filters are processes that identify to what extent a specific token is relevant to a query. Filters are often used to select a subset of a database by selecting only those records that contain a certain value, certain range of values, or only those records that do not have a certain value.

Indexing identifies information about a token relative to a particular representation. Indexing uses the vocabulary associated to a particular application domain and ties the words in that vocabulary to relevant tokens.

Extraction maps a token into a particular representation standard, like a template. This template identifies the significant terms, names, and associations within that token.

Retrieval identifies to what degree a set of tokens are relevant to a query. Any database query is a simple form of retrieval.

Routing identifies which of a set of queries a token satisfies. Automatic e-mail routing systems use user profiles to decide who receives a particular mail message.

Query processing translates general queries for a particular representation. Many database applications use query processing to translate user requests into a formal query language that the database system can understand.

Data visualization graphically represents the structure that exists among a set of tokens. Data visualization is generally used for exploring or navigating through complex data sets.

Many applications of knowledge discovery have been identified in professional and research domains. Some, like fraud detection and economic forecasting, are exciting and occasionally make the evening news. Other applications, like intelligent error detection and analyzing buying patterns, are less glamorous, even though they may be more valuable.

Error detection is an important application that includes discovering ways to detect errors in data based on the expected relationships among various data values and historical trends. As computer systems collect more and more information, catching errors becomes more important. Errors can be introduced in three different ways. First, random errors can result from data being entered wrong or transmission errors. Second, systematic errors can be produced by faulty sensors or other equipment. Third, are intentional errors, sometimes fraud, that are usually done specifically to avoid detection. Obviously, any business that involves financial transactions is acutely sensitive to errors.

Other applications of knowledge discovery include user profiling, shopping cart analysis, trend analysis, and dependency analysis. User profilers learn about a particular user's preferences or habits and then try to adapt the system to accommodate for them. Some profilers also make recommendations, like approving or rejecting credit applications. Shopping cart analysis looks at what people buy and tries to directly market additional goods or services. Profiling and shopping cart analysis can be powerful tools for improving customer support. Trend analysis uses historical data patterns to find anomalies in current data or to predict future values. Dependency analysis looks for correlations in data that can be used for error detection or prediction.

Despite the variety of applications for knowledge discovery, many of the processing steps are the same. Figure 4 shows the six steps of knowledge discovery as listed by Fayyad (1996). The process begins at the bottom of the mountain with raw data. Some of this raw data is selected through queries or filters to become the selected data. The selected data is chosen because it is applicable to the task. The selected data can then be preprocessed to remove outliers and records with missing fields, if appropriate. After preprocessing, the data may need to be transformed into normalized units (monetary units, weights, and measures), fixed value ranges, or quantified so that it can be processed by the analysis software. Pattern recognition algorithms that do classification, clustering, or regression can then be applied to the transformed

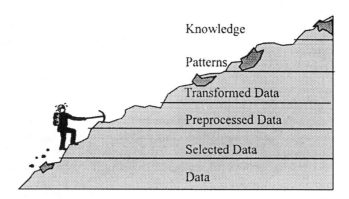

FIGURE 4 Steps of knowledge discovery.

data to find and characterize data patterns. This pattern recognition step is generally called *data mining*. Finally, the patterns that are identified as interesting by the data-mining step are analyzed in the context of a particular application or ontology to decide if they are significant enough to contribute to existing knowledge. Progressively higher steps in this process usually require progressively more human intervention. The final step of interpreting the patterns found during data mining is often entirely manual.

The algorithms used for selecting, preprocessing, and transforming data are usually fairly simple and application specific. The data-mining step, where patterns are identified, is usually the most challenging and interesting step. Data mining can include algorithms for clustering, classifying, or regressing the transformed data.

Data mining uses many algorithms, some developed 20 or more years ago under the label of artificial intelligence or general statistical analysis. The advantage of using these tried and true algorithms is that their strengths and weaknesses are well understood. Many challenges still exist for data mining, including:

- Scalability for large databases — The time required for some algorithms to execute increases exponentially or geometrically as the number of data elements increases. Even with faster computers, this is a real problem for very large data sets.

- Compounding noise and error — As shown above, knowledge management is a series of transformations and other functions on data. Each of these processes can reduce precision, introduce errors and noise.

- Heterogeneous data — When disparate databases are put together they may have different levels of precision in numeric fields or use different vocabularies in text fields. These problems may not be obvious.

- Changing data ranges — Maximum and minimum data values can change dramatically over time. Five years ago few would have believed that the Dow Jones Industrial Average would go above 9000 or that daily trading volumes on the New York Stock Exchange would exceed one billion shares per day.

- Changing clusters — Statistical clusters used in classification can change over time. Some classes may split into multiple classes, or multiple classes may merge into a single class.

13.2 Organizing Information for Knowledge Discovery

Knowledge discovery requires a frame of reference. Knowledge, as defined earlier, is a collection of information that is timely and relevant in some workflow. Retrieving, or grouping, information presupposes that some categories for that information exist. These categories may change from task to task, but for any given task there is at least one set of categories. An important part of designing a knowledge management system is defining these categories, their relationships with certain tasks, and their relationships with other categories.

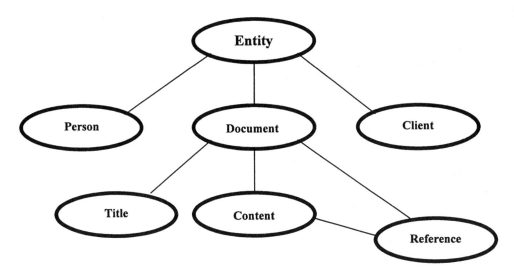

FIGURE 5 Ontology diagram.

A *conceptualization* is a set of objects with defined relationships among them. Conceptualizations can be used to organize the knowledge of organizations. An organization's conceptualization of knowledge manifests itself in all of that organization's work products, even though the organization probably doesn't have an explicit understanding of its own conceptualization. An explicit conceptualization model, comprised of objects, their definitions, and relationships among the objects is called an *ontology* (Gruber, 1993).

In practical terms, ontologies, like the example shown in Figure 5, are a set of objects, definitions of the properties of each object, and the relationships among those objects. Some researchers are developing very high-level ontologies that could be used to organize broad general information (Noy, 1997). In practice, however, ontologies are usually built for specific application domains. Ontologies are used widely in many fields, although they may be called by different names:

- *Genetic graphs* — pedagogy design
- *Semantic networks* — artificial intelligence
- *Relational graphs* — pattern recognition
- *State diagrams* — computer science

Within a particular ontology a well-defined terminology, or *taxonomy*, is used to describe the classes of objects, their properties, and relationships. Figure 5 shows a specific ontology for organizing text, images, and audio content into a multimedia repository to facilitate reuse and track licensing. This ontology defines a taxonomy with seven classes of objects. The top-level (most abstract) object is called an entity.

Entity has three subclasses: person, document, and client. Document also has three subclasses: title, content, and reference. Title is simply the title or name of a document. Content can be any text file, image, or audio clip stored in the repository. Reference contains information about the source of the content or document. A document can consist of a title, any number of content objects, and a reference. Person is an abstract class that can store data on authors, artists, or owners of any entity. A client is an entity that can license documents or content.

Ontology design is a difficult, yet critical, step in the implementation of a knowledge management system. Once an ontology is in use, it can be very difficult to change. The challenge is to build flexible and extensible ontologies, because the system will need to change over time to handle new situations, products, and technologies. Because change is inevitable and things will be forgotten, most research favors top-down design. Top-down design starts at the topmost abstract and general levels of the ontology and works its way down to the more specific details. One benefit of top-down design is that general objects can be defined to cover "what if" scenarios, even if they are not totally implemented at first.

Bottom-up and middle-out design begins with more concrete objects and then moves up or out to complete the design. Some researchers are trying to determine to what extent ontology design can be automated from analyzing free-text or database structure, but this work is still preliminary and a long way from being general purpose (Noy, 1997).

Several languages have been developed to encode ontology design and share ontology structures across different applications. The knowledge interchange format (KIF) was developed by Genesereth and Fikes (Genesereth, 1992) to communicate ontology and knowledge between applications. KIF uses a LISP-like predicate calculus to encode object relationships and properties. It is intended to be an intermediate format only, not to be used by programs internally. Similarly, the knowledge query and manipulation language (KQML) was developed by Finin and Wiederhold (Labrou and Finin, 1997) to give software agents a LISP-like language to communicate their knowledge and status with other agents.

Another such language is the ontology markup language (OML), based in large part on the SHOE (Simple HTML ontology extensions) initiative at the University of Maryland at College Park (Luke, 1996). OML was developed by the WAVE (web analysis and visualization environment) project at Washington State University and is a dialect of the extensible markup language (XML), which has been adopted by the Worldwide Web Consortium (W3C). XML is a somewhat simplified version of the standard generalized markup language (SGML). These simplifications were made so that Internet browsers and editors could be built to support XML, which is much more powerful and flexible than the current mainstay of the Web, HTML.

The conceptual knowledge markup language (CKML) is an extension of OML based on the conceptual knowledge processing work of Rudolf Wille at the Technische Hochschule in Darmstadt. OML supports the definition of ontologies and objects. CKML adds *attributes*, *conceptual scales*, and *conceptual views* from the field of conceptual knowledge processing. Attributes are used to define properties of objects. Conceptual scales are processes that can extract information of interest from an ontology because they understand the ontologies schema. Conceptual views are user interest profiles used to route the output from the conceptual scales (Wille, 1982).

Figure 6 shows the ontology from Figure 5 implemented in OML. The seven objects in this ontology are named in OML with the OBJECT TYPE tag. Ontologies require a formal *schema*, which defines the usage of field names, or tags, like OBJECT TYPE. Schemas specify the semantic details of an ontology, like the taxonomy, controlled vocabulary, and variable ranges.

Brief Overview of the OML Schema

All tags are delimited by '<' and '>', like <tag>. Most tags have an opening tag, <tag>, and a closing tag, </tag>. If the tag and its arguments all fit on one line, then for brevity a forward slash, '/', can be used alone at the end of the tag instead of the closing tag.

```
OML tags -
<OML>
    OLM file's contents
</OML>


Ontology tags -
<ONTOLOGY NAME="ontology name" VERSION="version number">
    ontology's body
</ONTOLOGY>


Comment tags for comments that are not processed -
<COMMENT>
    text description of ontology (this text is not processed)
</COMMENT>
```

```
<OML>
   <ONTOLOGY ID="Content Description" VERSION="1.0">
      <COMMENT>
         Ontology for managing reusable content (text, charts, images, audio)
      </COMMENT>

      <OBJECT TYPE="ENTITY"/>
      <OBJECT TYPE="PERSON"/>
      <OBJECT TYPE="CLIENT"/>
      <OBJECT TYPE="DOCUMENT"/>
      <OBJECT TYPE="REFERENCE"/>
      <OBJECT TYPE="TITLE"/>
      <OBJECT TYPE="CONTENT"/>

      <SUBTYPE SPECIFIC="DOCUMENT" GENERIC="ENTITY"/>
      <SUBTYPE SPECIFIC="PERSON" GENERIC="ENTITY"/>
      <SUBTYPE SPECIFIC="CLIENT" GENERIC="ENTITY"/>
      <SUBTYPE SPECIFIC="TITLE" GENERIC="DOCUMENT"/>
      <SUBTYPE SPECIFIC="CONTENT" GENERIC="DOCUMENT"/>
      <SUBTYPE SPECIFIC="REFERENCE" GENERIC="DOCUMENT"/>
      <SUBTYPE SPECIFIC="REFERENCE" GENERIC="CONTENT"/>

      <RELATION TYPE="LICENSED-TO">
         <ARGUMENT TYPE="CONTENT" VALUETYPE="#CONTENT">
         <ARGUMENT TYPE="CLIENT" VALUETYPE="#CLIENT">
      </RELATION>
      <RELATION TYPE="LICENSED-ON">
         <ARGUMENT TYPE="CONTENT" VALUETYPE="#CONTENT">
         <ARGUMENT TYPE="Date" VALUETYPE="Date">
      </RELATION>
      <RELATION TYPE="AUTHORED BY">
         <ARGUMENT TYPE="CONTENT" VALUETYPE="#CONTENT">
         <ARGUMENT TYPE="PERSON" VALUETYPE="String">
      </RELATION>

   </ONTOLOGY>
</OLM>
```

FIGURE 6   OML example.

```
Object tags for declaring objects in the ontology -
<OBJECT TYPE="name of object"
   <FUNCTION TYPE="function name">

      ...
</OBJECT>
```

```
SUBTYPE tag for linking general and specific objects to build the ontology
hierarchy -
<SUBTYPE SPECIFIC = "specific object" GENERIC = "generic object"/>
```

```
Define a relationship schema among objects or schemas in the ontology -
<RELATION TYPE = "function name">
   <ARGUMENT TYPE  = "argument type" VALUETYPE = "data type">
</RELATION>
```

The goal of the data mining and pattern analysis steps of knowledge discovery is to find data patterns that are interesting and significant based on an ontology. The information and ontology for a particular application must be analyzed to determine what portion of the information should be mined. Once mining has found interesting patterns, these patterns must be analyzed to determine how significant they are.

These analysis tasks can either be manual, automatic, or partially automatic. Populating an ontology manually may be necessary for small-volume, high-value knowledge management systems. Although this manual process is very expensive and slow, it may be necessary to ensure high quality. Automatic processing, using static classifiers or machine learning algorithms, can be a fast and cost-effective way to keep up with the rapid flow of information. The quality of the patterns and knowledge that are found depends on the limitations of the algorithms used.

For most applications, a high degree of automation is necessary to make the process cost effective and fast enough for business use. Researchers at a pharmaceutical company might use an automatic system to search a database of drugs, mechanisms of action, and side effects to look for new applications of existing medications. This would be a good use of technology. A local pharmacist, however, should not be experimenting with novel uses for drugs. They should be using a carefully crafted and tested database of side effects and interactions.

The next few sections will discuss some of the basic algorithms and strategies used to automate the data mining process for different types of information.

13.3   Data Mining in Fielded Data

Most references to knowledge discovery, or any of its constituent processes like data cleansing or data mining, refer to the processing and analysis of fielded data in a database. This section discusses some of the common analysis techniques and machine learning algorithms used in data mining for fielded data. Most of these techniques have been covered extensively in technical books and journals, so they won't be presented here in detail. The discussion will focus instead on the differences among these techniques.

Before evaluating methods for use in a particular application, it is important to understand the limitations of the information being used. Understanding the limitations of the information is critical to setting reasonable goals and expectations for a project. Poor information will produce poor results even with the best algorithms, while even simple algorithms can often produce good results with good information. Some information properties to pay attention to are:

- Information quality
- Information volume
- Information precision

Information quality includes issues like error rate and completeness. Error rate is the percentage of erroneous records in a database. Many data analysis algorithms can tolerate some errors, but it is important to have some estimate of the error rate. Completeness is the percentage of database records that are not missing any data. Missing data can be a serious obstacle for data mining and it is not always easy to detect. A local police department was experimenting with data visualization techniques to see if they could find patterns in their records on juvenile offenders. When they started using the offender's age in this analysis, they noticed that a large number of these juveniles were 97 years old. When they looked further, they realized that the arresting officer did not always fill in the date-of-birth field and that it was being treated as 1900 by the age calculation, which then reported an age of 97 in 1997.

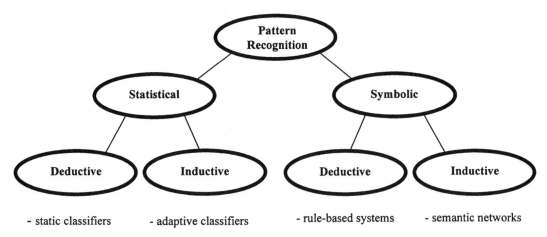

FIGURE 7   Pattern recognition algorithms.

Information volume in a data set must be sufficient for the analysis method that will be used. Most statistical methods have some way to estimate how many training examples are required to produce reliable results. The information must be examined to make sure that it is appropriate for the proposed application. About two years ago I joined a news group on the Internet that was discussing a very contentious subject. After about a month I realized that there were two diametrically opposed groups and that all their messages simply railed on previous messages, calling members of the other group names. I expected the two opposing groups, but I wasn't expecting there to be absolutely no useful information, despite the dozens of messages that were submitted daily. After lurking for a month, I posed a question asking for some statistical information. My question went unanswered for three weeks, at which time I resigned from the group. This is an example of lots of data but no useful information.

When setting goals for knowledge discovery, make sure that the information has sufficient temporal and value precision to generate useful results. Temporal precision or sampling frequency is the number of data samples recorded for a unit of time. The Nyquist Criterion is one metric used to measure sufficient temporal resolution. The Nyquist Criterion states that to accurately reproduce a signal at a particular frequency the sampling frequency must be at least twice that frequency. For example, audio CDs accurately reproduce sounds at frequencies up to 22 kHz. The achieve this, the audio signal is sampled at 44 kHz. In data mining terms, to predict a monthly sales volume from a historical series, the data should be collected every two weeks. Certainly this will work better than trying to predict weekly volumes from a time series of monthly data.

Value precision is much like temporal precision in that the value precision of a data set limits the precision allowed for any estimates or predictions. In data analysis 1,000,000 divided by 52 usually does not equal 19,230.76923077. Precision comes from measurement, not calculation. If the value 1,000,000 is rounded to the nearest 1000 before being entered in the database, then the real value could have been anything from 999,500 to 1,000,499. So the real answer could be anything between 19,221.15384615 and 19,240.36538462. Strictly speaking, the most accurate answer is 19,200 with the first three digits being significant. Carrying more precision than that into future calculations only propagates the myth that more precision actually exists. This means that if the application is monitoring for fluctuations of 100 or more, then this data can be used. However, if the application is monitoring for fluctuations of 99 or less, more precise data will be needed to produce reliable results.

Pattern recognition algorithms are used to find patterns in information. Some of these algorithms use statistical methods and others use symbolic methods (Figure 7). The best algorithms for any application depend on the information format and the application goals. Many of the pattern recognition methods discussed in this section have been used widely for many years. Their properties, strengths, and weaknesses have been published in books and journals. They have been applied to a wide variety of applications including fielded data, free text, as well as image and audio information.

Statistical methods measure important characteristics (or features) of the information. These feature values are plotted into an *n*-dimensional feature space, where *n* is the number of features. These plotted values are then segregated by decision boundaries to form classes or clusters. The algorithms that fall into this statistical group, including neural networks, histogram learning (Bock, 1993), and decision trees (Quinlan, 1986), are all different ways of building decision boundaries with the feature space.

Symbolic methods are used to find structure in information. Instead of calculating features, like statistical methods, symbolic methods look for pattern primitives using pattern description languages. Pattern primitives can be parts of speech in text, pixel patterns in an image, or a particular pattern of values in a time series. A symbolic method uses a *grammar* to govern how patterns can be put together to form *sentences* (higher order structures of pattern primitives). Recognition is done by parsing sentences looking for groups of patterns that fit into the application's grammar.

Within both statistical and symbolic groups, some methods are *deductive* and others are *inductive*. Deduction is a process that starts with a general ontology and matches it to specific examples. Deductive methods include rule-base and case-base reasoning systems. Induction (or concept learning) is a process that uses exemplars to build a general ontology that describes the information patterns. Decision trees, neural networks, and other machine learning techniques are all inductive.

Static Classifiers

Static classification systems use classical statistical methods, like Bayesian classifiers or band-pass filters, to classify information examples. These static systems are usually derived from empirical evidence of how best to delineate classes. They are fine for static environments where data ranges and the classes themselves are static.

Adaptive Classifiers

Adaptive classifiers train on data examples to determine the best way to delineate classes within the application domain. As discussed earlier, this training can be either supervised or unsupervised. Adaptive classifiers include neural networks, histogram learning, regression techniques, and decision trees. The main difference among these techniques is the decision boundary complexity they can support. More complex decision boundaries can more precisely define a class, but they require much more training than simpler methods.

Rule-based Systems

Rule-based or case-based reasoning systems (Leake, 1996) begin with a specific example and try to match it to one of the rules or cases in its knowledge base. Inference engines are used to find patterns in the example data that match a known rule to some extent. Once a rule's preconditions have been met, or enough evidence has been collected to select a matching case, that rule (or case) executes, telling the system what to do next.

Semantic Networks

Semantic networks (Michalski, 1983; Kodratoff, 1990), or S-nets, are very much like ontologies in that they are objects interlinked by their relationships. S-nets are used widely in natural language processing or image scene analysis to represent the constituent objects of an application and the constraints that these objects' properties and their interrelationships place on the system.

Inductive methods require training with representative examples to learn how to characterize each new pattern class. These training examples must be representative of the class (or classes) that the pattern recognition system is being used to differentiate. A user *supervises* the training process by showing examples to the system one class at a time. In this way, the user designates the classes that the system should learn.

Another approach to training is *unsupervised*. In unsupervised training, examples are shown to the system in no particular order and the system finds clusters of feature values in feature space or similar symbolic structures that it can group together into its own class definition. If the application is expected to classify specific classes, unsupervised training is probably not a good approach, because the clusters found by the system may not correspond to the desired classes. If, however, the application is trying to detect change or predict trends, then clustering is a good way to establish a baseline for comparison.

Aside from supervised and unsupervised training, the third type of training is called *reinforcement learning*. Supervised training is very effective if the classes of the training samples are known. In some cases, like process analysis and strategy learning, the class of each training example may not be known. Consider a game like chess. It is easy to classify winning board configurations as good and losing configurations as bad, but many of the intermediate configurations are more difficult to classify. In situations like this, reinforcement learning accumulates a series of unclassified examples leading to a known state or goal. Once a known state is found, the series of examples leading to it are updated accordingly. Two methods of reinforcement learning are collective learning (Bock, 1993) and Q-learning (Mitchell, 1997).

Most knowledge discovery applications are hybrids, involving more than one strategy. Using different methods at different stages in the discovery process allows the system to take advantage of the strengths and weaknesses of each approach.

13.4 Data Mining in Free Text

Information that has already been organized, indexed, and stored in a fielded database is relatively easy to search and extract. Vast quantities of knowledge are, however, stored in the free text of books, encyclopedias, manuals, notes, newspapers, and journals. Computers can search text for a word or group of words, but this simple approach to locating information is not practical, because the volume of data is too large, words are not always spelled the same, vocabulary changes make it difficult to know which words to use, and it is difficult to extract the meaning from a body of free text.

To address this problem, algorithms have been developed to:

- Select words from a text that should be used for indexing;
- Look for close matches to personal names, company names, product names, or places;
- Extract data from formatted tables or forms; and
- Search for words that regularly appear in the same context and therefore may be related.

Finding words in large free-text collections that are good for indexing is a critical step in making these collections more accessible. One simple approach, is to develop an authority file, which is a list of important words (and phrases) within a particular domain or area of expertise. Authority files are used widely, but they require significant domain knowledge and need to be maintained as vocabulary changes.

Back in the early 1970s, Sparck Jones discovered empirically that the most effective terms for searching large text collections are those that occur least frequently (Sparck Jones, 1972). Terms can be weighted as good indexes based on their inverse document frequency. This seems intuitive because frequently used terms that appear in almost every document cannot be used to help differentiate one document from another. Equation 1 is used to calculate the weight (ω) for a term that appears in n of N documents in a collection. Greater values of ω indicate words that are better indexes.

$$\omega = \log \frac{N}{n} \tag{1}$$

Equation 1 bases its weight simply on term frequency without any way to include feedback from a user. For applications where user feedback is available to help judge the importance of terms, this additional help should not be ignored. Robertson and Sparck Jones developed Equation 2 to calculate term weights based on term frequency and a user's judgment of relevance for that document. (Robertson and Sparck Jones, 1976) In Equation 2, p is the probability that a term appears in a document judged to be relevant. q is the probability that a term appears in a document that is judged not relevant by a user.

$$\omega = \log \frac{p(1-q)}{q(1-p)} \qquad (2)$$

So, if the collection of documents being queried has N documents, and R of them are designated as relevant to a query, then N-R are not relevant. If r of the relevant documents contain the query term and n out of all N documents contain the query term, then substituting these variables for p and q in Equation 2 produces Equation 3.

$$\omega = \log \frac{r(N - R - n + r)}{(R - r)(n - r)} \qquad (3)$$

In cases where the number of relevant documents might be very small (or zero) Equation 4 is preferred, because it behaves better for small values of R and avoids the divide-by-zero problem.

$$\omega = \log \frac{(r+0.5)(N - R - n + r + 0.5)}{(R - r + 0.5)(n - r + 0.5)} \qquad (4)$$

This algorithm is just one example of the many tools being applied to knowledge discovery in free text. Fuzzy name matching, parsing to determine parts of speech, and statistical models that characterize term collocation in text are also widely used to help index or extract information from free text (Robertson and Walker, 1997)

13.5   Data Mining in Multimedia

Although most knowledge discovery is being done with fielded data and free text, some organizations have vast multimedia (image, audio, or video) assets that can be made more valuable by data mining. Database systems have long supported the ability to search for keywords in text and image records. Only recently, however, have image and signal analysis systems offered some limited ability to actually "look" at the images or "listen" to the audio in a database. This process is referred to as *content-based query*. These systems have the ability to retrieve images that "look like" example images or search for images with particular visual characteristics.

Keywords can describe image contents at both syntactic and semantic levels. Syntactic keywords are based on low-level characteristics of an image. Examples of syntactic keywords include "rough texture," "horizontal edges," "smooth color transition," and "large patches of dark green." Semantic keywords are based on higher-level image information. Semantic keywords often require more information than is available in a single image. For this reason, semantic keywords must usually be manually entered. Examples of semantic keywords include "this is a picture of Aunt Sarah" or "this is a hotel in Nice, France" (Bock, 1993). Some research has shown that semantic keywords can be extracted from the document text around the image and from other associated documents, particularly in Web pages where related documents are linked together. This work uses the local context of an image, not the image content itself, to derive keywords (Harmandas, 1997).

Applications that would benefit from content-based query include:

- Customers searching retail catalogs for a particular color or pattern of material
- Publishers or multimedia producers searching a digital image library

- Researchers searching trademark database for similar logos
- Remote sensing systems looking for abnormalities in satellite or aerial imagery
- Medical researchers searching for similar looking tissue samples

Accumulating a digital image library is an expensive undertaking because of the fees for photographers, graphic artists, rights licensing, and the manual processing required to attach keywords to each image so it can be retrieved later. The return on this investment comes when the images are reused, saving the time and cost of developing or licensing new content. Image analysis systems are being used to help automate the process of linking keywords to images or finding images in the database that "look like" a sample image. This automation can both speed the process and reduce the cost of incorporating new content. The technology behind any particular image analysis system will determine what kinds of keywords it supports and how domain specific it is.

Content databases, those that store text, images, video, and/or audio, are generally indexed with a controlled list of keywords (sometimes called a controlled vocabulary or authority file). When a new piece of content is added to the database, an operator manually looks at the content and assigns some appropriate keywords from the controlled list. The list of keywords must be controlled so that all users of the database work with the same vocabulary. In the future, when someone wants to access things in this database they can build queries from the keyword list to find the things they are looking for.

Ontologies for organizing content include schemas that incorporate knowledge about the relationships among the keywords. These ontologies are used to evaluate how well a particular piece of content in the database matches the user's query and returns likely candidates in a ranked list, with the most likely matches at the top of the list.

Most keywords describe semantic characteristics. In image databases, for example, the keywords might describe the subject of the image, information about the setting of the image, or administrative information about the image.

Current image analysis systems operate at the syntactic level, looking at details below the semantic level. Most of the syntactic level characteristics are based on color, texture, or simple shapes.

Syntactic characteristics are useful for finding images based on color, texture, and simple shapes. They can also help prune the list of images retrieved. For example, "Find photos of flowers, but only red flowers." The color of the flower in the image may have been input as a keyword, but if not, an image analysis system could look for the desired color. Pruning lists of candidate images is even more important when the database server is being accessed over a wide-area network or the Internet. For these networks it is very important to minimize the volume of traffic to reduce network congestion and get the results back to the user faster.

Classification is the process of extracting a feature value or values from an image and then comparing those values with the known list of classes. The class that is statistically closest to the image's feature value is selected. In the case of an image database search, any database image that has feature values close to those of the sample image are returned in a candidate list to the user. Since the comparison is being done statistically, each comparison produces a confidence value, which is usually represented as a value from 0 to 100%. This confidence value indicates how close the database image is to the sample image. Most systems allow the user to specify how many images should be returned for a query.

Classification is only part of the general image analysis process, described in Figure 8. While the terminology varies, virtually all image analysis processes can be mapped into this general list of steps. A more detailed discussion of each step follows the list.

The input image can either be a query (sample image used to search the database for a similar image) or a new image being added to the database. Some features require transforming the input image prior to feature extraction. Typical transformations include:

- Converting an RGB (red, green, blue) color image into HSV (hue, saturation, value) so that hue or saturation values can be extracted
- Reducing the resolution of the image so that only large-scale characteristics are analyzed

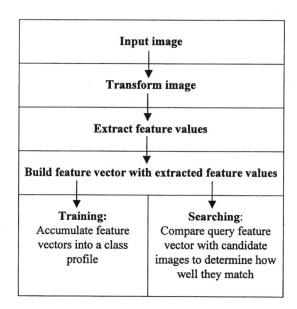

FIGURE 8   Basic steps in the image analysis process.

- Transforming the image into a different domain (e.g., Fourier, wavelet, or Hadamard transforms)
- Filtering the image to enhance edges or other features

The transformations used, if any, are usually coordinated with the feature extraction in the next step.

Feature extraction is the heart of any classification system. A great deal of research has been done to find features that can effectively characterize textures, edges, color patterns, and other image components. A feature value is calculated from the pixel values of a transformed image. Some features are based on an entire image taken as a whole, while other feature values are based on small regions of an image. Some popular features are based on the dominant colors, hues, or luminance levels in an image. Texture features tend to be filters that measure the activity, luminance gradients, or variance of small regions of an image. Selecting the features that are used for a specific application usually depends on the classification system that is used. Neural networks, for example, try to develop custom features for a specific application, which is why they require a large number of training examples. Most of the systems discussed in this chapter use a set of selectable features that can be quickly tuned for any application by an experienced user.

One key decision that must be made prior to feature selection is whether the application is looking for whole images or specific parts of images. For example, consider a system that has access to a large collection of medical images. The system being developed must scan heart images and look for evidence of aneurysms. The first step is to identify whole images in the database that are heart images, taken at the right aspect to see an aneurysm. Once an appropriate image is located, the image is scanned to see if an aneurysm exists. Finding the heart images in the library uses features that characterize the whole image as a single entity. The second step uses more localized features to look for the aneurysm within the image (Figure 9).

A single feature may be useful for identifying a simple texture, but more complex searches require multiple features to effectively characterize the query. After the required feature values have been extracted, they are combined into a feature vector as shown in Figure 10. This feature vector can just be a simple concatenation of the feature values. The resolution (number of bits) allocated for each feature value in the feature vector depends on the required resolution or priority of that feature. The feature vector is what represents the image for searches. A sample image or texture can be selected for a query. This sample can be used to generate a feature vector that is used to compare against the feature vectors of all the images in the database.

Adding a new image to the database requires calculating all feature values for that image and then linking those values to the image in the database so they can be used for searches. Searching the database

Healthy hearts   Aneurysms

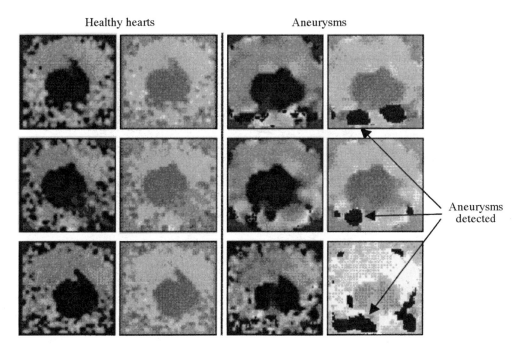

Aneurysms
detected

FIGURE 9   Heart images. (Images courtesy of the University of Ulm — Institute of Nuclear Medicine.)

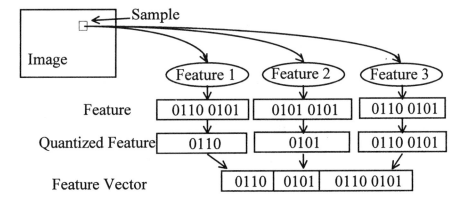

FIGURE 10   Creating a feature vector.

for images that look like a sample image is the process of generating this feature vector for the sample image and then comparing it against the same feature vector for each image in the database.

The ability to search image databases using features extracted directly from the image has wide-reaching application. Some applications that have already been implemented include:

- Fingerprint identification
- Searching image catalogs
- Trademark or logo matching
- Military uses like target identification
- Face matching for identification
- Medical image analysis

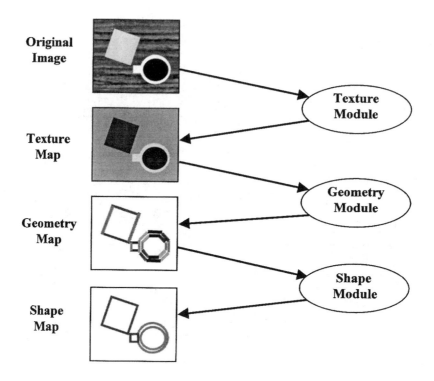

FIGURE 11   ALISA System modules.

The more controlled and regular the images are throughout the database, the more likely a current content-based query system will succeed. Most of these application areas have the following characteristics that help make them successful:

- Homogeneous image format (same resolution and color depth)
- Homogeneous subject perspective (same scale and orientation of the subject of the image)

Homogeneous image formats are important for applications that use texture and color features. If all the images do not have the same resolution (size in pixels) and color depth (number of bits used to represent each pixel) it is very difficult or impossible to compare them. It is also best if all of the images are compressed with the same technique and stored in the same color space. Loss-compression techniques, like JPEG, will affect an image's colors and textures, so if the images must be compressed, they should be compressed to the same degree and with the same technique throughout the database. JPEG, for example, uses the YUV color model that best preserves the luminance (Y) information, sometimes at the expense of the color information (UV).

Homogeneous subject perspectives are important for applications that use shape features. Take, for example, a system that recognizes faces. If it is trained to recognize one person's face only from the front, it probably would not be able to recognize the person from the back or top and probably not from the side either. Simple face recognition systems require a face image from a predictable aspect. These systems key on common facial characteristics to uniquely identify each person's face.

The ALISA system that was used to classify the heart images in Figure 9 is an image analysis system organized as layers of modules so that the modules in each successive layer look at the output of the module (or modules) in the previous layer as shown in Figure 11. The modules of each successive layer then classify more conceptually complex objects then their predecessors.

The ALISA system uses vector quantization to build a statistical representation of image data from a training set. ALISA uses texture features to characterize patterns in training images. These features are used to build a statistical representation of a class. ALISA then uses this class representation to determine if any test images are similar in part or in whole to the training set. ALISA learns by example and can be quickly trained to recognize new textures without reprogramming. (Bock, 1993; Howard, 1995).

Vector quantization is a machine-learning paradigm that learns which feature values are expected or unexpected for a particular class of images by accumulating the frequencies at which feature values occur in example images.

The current ALISA system (shown in Figure 11) has three modules, a texture module (Bock, 1993), a geometry module (Howard, 1995), and a shape module (Becker and Bock, 1998). The texture module uses features that characterize subsymbolic textures in an image. These texture features include measurements of roughness, wavy patterns, and the directionality of any intensity gradients. The ALISA texture module classifies pixel patterns in an image into known texture classes. The texture module generates an output image, called a *texture map*, in which each pixel's value corresponds to the texture class of the area surrounding that pixel in the original image. This means that areas of the texture map that have the same pixel value have the same texture in the corresponding area in the original image. The boundaries between these solid areas are the edges in the image (Bock, 1993).

The geometry module can be trained to classify a wide variety of two-dimensional patterns in the texture map. Edges can be classified based on their orientation (i.e., vertical, horizontal, up-slant, and down-slant) or other relationships like junctions or simple shapes. The geometry module builds an output image, called the *geometry map*, in which each pixel's value represents that region's geometric configuration (Howard, 1995).

The shape module accepts the geometry map from the geometry module as input and generates a *shape map* as output. Each pixel's value in the shape map indicates the shape class selected for that pixel's segment. The shape module expects the input geometry map to contain edges classified as any of the following four classes: horizontal, vertical, up-slant, and down-slant. The first processing step in the shape module is to segment the input geometry map. The segmentation process finds continuous regions of any one of these four canonical geometry classes.

A radial feature token (RFT) is then applied to each segment. The RFT measures the distance between the center of each segment and its surrounding segments, as shown in Figure 12. As with the other ALISA modules, the shape module must be trained to recognize particular shapes before it can be used in an application. While the shape module is being trained to recognize a new shape class, the feature vectors generated by the RFT are accumulated in the shape module's vector list for that shape. This list of feature vectors constitutes the definition of a shape for the shape module. When the shape module classifies shapes, the feature vectors generated by the RFT are then compared with the feature vectors learned during training to find the closest match.

After the shape module has classified each segment, groups of segments are analyzed with a parsing algorithm to remove small shapes so that larger shapes can be identified. The parsing algorithm takes advantage of the relative confidence of shape classifications to group segments into shapes.

Video analysis is on the frontier of content-based query. The ability to search through video archives for video of sports events, famous actors, chase scenes, or dance routines is beyond current technology, but not by much. This capability would greatly reduce the barriers to research tape archives of court cases, old films, or television news broadcasts.

For organizations with large investments in multimedia content, knowledge discovery tools that help organize, index, and retrieve those assets are an important part of controlling maintenance costs. These tools must also help integrate multimedia libraries into workflow processes that have historically handled only text and scheduling information.

13.6   Knowledge Discovery in Workflow Systems

Workflow systems are becoming more prevalent in the workplace, driven by the need for handling more information for distributed project teams with less administrative support. New technologies, like the

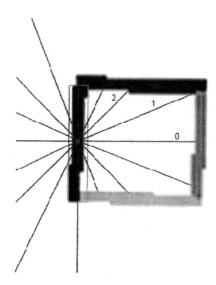

FIGURE 12   Radial feature token (RFT) and its feature vector.

World-Wide Web, CORBA, COM, and XML have been developed specifically to support highly distributed systems and information sharing.

The goals of knowledge discovery in workflow systems are to:

- Adapt to each user's specific needs, making that user more productive
- Find information that is relevant in the context of the workflow
- Detect patterns in the workflow that indicate problems, like bottlenecks or error sources

To adapt to users' particular needs, the knowledge discovery process must monitor each user's interaction with the system and build a profile of that user. Knowledge about the system's content must also be accumulated, either by direct user feedback or automated data mining algorithms. A user's profile can then be matched against the content profile to help improve queries based on the information needs or usage patterns of a specific user. Three algorithms are presented here that have been developed to adaptively customize or improve proformance:

- Rocchio's algorithm
- Collaborative filtering
- State transition matrices (STMs)

Rocchio's algorithm iteratively improves a user's profile based on that user's feedback about previously seen content. Collaborative filtering groups users into peer groups of others with similar preferences. Content that was popular within a user's peer group is then recommended to the user. State transition matrices (STMs) focus on the content and the workflow of the system. The STM learns which content is used in which order in particular scenarios. After some training, the STM can use this experience to predict which content might also be interesting to a user.

A video store will be used to illustrate profiling. Imagine someone trying to find a movie that he would really like and that he is willing to spend some time sampling movies to refine his selection. To help their customers, this imaginary video store has rated all of its movies on a scale from 1 to 10 on the following criteria: violence, believability, and humor. Comedies would have high scores for humor and melodramas would have low scores for humor. Science fiction movies would have low scores for believability and documentaries would have high scores. Figure 13 shows the ratings for eight sample movies, named A through H.

Movie	A	B	C	D	E	F	G	H
Violence	3	5	1	7	2	10	1	2
Believability	5	4	5	4	6	2	10	8
Humor	1	7	7	2	9	2	3	6

FIGURE 13 Sample movie ratings.

Rocchio (Rocchio, 1971; Singhal et al., 1997) developed an algorithm that uses the ratings of content, movies in this example, and a user's feedback to iteratively improve a profile for an optimal selection. After a user has seen all eight of the movies in Figure 13 and decided whether or not he liked them, Equation 5 can be used to determine the distance between the center of mass of the movies he liked and those he didn't. The larger this distance is, the more likely this algorithm is to be able to do a good job of selecting appropriate movies.

$$\vec{Q}_{optimal} = \frac{1}{R} \sum_{D \in Rel} \vec{D} - \frac{1}{N-R} \sum_{D \notin Rel} \vec{D} \tag{5}$$

For purposes of this example, say that someone is looking for believable and funny movies that don't have a lot of violence. That is to say that he liked movies C, E, and H, and disliked the others.

In Equation 5,

- N is the total number of movies (N = 8)
- R is the number of relevance (or liked) items (R = 3)

- $\sum\limits_{D \in Rel} \vec{D}$ is the sum of vectors from the relevant (or liked) items

- $\sum\limits_{D \notin Rel} \vec{D}$ is the sum of vectors from the not relevant (or disliked) items

Selecting a static user profile for our movie rental example would be quite simple. If the user likes nonviolent, believable, funny movies, then his profile should be [2, 6, 9], an exact match for movie E. As time goes on and he sees other movies, his preferences will change and his profile should track that change. Equation 6 is an extension of Equation 5 that uses the current profile ($Q_{current}$) and the centers of mass from the liked and disliked movies to calculate a new profile (Q_{next}). The weights α, β, and γ can be adjusted to determine how much each term contributes to the new profile. Relatively large values for α will keep the new profile close to the current profile, meaning that profile changes will be gradual. Relatively small values for α mean that recent movie evaluations will be used to set the new query, allowing the profile to change more rapidly. The relative values of β and γ determine whether the new profile is more affected by likes or dislikes.

$$\vec{Q}_{next} = \alpha \vec{Q}_{current} + \beta \frac{1}{R} \sum_{D \in Rel} \vec{D} - \gamma \frac{1}{N-R} \sum_{D \notin Rel} \vec{D} \tag{6}$$

Although Rocchio's algorithm does support adaptive profiling of users' preferences, it also has some significant limitations. For example, each user must have seen all of the movies being used to generate his profile and all of the movies had to be subjectively rated in several content areas. Collaborative filtering addresses this second issue by using feedback from many people to build peer groups. Each person's particular preferences are first used to identify his peer group. The filter can then recommend other movies that his peer group liked.

Movies

	A	B	C	D	E	F	G	H
Cheryl	+	-	+	-		0	+	+
Lloyd	+	0	+					-
Thelma	0	-	+	+	0	+		+
Dave	-	0	-					
Al	+	+	+	+	-	-	-	+
Cindy		0	-			+		-
Carl	+	-	+	-	-			
Kathy	-	+	-	+			-	-
Arnie	-	+	-	-	+		+	-
Carol	+	-	+	-	-		+	+
Steve	+	-	+	+	+		+	+

FIGURE 14 Collaborative filter example.

Returning to the video store example again, Figure 14 shows a chart of the eight movies A through H and a list of customers. Each customer evaluates each movie they have seen with a '+' for those they liked, '0' for those that were indifferent, and '-' for those they disliked. A blank space means that that person hasn't seen that movie.

The collaborative filtering algorithm groups customers into peer groups. According to Figure 14, Cheryl, Carl, and Carol have the same preferences, ignoring '0's and blanks. This constraint on grouping can be relaxed to allow groups with one or more differences to also exist. The filter can be used to recommend movies G and H for Carl because others in his group liked it. Likewise, the filter could warn Cheryl that people in her group don't seem to like E. Of course, this is a very simple example, but it does illustrate the basic operation of collaborative filters.

Both Rocchio's algorithm and collaborative filtering are effective ways to develop user profiles based on feedback from individuals or groups of users. These methods do not, however, take into account the sequential nature of many workflow systems, where the next transaction or selection is based on the current selection. In the case of our video store example, video stores have found that people who rent one movie are more likely to rent a sequel or another movie with the same star.

A state transition matrix (STM) can be used to learn the usual sequential order among a collection of content (Bock, 1993). In most workflow scenarios or information searches among hyperlinked content, a user is not the first person to travel a path through the maze of content. An STM can accumulate information about which direction each predecessor went and then recommend next steps to subsequent users.

Consider an example database of eight reference tapes in a company library. The STM shown in Figure 15, shows that if a person watches Tape A (current item), returns it to the library, and checks out another tape, that next tape will be Tape B 90% of the time, Tape E 4% of the time, and Tape G 6% of the time. Each time someone checks out a tape, the STM is updated. In this example, if Tape A had been checked out 50 times, then Tape B was checked out next 45 times. If the current user returns Tape A and checks out Tape B, then the new STM value for Tape B becomes 90.2% (46/51 = .902). The percentages for Tape E and G are decreased using the same calculation so that each column sums up to 100%.

This same approach would work well for research applications where a researcher is following a chain of bibliographic references or legal citations. Each document may have dozens of references, making an exhaustive search too time-consuming and expensive. Having information about the most popular related materials can help make the problem more tractable.

This simplistic STM approach can be made more powerful by using user feedback to update the STM, instead of just the fact that the next item was selected. Most other people may have selected the first citation on the list, so the STM recommends it most, and even more people follow. It would be nice to know how relevant they thought the document was. In some applications it might be possible to solicit

Current Item

Tapes	A	B	C	D	E	F	G	H
A	•	0	95	0	0	0	5	0
B	90	•	0	0	33	0	0	0
C	0	0	•	0	33	0	0	8
D	0	38	3	•	34	0	0	0
E	4	48	0	1	•	0	0	92
F	0	0	0	0	0	•	95	0
G	6	0	2	0	0	99	•	0
H	0	14	0	99	0	1	0	•

FIGURE 15   STM example (cell values are percentages).

explicit user feedback about a document's relevance. A "like/don't like" button or simple rating scheme would work if users are willing to cooperate.

Other, more subtle, ways can also be used to implicitly measure a user's opinion. The length of time a user looks at an electronic document can be measured and used as a relevance metric because users will normally look at information relevant to their current task longer than information that is not. Subsequent links after a document is viewed can also be analyzed and used as feedback. If a user links to a document and then immediately returns to its predecessor, this document was probably not very useful. If, however, the user continues to link down to more detailed subdocuments, even though they spent very little time on the initial document, then this document could be considered very useful.

Another way to make the STM more aware of the workflow context is to change the "current item" column headings into a more general "current state" description. The current item is now only part of the current state, which can also include a more general task designation. In other words, the next document you want depends not only on the current document but also on the task you are performing. Of course, these additional rows will increase the size of the STM dramatically. Even in the simple video tape example the number of cells in Figure 15 is the square of the number of tapes.

Knowledge discovery in workflow systems is and will continue to be an active area of research and development, as companies strive to make their workers more productive. Despite ever-increasing and broadening information sources, workflow systems that can adapt to a company's utilization patterns or a user's interest profile will be critical.

13.7   Measuring the Success of Knowledge Discovery

What does "successful" knowledge discovery mean? Is it performance, usability, return on investment? How can these things be accurately measured to determine if they are successful? How long does the evaluation take? Six months? One year? Five years?

Gross measures like revenue or profit margin are easy to determine for an organization, because they measure money flow in and out of the organization. If, however, problems exist within an organization, they don't help to identify the problem or suggest possible solutions. Knowledge discovery, as part of a larger knowledge management system, operates within an organization. Certainly the organization can be very complex and distributed, but the interface between the knowledge discovery system and its users can always be defined. Just as revenue and profit are measured at the interface between an organization and the outside world, the success of a knowledge discovery system can also be measured in much the same way.

When organizations are evaluated the criteria include:

- The experience of the management team
- The quality of the people

- Utilization levels of the staff
- Feedback from their customers

By making an analogy between a commercial organization and a knowledge discovery system, some very useful metrics become apparent. The ontology of the knowledge discovery system can be viewed as the org. chart for our evaluation. From the ontology shown in Figure 5, the entity object is the senior management, person, document, and client are the middle managers, and the rest are staff.

To evaluate the experience of the management team, look at the history of using Entity as the top level in previous ontologies. Were these ontologies built top-down like this one, or were they bottom-up (a situation that entity has no experience with). If entity has worked successfully in similar ontologies before, then it is probably a good place to start.

The quality of the people becomes the quality of the discovered knowledge in this analogy. Usually the best way to measure the quality of people is to measure the demand for them. This links the quality of the people with the staff utilization as well. People who are current on technology or best business practices, easy to work with, and effective are generally in high demand. Likewise, objects within the ontology that have current and complete information, easy interfaces, and can be accessed effectively as part of a workflow will be in high demand.

Measuring the utilization (or demand) for the knowledge within a particular object's domain can be done objectively by counting the number of accesses. This is much easier than trying to develop some subjective way to evaluate the quality of knowledge within an object. Objects that have useful information that is valuable to people and can be accessed effectively will be used. So just meter the usage. Objects that are out-of-date or difficult to work with will not be used as often. By monitoring usage, these trouble spots can be isolated and addressed.

When measuring utilization, be careful to measure access to real knowledge elements and not just accesses to a table of contents. Companies that advertise on the Web discovered early on that just counting the number of times someone downloaded their banner ad was not a good way to measure the success of an ad. Now companies meter these ads by counting the number of "click throughs" where the viewer actually selects the ad to get more information. When metering knowledge access, be careful to count only "click throughs."

The final criterion is feedback from users. Explicit feedback can be gathered in applications where the users are willing to provide it. Explicit feedback is most useful in cases where the implicit feedback is not specific enough to isolate problems.

Making this analogy between a commercial organization and an ontology is also a useful way to communicate the results to an organization's management because they will readily see the relationships and understand the vocabulary.

13.8   Summary

At this point knowledge discovery is at the hunter-gatherer stage of its evolution. As in hunter-gatherer societies, more than 90% of the food (or knowledge) consumed by the group will be collected by the gatherers. Gatherers collect the information that is readily available within an organization, like: forms, policy manuals, report templates, project proposals, source code modules, and customer contacts. They gather this valuable information and organize it so that others can access it and use it more easily. The hunters are the researchers looking for new information (red meat) for the organization. Although hunting is more glamorous and exciting than gathering, it only contributes a small percentage to the organization's daily information diet. That's not to say that organizations don't need hunters. The research they laboriously stalk, club over the head, and drag home is a critical part of the group's survival as well.

The challenge for knowledge discovery is to make information and its supporting technologies work for us — making us more productive — instead of us spending our productive hours working to find or recreate information. Making information work for us means there must be quick and intuitive ways to locate important information and understand the importance of the information in a work context.

Developing effective knowledge discovery must be part of everyone's job, not a special committee or task force. Because of the cultural changes needed to adopt these technologies, everyone must be a stakeholder in the development, implementation, and maintenance of the system. This way everyone will be more motivated to work through the often difficult process of adopting the new technology.

Work in content-based query technology will continue to extract more information from images, audio, and video. This information will be used to identify people, famous buildings, or places in photographs. Medical images, like X-rays, will be double-checked by image analysis systems that have been trained to recognize abnormalities. Integrating natural language processing (NLP) capabilities with content-based query systems also offers some exciting potential. Some systems already offer the ability to combine keyword searches with content-based searches. But these capabilities will continue to evolve. NLP technology allows searches to find image records with keywords that are close in meaning to the keyword used in the query. Likewise, content-based queries search for images that "look like" a sample image.

As research into the use of intelligent systems progresses, knowledge discovery systems will evolve from the hunter-gatherer stage into the agricultural stage. Intelligent systems and enterprise-wide collaborative systems will then grow ideas from conceptual seeds into complete designs and reports.

13.9 Abbreviations

ALISA	Adaptive learning image and signal analysis
CKML	Conceptual knowledge markup language
COM	Component object model
CORBA	Common object request broker architecture
HTML	Hyper-text markup language
JPEG	Joint photographic expert group
KIF	Knowledge interchange format
KQML	Knowledge query and manipulation language
NLP	Natural language processing
OML	Ontology markup language
RFT	Radial feature token
SGML	Standard generalized markup language
SHOE	Simple HTML ontology extensions
SQL	Structured query language
STM	State transition matrix
W3C	World-Wide Web consortium
WAVE	Web analysis and visualization environment
XML	eXtensible markup language

Acknowledgments

I would like to thank my colleagues Cheryl Howard and Steve Finch for their help in preparing this chapter. I would also like to thank my wife, Linda, and children, Adam and Melissa, for their continuing support.

References

Allee, V., *The Knowledge Evolution: Expanding Organizational Intelligence*, Butterworth-Heinemann Publishers, 1997.

Barford, N., *Experimental Measurements: Precision, Error and Truth*, Addison-Wesley Publishing Company, 1967.

Becker, G. and P. Bock, Collective learning systems II: The ALISA image analysis system, *PC AI Magazine*, January 1993, pp. 42-44.

Becker, G. and P. Bock, The ALISA shape module: Shape classification using a radial feature token," Proceedings of the 4th International Workshop on Multistrategy Learning (MSL-98), June 1998, pp. 23-37.

Bock, P., *The Emergence of Artificial Cognition: An Introduction to Collective Learning,* World Scientific Publishing Company, 1993.

Davenport, T. and L. Prusak, *Working Knowledge: How Organizations Manage What They Know,* Harvard Business School Press, 1998.

Domingos, P., Data mining with RISE and CWS, Proceedings of the 4th International Workshop on Multistrategy Learning (MSL-98), June 1998, pp. 1-12.

Fayyad, U. M. et al. (editors), *Advances in Knowledge Discovery and Data Mining,* AAAI Press, 1996.

Fu, K. S., *Syntactic Pattern Recognition and Applications,* Prentice-Hall, 1982.

Genesereth, M.R. and R. E. Fikes, Knowledge Interchange Format, Version 0.3, Reference Manual, Knowledge Systems Laboratory, Stanford University.

Gruber, T. R., Toward principles for the design of ontologies used for knowledge sharing, KSL-93-04, Knowledge Systems Laboratory, Stanford University, 1993.

Harmandas, V., M. Sanderson, and M. D. Dunlop, Image retrieval by hypertext links, *Proceedings of the 20th Annual International ACM SIGIR Conference on Research and Development in Information Retrieval,* ACM Press, July 1997, p. 296.

Hofstadter, D. et al., *Fluid Concepts and Creative Analogies,* Basic Books, 1995.

Howard, C., *An Adaptive Learning Approach to Acquiring Geometry Concepts in Images,* Doctoral dissertation for the School of Engineering and Applied Science of The George Washington University, 1997.

Kodratoff, Y. and R. Michalski, *Machine Learning: An Artificial Intelligence Approach,* Volume III, Morgan Kaufmann Publishers, 1990.

Labrou, Y. and T. Finin, A proposal for a new KQML specification, Computer Science and Electrical Engineering Department, University of Maryland College Park, Feb. 3, 1997.

Leake, D. (editor), *Case-Based Reasoning: Experiences, Lessons, and Future Directions,* AAAI Press, 1996.

Luke, S., L. Spector, and D. Rager, Ontology-based knowledge discovery on the World-Wide Web, Proceedings of the Workshop on Internet-based Information Systems, AAAI-96 (Portland, Oregon), 1996.

Michalski, R., J. Carbonell, and T. Mitchell, *Machine Learning: An Artificial Intelligence Approach,* Volume I, Morgan Kaufmann Publishers, 1983.

Mitchell, T., *Machine Learning,* McGraw-Hill, 1997.

Mulder, J., A. Mackworth, and W. Havens, Knowledge structuring and constraint satisfaction: the Mapsee approach, *Journal of Pattern Analysis and Machine Intelligence,* IEEE Computer Society, Vol. 10, No. 6, November 1988, p. 866.

Nonaka, I. and H. Takeuchi, *The Knowledge-Creating Company: How Japanese Companies Create the Dynamics of Innovation,* Oxford University Press, 1995.

Noy, N.F. and C.D. Hafner, The state of the art of ontology design, American Association of Artificial Intelligence, *AI Magazine,* Vol. 18 No. 3, Fall 1997, pp. 53-74.

Quinlan, J. R., Induction of decision trees, readings in machine learning (edited by J. Shavlik and T. Dietterich),1986, pp. 57-69.

Robertson, S. and K. Sparck Jones, Relevance weighting of search terms, *Journal of the American Society for Information Science,* Vol. 27, 1976. pp.129-146.

Robertson, S. and S. Walker, On relevance weights with little relevance information, *Proceedings of the 20th Annual International ACM SIGIR Conference on Research and Development in Information Retrieval,* ACM Press, July 1997, pp. 16-24.

Rocchio, J., Relevance feedback in information retrieval, in the SMART retrieval system, *Experiments in Automatic Document Processing,* Prentice Hall, 1971, pp. 313-23.

Roos, J. et al., *Intellectual Capital,* New York University Press, 1998.

Shen, W.-M., *Autonomous Learning from the Environment,* Computer Science Press, an imprint of W. H. Freeman and Company, 1994.

Singhal, A., M. Mitra, and C. Buckley, Learning routing queries in a query zone, *Proceedings of the 20th Annual International ACM SIGIR Conference on Research and Development in Information Retrieval,* ACM Press, July 1997, pp. 25-32.

Sparck Jones, K., A statistical interpretation of term specificity and its application in retrieval, Journal of Documentation, 1972, Vol. 28, pp. 11-21.

Tou, J. T. and R. C. Gonzalez, *Pattern Recognition Principles,* Addison-Wesley Publishing Company, 1974.

Wille, R., Restructuring lattice theory: An approach based on hierarchies of concepts, Ordered Sets (J. Sowa, editor), pp 445-470, Reidel, Dordrecht-Boston, 1982.

14

Knowledge-Based Systems: A Look at Rule-Based Systems

14.1 Knowledge-Based Systems for Knowledge Management

A key aspect of knowledge management is the capture and dissemination of knowledge — as distinct from the dissemination of data. There are many technologies for the dissemination of data, ranging from printed manuals to large databases and the Internet. However, providing access to data alone assumes that the user already has the knowledge to understand and utilize the data or that the user can read the data and understand it well enough to convert it to knowledge. The first assumes that the user is already trained in the subject area, the second assumes that the user can train themselves. These assumptions are frequently not valid. No one can be trained in all areas, particularly with the increasing complexity of decision-making. Likewise, in most cases, people do not have time to learn a new subject area when they simply need an answer to a specific question. This is why there are expert consultants or specialists who can answer specific questions.

The goal of expert systems is to capture a portion of the expert's decision-making knowledge, codify it in a way that preserves it, and allow its effective dissemination to users — without the users needing to learn the subject area. Rather than just presenting the user with ways to access data, the expert system emulates the interaction that the user might have with a human expert in the subject area. The expert system asks the user questions, comes to recommendations, and can explain the logic for the decision. As with a human expert, the user can either just take the system's answer or can interrogate the expert to learn how and why the decision was reached.

This ability to capture and disseminate knowledge should be part of any comprehensive program of knowledge management. As described in this chapter, expert systems are not the appropriate solution to all problems, but they are an excellent solution to many, and for those areas where they work, they should be used.

The term *knowledge-based expert systems* has been used to describe a wide range of approaches to delivering knowledge in computer programs. This chapter is limited to those that have the knowledge in the program represented as rules that describe pieces of a logical process that can be maintained independent of the rest of the program. These rules are processed by an inference engine that can utilize combinations of the rules to reach a conclusion. This is largely equivalent to what have generally been called expert systems. The definition excludes programs that simply use the term *knowledge-base* as an alternative for a database, which can be searched for relevant items matching a query.

14.2   History

Even the earliest expert systems proved very effective at handling problems not easily solved by other means. The promise of this new problem-solving technology led to a large number of commercial ventures in the expert system field. The mid-1980s saw an explosion of companies in the expert system market. Trade shows such as that of the American Association of Artificial Intelligence (AAAI) were major events for industry and the media. Unlike many computer science events of the time, expert systems and artificial intelligence also captured the attention of the general public due to sci-fi films, especially Stanley Kubrick's enormously popular *2001*. A popular expectation for the future of AI was a true thinking machine. Unfortunately, such goals proved totally unrealistic. Many companies set goals that could not be met. This resulted in an industry shakeout in the early 1990s and many of the expert system companies went out of business.

Unfortunately, this was incorrectly perceived as indicating that the technology had failed. Actually it was unrealistic goals that were beyond what the technology could do that had failed. This should not have been much of a surprise, especially with hindsight, but it fostered a widespread impression that expert systems were a failed technology. In reality, companies that had used expert system technology appropriately had been extremely successful. DuPont was one of the leaders in the implementation and use of expert systems under the inspired guidance of Dr. Ed Mahler. DuPont made extensive use of small expert systems throughout the company to disseminate problem-solving knowledge on relatively simple problems. According to Dr. Mahler, DuPont's savings due to the use of expert systems was over $1 billion.

What separated the expert system successes from the failures was the nature of the problems they attempted to solve. Expert systems worked well for specific, well-defined, well-understood problems. They did not work well for problems requiring deep insight, original creativity, or anything approximating true intelligence. Despite the successes in certain industries, interest in expert systems gradually declined and was partially replaced by research in other AI technologies such as neural nets.

14.3   Future — The Web

Expert system technology has been quietly waiting for the last half decade. Now the World Wide Web has given it new life. Web sites have become one of the main communication and marketing tools of a company. The company site has become the virtual representative. Most Web sites attempt to give the user enough information to understand and decide on a product or the company and provide support to existing customers. This has meant presenting information and relying on the user to read it, understand it, and assimilate it into the "knowledge" needed to make a decision. Expert systems that can be run from Web sites enable the direct delivery of knowledge, rather than data, to the user. The types of problems that expert systems work particularly well for are exactly the types of problems that need to be handled on the Web.

The Web also solves the maintenance and distribution issues that have always been a problem. A single copy of the expert system is maintained on the Web server and can be accessed worldwide 24 hours a day. Web-based expert systems are already in use in government and industry. This paradigm shift in how people access companies is leading to a tremendous new opportunity for expert systems.

Expert systems can be embedded into existing Web sites. The end user often may not even realize that they are communicating with an expert system. They will simply answer some questions and get an answer. For example, consider ways to do tech support on a Web site. The common technique is to use a FAQ section, which lists symptoms and fixes for a wide range of problems that commonly occur. The user is expected to read through the list and find one that might be relevant to the problem they are having. This can take a long time and even when a possible fix is found there is no indication if another solution might be more likely. Some more advanced systems allow a keyword search to try to find topics more quickly, but this relies on using exactly the same terminology to describe the problem and also has the problem of not indicating ranking. Such keyword systems work well when used by an experienced help desk operator (which is whom they were originally designed for), but they are not as effective when they are made widely available on the Web and accessed by novice users.

The expert system approach prompts the user with questions, rather than relying on them to ask the correct questions. Based on the data that the user provides, the best solution is found and presented. If there is more than one solution, they can be ordered by likelihood.

Every company has information that it wishes to explain to its customers. This may include product information, company policy, and regulatory information. Traditionally there have been two ways to do this — media (traditionally print, but now Web pages) that explains the subject and educates the user, and support personnel that can talk one-on-one with a customer and give them the answers that they need.

Education is difficult. There is no way to know that the user actually grasps the concepts or understands enough to make an informed decision. For complex areas, such as regulatory legal requirements, it is usually impossible to quickly educate a novice user to the level of being able to make a reliably good decision. Support personnel are expensive, require training, and may not all give the same answers. A better solution that more and more companies are using is to integrate expert systems on their Web site. The expert system emulates the interaction that the user would have with a support person, and produces consistent answers 24 hours a day.

Another technology that is often used on Web sites is database queries. It is the ability of expert systems to use confidence values to select a best solution that differentiates them from these database searches. For example, product selection can be handled by both database and expert system techniques. Suppose the system will ask three questions, and based on the input recommend a product. The database approach is to search for all products that meet ALL of the three criteria — any mismatch and that product is eliminated. If no product was a perfect match, the database would produce no recommendation. The expert system approach would use each answer to increase or decrease the confidence of a product being appropriate individually. The expert system would recommend the product that received the highest confidence value and was the best match to the customer's needs, perhaps with an explanation of what criteria could not be met. The expert system approach is a closer emulation of the interaction a customer would have with a human salesperson. A human would not tell a customer "There is nothing that matches what you want — go away." Instead they would say "We can't meet all of your requests, but here is the best fit.." This is what the expert system does, and it is a much more effective virtual sales representative.

The Web is a perfect medium for expert system dissemination, and expert systems are very well suited to answering the types of questions that are needed on Web sites. This match is leading to a new wave of system development and utilization.

14.4 Concepts and Terminology

Knowledge-based expert systems can be built for a wide range of decision-making problems. A specific decision-making task is made up of facts and "rules of thumb" used to make a decision. These "rules of thumb" are referred to as *heuristics.*

A heuristic is simply a step in the decision-making process. For example, an expert system to diagnose a printer might incorporate heuristics such as "If the green light is lit, the printer has power" or "When the ink gets low, the printing becomes gray." Not all steps are used in every decision, but the combination of the relevant ones for a particular situation will produce the appropriate conclusion.

The process of building an expert system is to convert the heuristics into **rules**. Heuristics are quite free-form. They do not have fixed syntax and are just an explanation of a decision-making step. Rules on the other hand, have a defined syntax. There are many types of detailed rule syntax, but typically are IF/THEN production rules. IF/THEN rules have an IF statement that can be evaluated to TRUE or FALSE based on data for the particular case. When the IF statement is determined to be TRUE, the THEN portion of the rule is considered to be TRUE and added to the data the system can use to test other rules or present as a conclusion.

The conversion of heuristics to rules can be a simple matter of syntax or may be quite complex. In the above case, "If the green light is lit, the printer has power," the heuristic is already stated in an IF/THEN form. In this case conversion to an IF/THEN rule is easy.

IF
    The green light is lit
THEN
    The printer has power

Rules can be built in many ways. There have been dozens of different *development tools* to assist in building and maintaining the rules. These have varied from directly writing rules in text editors to advanced developer interfaces using tree diagrams allowing the developer to visualize the logic of large sections of a project.

The learning curve and complexity of syntax of these tools have varied greatly. Many of the early tools were written in the LISP computer language and the rule syntax reflected the powerful, but not particularly easy to read, syntax of that language. Later tools used a more English-like syntax making rules far easier to write and maintain. Some development tools were made for specific vertical industries such as process control and sacrificed general utility for special features unique to the individual industry.

Once the rules of a system are defined, they are run with an *inference engine*. The engine is a program that runs the rules. It makes use of the rules to:

1. Determine what rules are relevant to reaching a conclusion.
2. For the relevant rules, determine what data is needed to tell if the IF statement is true.
3. Determine if that data can be derived from other rules or must be asked of the user or some other source.
4. Use the appropriate rules to derive needed data where possible.
5. Ask the user for data via the system interface.
6. Present conclusions and recommendations via the system interface.
7. Answer user questions about how conclusions were reached.

A common misconception when looking at the IF/THEN rules of an expert system is to think that they are the same as the IF/THEN statements in a language such as BASIC or C. There are fundamental differences. IF/THEN statements in BASIC are simply procedural operations. The BASIC program just goes through them in order. One BASIC rule has no "knowledge" that it could use another rule to derive data it needs or the ability to invoke that rule. The BASIC program has no automatic knowledge of what rules fired or how conclusions were reached. All of these functions are provided automatically by the inference engine. Of course, they could be programmed into the BASIC program, but that would in effect be writing a simple inference engine in BASIC. In an expert system, rules should not be viewed as executable pieces of code, but as individual statements relevant to reaching a conclusion that may (or may not) be needed and used.

There are many aspects and concepts involved in various inference engines, but two of the most fundamental are backward and forward chaining. Unfortunately, these terms have been used to mean somewhat differing things over the years — largely for marketing purposes. A standard distinction is that *backward chaining* is goal driven and *forward chaining* is data driven.

In backward chaining the system has a goal that it is trying to reach or determine to be true or false. Typically, this goal is one of the possible conclusions of the system. The inference engine determines which rules are relevant to this goal. Those rules are examined to see whether their IF statement can be determined to be true. If there is not enough data to do this, the needed data becomes the new goal and the process starts over recursively, now trying to determine information on the new goal. It is this recursive process that enables the expert system to make effective use of the rules. For example, in an expert system to diagnose printer problems there could be the rule:

IF
  The printer is not receiving power
THEN
  Check the power cord connections

When the inference engine is trying to determine if the power cord connections are possibly of concern, it would use this rule. To determine whether the IF part of the rule is true, it needs to know if the printer is receiving power. "Receiving power" becomes the new top goal for the backward chaining. Somewhere else in the set of rules there might be a rule:

IF
  The green light is on
THEN
  The printer is receiving power

This rule would be used by the inference engine to satisfy the current goal. Since the system does not know if the green light is on, this becomes the new top goal. In this case, there might not be any other rule to derive information on the light. The system would have to ask this of the user directly via the system interface. Suppose the user says the light is *not* on. This means the rule is false and cannot be used to add to the data in the system; it does not mean that the printer is not receiving power — that would require a different rule. Since the "green light" goal was satisfied (by asking the user), that goal is now dropped and the previous top goal "receiving power" again becomes the top goal.

Another rule in the system is:

IF
  The printer is making noise
THEN
  The printer is receiving power

The first rule to satisfy the "receiving power" goal was not true and could not be used, so the next relevant rule is tested. "Making noise" becomes the top goal. Again, there is no rule to derive information on "making noise," so it is asked of the user via the system interface. This time the user says, yes it is making noise. This satisfies the top goal of "making noise." This rule is true and the system now knows that the printer is receiving power. The system has now satisfied the "receiving power" goal. The "power cord connections" goal now again becomes the top goal. In this case, the user input and data derived from rules indicate that the original rule:

IF
  The printer is not receiving power
THEN
  Check the power cord connections

is false and cannot be used. The system would then go on to other rules relevant to "power cord connections" or other goals in the system. When all goals have been examined, the system results are presented to the user.

An important fact to remember is that the inference engine finds and invokes the appropriate rules automatically. The developer should not need to explicitly link the rules together to build the system.

A forward chaining system is data driven. The rules are tested in order based on the available data. When data is not already available in the system, it is asked of the user. This results in a faster execution, but lacks the ability to have the system automatically combine rules to derive information.

In most cases, backward chaining is the easier and preferred method to use. Forward chaining is useful for expert systems that have very high throughput and that receive all data from automated sources such as databases. If the system asks questions of the user, the time spent reading and answering the questions will far outweigh the speed improvements of forward chaining.

One of the most effective aspects of an expert system is the ability to rank more than one recommendation based on a *confidence value*. In many real world problems, there is not a single, specific, absolutely correct recommendation for a set of input data. The input may indicate that two or more recommendations are possible, though perhaps with different degrees of likelihood. This may be due to:

1. An inability to distinguish between the recommendations based on the input data.
2. Cases where the final decision is based on personal preference, emotion, or other factors that cannot be put into the expert system.
3. Situations where there are more than one appropriate solutions.

Expert systems handle this complexity through confidence values. There are many methodologies for defining and handling confidence values. The requirements of the individual problem determine which technique is best to use. The confidence value selected should allow the developer to build rules based on heuristics such as "When you see… that USUALLY means …, but occasionally it means …" The rules need to reflect that the IF condition is indicative of more than one conclusion, though with different degrees of confidence.

14.5   Methodology for Building Systems

The process of building an expert system is taking the heuristics that describe how to solve a decision-making problem and converting them to the rule form needed by the inference engine. Usually then the system needs to be integrated into an existing MIS structure to obtain data, communicate etc. There are three main areas when building an expert system: selecting a problem, selecting a development tool, and building the rules.

Selecting a Problem

Some companies have been very successful in their use of expert systems and others have not. What typically separates the successes from the failures is the problems that were selected. Expert systems work very well to solve many problems, but they are not an appropriate solution for all problems. There are a variety of factors that determine how feasible, practical, and desirable an expert system solution will be to a particular problem. Setting unrealistic goals for an expert system project will doom the project from the start, but using the technology in an appropriate way will enable a wide range of highly valuable applications to be fielded.

Does the Knowledge Exist?

Expert systems are knowledge capture and delivery tools — they are not knowledge creation tools. The heuristic knowledge must exist somewhere. This may be only in the head on one person, or it may be well documented, but it must be somewhere. A good first test is "Would it be possible to teach someone how to solve this problem?" If the answer is "no," it is very unlikely that an expert system can be built to solve the problem.

If the knowledge does not exist, the project becomes a research project to find the knowledge — a far more difficult task than converting existing knowledge to rules. Technologies such as data mining and neural nets can, in some cases, uncover interesting relationships within large bodies of data. Finding and understanding the underlying causes for these relationships can eventually lead to heuristics that can be used to build an expert system; however, these technologies do not build rule-based systems automatically.

In addition to the knowledge existing, it must be heuristic knowledge that predicts an outcome or solution. Many problems have been studied and the causes for an event can be analyzed after-the-fact, but not predicted. Processes that are highly dependent on random or chaotic factors cannot be predicted. This is why there are no expert systems to predict the weather or stock market except in very specific, limited situations.

How Big Is the Problem?

This measure of the size of the problem is the number of rules, not its significance or importance. Expert systems work best for small, well-defined problems. They can handle large problems, but the complexity of the system often increases rapidly. One thing to consider is "How many possible solutions are there to the problem?" An expert system to select among a few options has only a few possible solutions. An expert system to diagnose the failure of a component has as many possible solutions as there are things that can go wrong with the component — probably less than 100. An expert system to configure the components that make up a complex system could have thousands of possible outcomes. The more the possible outcomes, the larger and more complex the system. Problems with small numbers of possible outcomes are usually easier to build, are much more likely to succeed, and require fewer rules. In some cases, a large problem can be divided into several systems that are more easily developed.

Does the Problem Require Insight or Novel Solutions?

Expert systems do not think — they simply apply the rules that are in the system to data in a specific situation and produce the results. Expert systems are not good solutions to problems requiring human insight or novel solutions. For example, it is possible to build an expert system to diagnose problems with a printer, but it is not possible to build an expert system that will design a better printer. The design problem has an infinite number of solutions. Creating a new design is a process of human insight — this cannot be converted into rules.

What Percent of Problems Is the System Expected to Solve?

Usually human experts cannot solve 100% of the problems in an area. Some problems must be referred to other specialists or may require additional study. The same is true of expert systems. If the goal is to have the system solve 100% of all possible problems, either:

1. The problem needs to be very well defined, small, and fully understood so that heuristics to solve 100% of the cases can be defined.
2. Heuristics should exist that allow some *adequate* solution to be found for all cases. (This can often occur in product selection problems.)

For most other problems, it is unlikely that an expert system can be built to solve 100% of the cases. This is not necessarily bad, it is just a reality that must be considered in system design. An expert system that can answer 80% of the questions in a complex area is still a highly valuable system. Remember, most human experts cannot answer 100% of the questions either.

Common Questions

When selecting an expert system project, select a system that answers commonly occurring questions. Common questions are the ones that have been most studied and should be best understood. There is immediate payback to fielding a system that can answer common questions. An expert system that can solve 80% of the total number of possible questions will probably answer a far higher percentage of the actual questions by volume. The volume is made up of the common questions, which are the ones most likely to be handled by the system. The unusual questions that the system cannot handle can be summarized and passed on to a human expert. When those problems become more fully understood, they can be added to the system.

Real Time

The speed of an expert system is rarely an issue. Current hardware and software allows expert systems to run at high speed. On any interactive system that asks the user questions and presents results for them to read, far more time will be spent in the user reading screens than in rule processing. Speed issues only become important for systems that run in batch mode with very large numbers of records or systems that are automatically watching rapidly changing data inputs. As a general rule of thumb, even on fast systems, processing a set of data with enough rules to handle complex decisions will take from one to a few seconds. Applications that require significantly faster rates will be more difficult and need to use special techniques to achieve the needed speed.

Learning

Expert systems do not learn. They do not become "smarter" automatically. There have been a variety of techniques that have been tried to make systems that automatically learn, but these are mostly at the academic and theoretical level. For most applications, "learning" is achieved by adding rules to cover new areas, but that is really system maintenance and enhancement, not learning.

Selecting a Development Tool

Once an appropriate problem has been selected for the expert system, the next step is to decide on a development tool to use. There are a variety of expert system development tools on the market and selecting the best one for a project can be an intimidating task. Various tools have specific strengths and weaknesses. The following are some of the factors to consider:

Proven History of Success

Look for a development tool with a good track record of success in building applications similar to the planned project. Look for tools that have proven ability to handle all required aspects of the project.

During the mid-1980s there were several dozen expert system development tools, but most went off the market in the industry shakeouts that occurred in the early 1990s. Many of the better known current tools are the ones that survived the shakeout. Unlike some other types of software, the tools with proven technology are often the best, since they have been tested and refined on a wide variety of problems. The newest tool may have good concepts, but refining a tool to meet real-world problems takes time.

Company support

Look for a development tool company that will support the development process. Complex application development often requires using and understanding advanced features of the tool. The company should provide training in using the tool, along with building a prototype and consulting should that be necessary. The company should have a history of having worked with users in similar problems.

Platforms

Make sure that the tool supports the intended development and delivery environments. At one time, virtually all tools were for LISP machines. That gradually shifted to UNIX and now most are aimed primarily for Windows. There are some tools that run on special environments such as mainframes. If the application requires running on a specific platform, make sure that it is supported.

Web-based delivery

Probably the best way to deliver expert systems to users is via the Web. This is the future of the industry. If the application is intended for Web delivery, check on support for this option. Also check on what is required of the user — Java, a plug-in, or can applications be run with just a standard browser? Make sure that system requirements match the capabilities appropriate to the anticipated users.

Learning Curve

Development tools have widely varying learning curves ranging from a few hours to almost a year. The longer times are due to the high degree of complexity of some of the most powerful tools. The very short times are for specialized tools that may not have enough flexibility to handle complex problems. Judging what is a good level of flexibility and power, without overkill, can be difficult. The expert system industry is full of arcane nomenclature and syntax. In general:

1. If you know how to use a tool, and it is suitable to the intended application — use it. It is usually more efficient to use a familiar tool, even if it is not ideal, than spending the time to learn a new tool. Unlike some types of software where there are common conventions of interface and design (e.g., paint programs), in expert system tools there are tremendous differences in design and approach. Knowledge of one tool is not usually much help in learning another.
2. Use the easiest tool that will meet the development needs along with reasonable expectations for application growth. Many tools can be used to build most typical applications, but some will be much faster and easier than others.
3. The fundamental concepts of the development tool in terms of how rules are built should be easy to understand. If the syntax or approach is too complex or specialized when doing simple problems, it will be very difficult to use for actual applications. Conversely, if everything the tool can do can be learned in a day, probably the tool is not flexible enough to handle complex applications.

Simple Things Should Be Easy

When Dr. Ed Mahler was at DuPont, his visionary and widespread use of expert systems is one of the greatest expert system success stories. He was instrumental in getting hundreds of experts in DuPont to build expert system applications. His ideal was an "inspection obvious" development tool — that is a tool where just looking at the tool is enough to understand how to use it. (His favorite example of an "inspection obvious" interface is a Coke machine.) While a highly powerful, inspection obvious, development tool has not yet been created, some tools have proven to deliver power without unnecessary complexity. One measure of a tool is that it should be easy to perform simple, common tasks. Different tools have different design objectives, and what may be easy in one tool may not be in another. Complex or advanced functions are likely not to be "inspection obvious," but common functions should be easy. Remember, it is typically the domain expert that will be building the system. They need to be able to learn the tool quickly and remember how to use it for system maintenance.

Interfaces

Expert systems require data to run a session. Normally, the user will be asked questions to obtain this data; however, if the data is available on-line from a database or sensor, it should be obtained automatically. This both reduces the number of questions that need to be asked and reduces the chance for an error on inputting the data. The potential availability of on-line data varies with the specific application, but the tool selected should be able to communicate with required data sources. In cases where the data source is uncommon and not likely to be supported by any expert system tool, look for those with open architecture that allow extending the interface capability of the tool.

Training and Consulting

Developer training is always a good idea when starting an expert system project. Training allows the developer to learn how to build various types of applications in a classroom setting. Without training often the first system the developer builds is the actual application. It is rare that someone masters any expert system tool on the first try. Look for tools that provide an appropriate amount of training. Consulting with a developer on the specific project is also highly desirable. Working with someone very familiar with the tool to establish the approach and architecture can save major problems later.

System Validation

Expert system applications require extensive testing before they can be fielded. There are typically many thousands of possible paths through the system, and validating that all are correct is a major task. Look for tools that offer automation features to assist in this important testing aspect.

Growth

Expert system projects tend to expand with time. The technology is very appealing and applicable to many problems. Make sure that the tool selected is capable of both the current project and reasonable growth in the future. If the basic project uses 100% of a tool's capability, extending the project will be impossible or will require changing tools later. Once expert systems start to be fielded in a company, people will find more areas where they can be used. Unless the original project requires a specialized tool, select a generic tool that can be used for a wide range of projects.

Building the Application

Once the problem and development tool are selected, the final step is actually building the application. This is where many projects fail due to not considering all of the factors involved.

Who Is Going to Build the System?

Basically there are three choices on who builds the system:

1. The person with the domain knowledge — the expert.
2. Someone inside the company that generally knows the domain technology, knows (or can learn) the tool, and can work with the expert.
3. A consultant "knowledge engineer," typically from the tool company that knows the tool very well, but may be unfamiliar with the technology.

Each option has pros and cons. Experience has shown that with the appropriate tools, experts can build excellent applications themselves. However, the experts are busy people and often cannot spend much more than a few days to learn the tool. This means learning basic rule representation and tool functions, but usually not advanced features that might be useful or desirable for the problem. The largest advantage the expert has is the ability to rapidly recognize errors and incompleteness in the logic and know-how to correct them. This is particularly important in cases where the heuristic knowledge is not documented.

Using external consultants provides immediate high skill level with the tools and greatly reduces the time needed to build the system. In cases where the heuristics are well documented, the lack of specific domain expertise is not as important. However, if the heuristics are not documented, this means bringing the knowledge engineer up to the expert level for this specific problem. This may be a difficult task. Another advantage of using knowledge engineers is that they also bring the important skills of user interface graphic design, Web-based delivery, and documentation that are important aspects of any fielded system.

Using internal staff trained in using the tool and able to work easily with both the expert and external consultants is often the best solution for many problems. This approach provides strong in-house expert system capability while reducing learning curves and enhancing the likelihood for success.

Not as Easy as It Looks

Actually converting the decision-making heuristics into a syntactically correct rule form can be difficult. This depends largely on how precisely the heuristics are known, and how the heuristics are stated. Well-documented steps stated in IF/THEN form, decision trees, decision tables, or some other form syntactically close to the way the rule is defined in the development tool are fairly easy to convert. More generalized knowledge that is not formally documented or that does not completely cover all cases is far more difficult.

Consider the apparently simple heuristic: "When the ink gets low, the printing becomes gray." It is not in IF/THEN form and only indicates two things that occur at the same time. This could become

IF
   The printing is gray
THEN
   The ink may be low

It was converted to " may be low," since there may be other reasons for gray printing. If low ink is the only possible reason for gray printing, the THEN should be "The ink is low."
An alternative is

IF
   The ink is low
THEN
   The printing is gray

This is another way to express the heuristic. The difference is now the IF part, which is the test part, would need to determine if the ink was low, rather than determining if the printing was gray. For a diagnostic system, where the user would be asked about print quality, the first form would be preferable — but both describe the heuristic. However, the two rules would be run differently by the inference engine.

As even this simple example shows, converting heuristic knowledge from the form that people express it, to the more syntactically rigid form that the computer needs is not always simple or straightforward. As problems become larger and the heuristics become more vague, it becomes increasingly difficult to convert them into rule form. This difficulty of converting vague human heuristics into precise statements for the computer is why expert systems become impractical for large problems requiring complex, imprecisely stated heuristics. The rule form of a heuristic may seem correct in one place, but the inference engine will use it in other places where needed and it must be appropriate in all places.

Resources

To plan an appropriate amount of time and resources needed, careful consideration needs to be given to how difficult the conversion of a particular set of heuristics to rules will be. In addition, resources for graphic design and system integration need to be considered. Are these resources available and can they be committed to the project? Is the expert available to provide the needed input and feedback on the system? Many expert systems have failed simply because the expert did not have time to work on them. When planning resources, remember that the validation and final "polishing" of an application is likely to take almost as long as inputting the basic logic.

One "Simple" Function Can Be the Show Stopper

The design specification for the project needs to be examined carefully by someone very familiar with the development tool. Functionality commonly found in standard programs like word processors and spreadsheets is not necessarily found in expert system tools. A desired "look and feel" or function may be unexpectedly difficult to do in a specific tool. Often the desired functionality can be achieved, but it may require far more effort in the system design to achieve it. Functionality that will be difficult to implement needs to be identified as early as possible.

Expert Acceptance

It is essential to get the expert's enthusiastic support for the project. Expert systems can sometimes be perceived as reducing the status or importance of the expert. This should never be the case. An expert system is only a means to capture and disseminate a small part of what the expert knows. Fielding an expert system should both make the expert more productive by not having to answer the routine questions and establish them as the Expert in the domain area, much like a journal article or other publication. Without the expert's support, the project will fail — and that support should not be taken for granted.

If the expert has not been exposed to expert system technology, they may have some of the common misconceptions. Building familiarity with the technology will usually correct this.

End User Acceptance

Careful consideration must be given to the actual needs and requirements of the end users of the application. Unless the application provides the information they need, in the form that they need and expect it, they will not use it. A fully developed application that is never used is just as much a failure as one that was never built — perhaps even more so due to waste of resources. Check with end users on what they need. Get intended user feedback on planned graphic design and the "look and feel" of the interface. Modify the project as required to get good user acceptance. Consider if the system is best fielded stand-alone or over the Web.

System Integration

Expert systems that require delivery over the Web or intranet, or that access data automatically, need to be integrated into the overall information system. This is often not an appropriate task for the system designer and will require coordination and additional resources.

User Interface Design

Virtually all expert system tools provide the ability to customize the look of the delivered application. This is particularly true in applications that are delivered over the Web. Applications that are to be run by a company's customers for tech support, assistance, or product selection are in effect virtual representatives of the company. It is very important that they be well designed and well integrated into the company Web site. These may require special talents and not be something that the expert or knowledge engineer can do. If this is the case, graphics designers either as consultants or from the company's Web design group need to be involved in the project.

Maintenance

After a system is fielded, it must be maintained. Depending on how rapidly the knowledge in the expert system changes, this may be an occasional or ongoing process. Typically, it should be done by the original system developer; however, there needs to be adequate documentation to allow others to also maintain the system.

14.6   Methodology Conclusion

Knowledge-based expert systems are a well proven and effective technique for capturing and delivering expert knowledge. They work well only for certain types of problems, but those problems are very common and the ability to deliver solutions for them has tremendous payback. There can be pitfalls to developing an expert system, but they are easily avoided with planning. The Web presents an unparalleled opportunity for companies to deliver knowledge to their customers and employees, and expert systems are the ideal way to do it.

14.7   A Few Case Studies

Component Failure Prediction for Diesel Engines

Canadian Pacific Railroad

Over 10 years ago, Canadian Pacific Railroad developed an expert system to predict component failure in diesel locomotive engines. The system is still in use and is estimated to have saved the company over $300,000,000.

The system bases the analysis on the type and concentration of metal impurities in the lubrication oil from the engine. The oil sample is analyzed in a spectrum analyzer to determine this data. The visual interpretation of these spectra is a skill that takes years to learn and is very difficult to teach.

The expert system developed at Canadian Pacific allowed a PC running the expert system to analyze the data automatically. All of the input required by the system is supplied directly from the spectrum analyzer. The system produces a printed report for the technician telling which components require service and which are likely to fail. The system is highly embedded and there is no direct user interaction with the expert system.

The system has been highly successful at CP. Over 10,000 samples have been run with 98% accuracy. Major cost savings have been realized through the detection and replacement of components before failure. In some cases, the detection of a single failing component has saved more money than the cost of the entire system.

Credit Analysis Advisor

Financial Proformas, Inc.

Financial Proformas develops and markets software products to support the needs of commercial lending institutions. Over one third of the top 100 commercial banks in the U.S. and Canada along with some of the largest industrial and financial companies in the world use FPI's FAST (financial analysis support techniques) software for credit analysis. This software provides a complete range of traditional analytical reports on both historical and proforma basis.

FPI has released an expert system module, FAST Advisor, to provide complex analysis of the traditional report data. This supports the company's philosophy of utilizing the latest technology to increase loan officers' productivity. The expert system not only provides an English language interpretation of the historical financial output generated by FAST, but also prepares the assumptions for FAST annual projections and produces output that links to the banks' word processing software. This link eliminates much of the tedious writing of analytical reports by generating standard financial statement reviews.

Through periodic updates, the knowledge base can be customized for a bank's current loan policy, U.S. and local economic forecasts, and interest rate projections. FAST Advisor consistently and reliably interprets the relationship of these variable factors and user-defined levels of sensitivity associated with a particular financial statement. The expert system gives credit analysts access to the expertise of more experienced advisors, thereby accelerating the training process and increasing productivity.

Diagnostics/Repair of Complex Machines

ElectroCom Automation L.P.

ElectroCom offers unique capabilities in their electronic performance support systems (EPSS) by providing their users with a fully integrated system for computer-based training and electronic documentation. Each system can be tailored to special needs. EPSS enhance user's productivity and quality of work, as well as reducing overall personnel support and training costs.

The computer-based, self-paced training covers all aspects of target system operations, processing, and repairs. Electronic documentation provides on-line, easily updated system support information. The expert system incorporates multimedia interfaces in a high-performance hypertext format that provides the best available expertise in system operation and processing. Alternative platforms and interfaces, including the latest in portable PC technology and digital video are also supported.

ElectroCom has fielded a fully integrated EPSS to the United States Postal Service designated TOPS (total organization productivity support). TOPS is a totally integrated system. Designed to facilitate on-the-spot maintenance and operation of the U.S. Postal Service's automated ZIP-code mail sorter, it provides a multimedia single source for interactive training, diagnostic assistance, and system support. TOPS dynamically integrates the expert system and fully hyper-linked on-line reference with an exceptional visual database. The system provides entry level personnel with all training and assistance that is required to operate and maintain large, complex letter sorting machinery.

"TOPS" won the Best Overall Achievement in the 1993 Nebraska Interactive Media Awards, presented at the Ziff Institute's Interactive '93 Conference.

Bearing Troubleshooting Advisor (BTA)

Electric Power Research Institute and Automation Technology Inc.

Bearing system failures constitute one of the leading causes of unplanned outages of power plants. Because of the direct loss of plant availability, determination of the causes of bearing failure, methods of effective repair, and proper maintenance procedures are of paramount importance.

To address these issues, the Electronic Power Research Institute (EPRI) produced the bearing troubleshooting advisor (BTA) software package. Under EPRI sponsorship, Automation Technology Inc. developed BTA as an integrated software product. The system provides a complete environment for bearing maintenance record keeping, failure troubleshooting, and reference information on journal bearings. BTA provides quick access to all pertinent bearing information and helps to diagnose bearing problems with greater accuracy. By using BTA and establishing proper bearing maintenance procedures, utilities can reduce the number of bearing failures and forced plant outages.

The expert system helps users determine the mode of bearing failure and possible cause. Through an interactive series of questions and answers, the expert system provides probable failure modes. While answering the questions, the user may also view an example picture of a particular symptom, to look for identifying characteristics. After all questions are completed, a detailed analysis is provided of possible modes of failure, causes, repair solutions, and supporting reference text. This information is then automatically stored in the database for future reference and retrieval.

Financial Options for Restricted Stock

MultiLogic

This expert system is designed specifically as an on-line portfolio planning tool for holders of restricted and control stock. It combines the power of the Web and expert systems to provide free advice for users to assess their options for trading or borrowing on their restricted shares as they seek to protect the overall value of their portfolios.

In April 1997, there were changes to the SEC rules 144 and 145, which regulate restricted stock trading, making restricted stock less restrictive. Inside-trading tracker CDA/Investment anticipated sales of restricted and control stocks to nearly double to $10.5 billion in the second quarter of 1997 as a result of the rule change.

Through the system RestrictedStock.com (http://www.restrictedstock.com), stock owners can privately check out their alternatives. It attests to the power of converging the intelligence of expert systems and the universality of the Internet. Because of the complexity of this often-confusing area and the many variables possible, building this application would have been a massive undertaking without expert systems technology.

Holders of restricted stock are typically company insiders — executives who have become wealthy on paper, usually as a result of an initial public offering. Previously, the restrictions included a minimum two-year waiting period before company insiders could cash in their investment. As of April 1997, the minimum waiting period is reduced to one year.

MultiLogic worked with restricted and control securities experts to develop the site. RestrictedStock.com is modeled after the interaction that these experts have with their clients. Site visitors simply answer a short series of questions to find their specific alternatives for trading their stock or using it as collateral. Because of the numerous possible variables, such as state of residence and how shares were acquired, MultiLogic employed expert systems technology that deals with complex variables far more effectively and flexibly than traditional software. Customers desiring to take action based on the system's recommendations are directly linked to brokers.

Distribution of Regulatory Information on the Web

Occupational Safety and Health Administration (OSHA), Department of Labor

Under the Occupational Safety and Health Administration (OSHA), small and large businesses are subject to thousands of complex regulatory guidelines and standards for the safety and well-being of their employees and their facilities. These highly technical rules and regulations vary with time and change as new levels of knowledge are acquired in each industry. Every day, OSHA officials field thousands of telephone calls about safety in the workplace and new health rules. These callers need prompt and accurate information so that new regulations and guidelines can become incorporated into their daily business operations.

In the past, OSHA could not efficiently respond to such requests. Many times a caller would be transferred to several OSHA departments before finding the appropriate information. In 1995, OSHA began efforts to create an integrated technology solution in partnership with business to help employers learn the required information and knowledge about OSHA regulations and standards.

This undertaking began with the very complex regulations governing the use of cadmium, a soft metallic element found primarily in zinc, copper, and lead ores and used in alloys, solders, batteries, nuclear reactor shields, and electroplating. The International Cadmium Association worked with OSHA to provide the information required to guard the safety and health of employees who work with and/or come in contact with the product. This effort was a springboard for a pilot project called the Interactive Expert Advisor, which led to the successful and acclaimed Asbestos Advisor project.

The Asbestos Advisor is a computer program developed in partnership with U.S. corporations to make OSHA rules and regulations easily available to employers. Through this program, OSHA Advisors guide users, small business owners, safety and health personnel in larger organizations, building owners, managers, and contractors through the decision logic of OSHA regulations, inquiring about the employer's worksite and any work performed at the site. The Asbestos Advisor includes a list of frequently asked questions (along with correct answers) and an instant pop-up glossary of terms. Many professionals, trade associations, trade press, and consultants helped OSHA make the Advisors an effective tool.

The Asbestos Advisor uses a variety of technologies: a rule-based expert system, hypertext markup language (HTML), and the Internet. Industry subject matter experts and a group of consultants created the logic in the expert systems; the Internet and HTML provided a simple and successful delivery vehicle for the information. Combining these qualities in a team and working with industry has proved to be a viable approach that will be continued on all future Advisors. In fact, OSHA believes that its involving the industry customer in the development of the solution is what made the Advisor a success.

Since the Asbestos Advisor program went on-line in October 1995, more than 13,000 users have accessed the Asbestos Advisor from the Department of Labor's electronic bulletin board and from OSHA's Web site. Affected trade associations are also distributing copies of the Asbestos Advisor to their members. Major corporations are also electronically redistributing the Advisor tool to thousands of sites that require the information. Industry and military users have reported that the Asbestos Advisor has been used by up to 80,000 businesses in the first year it was put on the Internet. In March 1996, the Asbestos Advisor was honored with the NPR Hammer Award, presented by Vice President Al Gore.

Public use and acceptance of the Asbestos Advisor attests to the success of OSHA's Advisor approach. Since the release of the final version of the Confined Spaces Standards Advisor six months ago, the "on-line" version has been queried more than 9000 times. OSHA plans additional advisors on Lead in Construction, and the Fire Safety Advisor. The Advisors are available to the public via the Internet on the OSHA Web address http://www.osha.gov/oshasoft/ or www.dol.gov/elaws

Suggested Readings

Awad, Elias M., *Building Expert Systems: Principles, Procedures and Applications,* West Publishing, 1996.

Ignizio, James P., *Introduction to Expert Systems, The Development and Implementation of Rule-based Expert Systems,* McGraw-Hill, 1991.

Liebowitz, Jay (ed.), *The Handbook of Applied Expert Systems,* CRC Press, 1997.

Turban, Efraim, *Decision Support and Expert Systems, Management Support Systems,* Macmillan, 1988.

Turban, Efraim, *Expert Systems and Applied Artificial Intelligence,* Macmillan, 1992.

References

Awad, Elias M., *Building Expert Systems: Principles, Procedures and Applications,* West Publishing, 1996.

Best IT Practices in the Federal Government — A Joint Project by the Chief Information Officers Council and the Industry Advisory Council, October 1997.

Ignizio, James P., *Introduction to Expert Systems, The Development and Implementation of Rule-based Expert Systems,* McGraw-Hill, 1991.

Liebowitz, Jay (ed.), *The Handbook of Applied Expert Systems,* CRC Press, 1997.

Toms, Larry, *Expert Systems, A Decade of Use for Used-Oil Analysis,* Technology Showcase, Integrated monitoring, Diagnostics and Failure Prevention, Proceedings of a Joint Conference, 1996.

Toms, Larry, *Machinery Oil Analysis, Methods, Automation and Benefits, A Guide fir Maintenance Managers and Supervisors,* 1995.

Turban, Efraim, *Decision Support and Expert Systems, Management Support Systems,* Macmillan Publishing Company, 1988.

Turban, Efraim, *Expert Systems and Applied Artificial Intelligence,* Macmillan, 1992.

Section V
Knowledge Management: Applications

15

Knowledge Management in Industry

Patricia S. Foy
PricewaterhouseCoopers

15.1 Background and Context

The concept of knowledge management (KM), which helps to turn raw data into workable, actionable information, has been around for a long time. However, advances in technology and communications tools are expanding the opportunities and new platforms for sharing, accessing, and creating information. Technology, in combination with changed business processes and worker incentives, is turning basic KM concepts into real-live programs in many corporations.

Knowledge creates business value only when it serves to meet business goals. In most companies, this means that it adds dollars, improves performance, or provides other measurable worth to the organization. In order to be successful, knowledge efforts need to be aligned with the strategy for the business.

15.2 Definitions, Principles, and Concepts

In 1998, knowledge management might be viewed as any one of the following, depending on whom you talk to:

- A buzzword
- Hype, a fad with no substance
- Information process redesign, an outgrowth of BPR (business process reengineering)
- Document retrieval
- Software
- Part of the knowledge economy
- Leveraged intellectual capital

In most businesses or corporations today knowledge management programs are, in many ways, an outgrowth of other previous business analysis and improvement efforts, such as TQM (total quality management), BPR (business process reengineering), continuous improvement and change management efforts. Knowledge management helps generate assets to an organization by people's transfer of knowledge, expertise, and innovation. Fundamentally, knowledge management is mainly about combining insight and information; it leverages off an organization's "intellectual capital" (people's knowledge and experience), in combination with accurate, qualitative, and pertinent data, to provide some sort of better product. This product could be better advice or counsel, a new or more effective way of doing things, or clearly a different level of understanding or expertise. Basically, knowledge management is all about facilitating learning, through sharing, into usable ideas, products, and processes. In the business world, the operative word here is the adjective — usable.

Some definitions of knowledge management offered by corporations and research firms include:

- KM is an emerging set of policies, organizational structures, procedures, applications, and technologies intended to improve the decision-making effectiveness of a group or enterprise.[1]
- KM is a discipline that promotes an integrated approach to identifying, capturing, retrieving, sharing, and evaluating an enterprise's information assets.[2]
- KM is codifying the knowledge your company creates and disseminating it to the people who need it — when they need it.[3]
- KM is the art of transforming information and intellectual assets into enduring value for an organization's clients and its people.[4]

Much has been written recently about the knowledge economy and the knowledge workers who contribute to it. Thomas A. Stewart says, "Knowledge is replacing matter and energy as the primary generator of wealth."[5] The common history lesson goes something like this: Sources of wealth in the Agricultural era (mid 19th century) were land; in the Industrial era (late 20th century) they were labor and capital, and now in the Knowledge era, physical assets are being replaced by intellectual assets such as knowledge and skills. This new knowledge economy means that there are a variety of new business models, which are staffed with less traditional full-time and part-time jobs, which require new skills and staffing requirements. These knowledge management efforts help focus the intellectual capital found through new work nets and virtual communities.

To effectively locate and use intellectual capital in business, knowledge management activities should be vigorously tied to changes in process, culture, and organization. They are also heavily reliant on the ongoing communication and sharing of ideas and expertise. Today, internally and externally, corporate organizations are trying to stand up and say that what people know and can learn is more valuable than any other resource. The source of value for companies in decision-making is discovery, gaining insight, understanding, and extracting value, not just accessing information.

15.3   KM Projects, Techniques, and Tools

So far in industry, KM activities and projects are being applied in certain key areas. These include:

- Content and information
- Technology
- People, including learning, culture, and personal responsibility
- Process, including measurement

---

[1]Gartner Group, KM Scenario, conference presentation, October, 1997.
[2]Gartner Group, KM Scenario, conference presentation, October, 1997.
[3]Forrester Research, *The Forrester Report,* "Managing Knowledge," November/December, 1997.
[4]Ellen Knapp, Coopers & Lybrand L.L.P., 1997.
[5]Thomas A. Stewart, "Getting Real About Brainpower," *Fortune,* November 27, 1995, p. 97.

There are a number of programs and initiatives under way in these areas in many business organizations. Service organizations, like consultancies, and firms that are heavily reliant on intellectual property, research, and exploration are probably the farthest ahead in trying to effectively implement and use KM programs. Professional services and consulting firms are the only organizations that exist as *pure players* in the knowledge economy, because knowledge is, in essence, their only product. The better they are at knowledge management in their own organizations, the more valuable they can be in helping their clients in this area.[6]

Content and Information

What information do people really need? How can the information that one person or expert knows be broadly communicated, shared, and saved for effective application, use, and innovation in another situation? How do we know what we know? In most organizations those fundamental questions have been around a long time. In an age of information overload, the scope of these questions becomes much more acute. Thus, capturing knowledge is the focus of many KM projects in business. Such projects may be as simple as making information that has been available only in print format easily accessible electronically; it might be codifying information that has been available only in a variety of documents, lists, or databases; it might be harvesting information from experts or best practices or lessons learned. Some might ask whether these projects really constitute knowledge management? KM opportunities can start with very basic tools; they are the starting steps to more encompassing initiatives.

The Institute of the Future has a chart that effectively illustrates the paradox of trying to turn data into knowledge. It considers the development of a repository of information, through changes in process, to the point where knowledge is created. Other charts that try to document knowledge management show the evolution of data to information, from information to knowledge, and, finally, from knowledge to wisdom. Key considerations in this conversion include the need for social brokering, increasing the timeliness and utility of information, and looking at content in the right context.

There is a real continuum of information — an information value chain — that leads to effective knowledge management. The value of information evolves from gathering and storing data — to accessing content — to awareness and reaction — to better evaluation in context — to interactive tailoring — and finally to integrated interaction. According to David Marshak of the Patricia Seybold Group, that ultimate relationship-based information convergence is where real knowledge management starts to take place.

Knowledge creation starts with the identification of a business-related need, where different, more timely, or better information can contribute to the value proposition of the company. In a services firm such as consulting, the value proposition to effectively compete in the marketplace is supplying better expertise and understanding to a customer situation. In the publishing industry, the proposition changes from just information gathering, printing, and distributing to better merchandising and tying in different media and services. In a high-tech or research-focused company, the value proposition of managed intellectual property and patents and their contribution to new products is a fundamental factor in moving quickly to market.

Many companies have started their KM initiatives with compilations of their products, experts, or patents. These are basic and key information sources that span the needs of many people. Knowledge inventories might include lists of skills and experts, as well as expert opinion; profiles of competitive firms, along with strengths, weaknesses, or samples of their work; reference and research databases and sources; market analysis, reports, and commentary; best practices databases; company policies and procedures, etc. In other words, an inventory might attempt to show a comprehensive picture of the knowledge resources of a company or organization, which could, in turn, be used as the basis for continued information mapping and content architectures.

---

[6]Ellen Knapp, Coopers & Lybrand L.L.P., 1997.

Technology

With levels of automation constantly increasing over the last twenty-five years, there are lots of changes in the way people are able to work. Technology greatly facilitates knowledge management. It is because of the technology capabilities available to us today that the knowledge management concept has taken on a heightened focus. Workers can get and do much more with information than has ever been possible before.

Knowledge management technology infrastructure is taking on a whole different scope than just traditional information retrieval, storage, and management. Although huge databases, repositories, and search tools are part of the technology used to support KM, the primary focus of knowledge management is not just getting data, but rather the proactive use of information and ideas. It is the relationships, the intuition, the insights, and the fundamental sharing of ideas that are the most telling and measurable worth of an "information experience."

A number of technology tools have been created or renamed to support KM activities. However, you can't buy knowledge management only through an investment in technology; technology just greatly helps the knowledge management process. Commercial software and tools vendors are making a wide variety of products available that purport to address some aspect of knowledge management. Tasks and activities that are supported include codification; information gathering, distribution, and retrieval; filtering; visualization; reasoning; workflow and collaboration; search and analysis. Some knowledge management products are really just the next version of document retrieval or storage; some are complicated and burdensome layers of data hierarchy and storage; some aim to allow easy indexing and categorization; and some are actually attempting to link performance and measurement of success to an information process. Examples of KM vendors by function include those that specialize in:

- Dissemination, e.g., TIBCO, Vitria, Level 8, Open Horizon, (middleware)
  Wayfarer, BackWeb, Diffusion, DataChannel, (delivery systems)
- Search & retrieval, e.g., Verity, Excalibur, Infonautics,
  Semio, AltaVista
- Categorization, e.g., Dataware, Intraspect, Fulcrum,
  Plumtree, GrapeVINE, (enterprise products)
  Inference, WinCite, Abuzz, KnowledgeX, (business function-specific)
- Unstructured text capture, e.g., Lotus Notes, Dataware, Intraspect
- Structured data capture, e.g., Oracle, Informix, IBM, Sybase, Microsoft[7]

Information technology practitioners must be aware that content is not solely software, tools, or sophisticated warehouses of information. There are clear differences between the consideration and the evaluation of content and that of software or hardware. Content is the domain of information, not technology, specialists. The MIS (management information systems) departments only play a support role in the world of knowledge management initiatives. By virtue of their name — MIS or IT (information technology) — they should not assume that they have primary responsibility for the choice or creation of knowledge and content. There truly is a difference between "I" (information) and "T" (technology). Technology infrastructure is very different than a knowledge infrastructure or content architecture. Knowledge management is not just another database, a new software program, or a methodology. Simply stated, KM is not the purview of technologists.

People

There is no doubt that as businesses try to build infocentric organizations, there are a variety of changes necessary in the skills and processes used to get at and use the information. This means taking the

---

[7]Forrester Research, *The Forrester Report,* "The Third Skill," December, 1997.

knowledge workers of an organization, making them information-savvy enough both to get information and to know when they need help to do so. Knowledge workers must be provided with the means to move into the sphere of knowledge management. In many companies this means that they must learn to be a part of a global knowledge network, a virtual community of practitioners.

Roles and skills are changing in many respects. Company culture and expectations, as well as the attitudes of people toward their own personal responsibility in growth and sharing, are important parts of these changes. In a learning organization, there is a focus on competency development and continuous learning. Remote or distance learning may also play a significant part in this. Cultural aspects of the work environment that must be built and nurtured include trust, teamwork, and behavior that supports other people at the same time that it supports the value proposition of the organization. Many components of a company's culture are grown over a period of years, not months, so knowledge management precepts may take some time to be fully acknowledged, let alone accepted and endorsed.

Information discovery and brokering have been part of many people's jobs in the past (e.g., librarians). This has usually been on a one-to-one basis or in a "Question?, then Response," mode. Effective knowledge management today must be carried out in different ways. For corporations looking to maximize new services in a global market, for example, one-to-one sharing is way too little and usually not possible. Sharing must be done across audiences of hundreds of people or to someone thousands of miles away. Knowledge must be supplied proactively, not reactively. *HR Magazine* describes the people at the center of knowledge management as technology experts, catalogers, archivists, guides, scouts, researchers, librarians, analysts, debriefers, content creators, and facilitators. Although this definition outlines part of what is necessary, there probably are significant differences in the type of skills that drive and define knowledge management activities and those that merely support some element of knowledge management. Value-added skills focus on advisory, analytic, business intelligence, communications, and customer service skills, leaving many searching, routing, cataloging, and support activities to be handled through technology.

Knowledge management practitioners must be able to function alone or as part of a team. KM activities for an information end user might range from timely information searching to best practices sharing and analysis. Industry studies show that 82% of people who use the Internet use it for research and finding information. The key to any information access tool, such as the Internet, is to save people time in getting to pertinent, quality information. "Expeditious convenience is what information seekers ...need."[8] If one posed a question to a typical end user as to their key requirements for getting information, the answer would probably sound something like, "I want it now; I want it my way; and I want you to know what I want." Some other realities of end user access relate to training, security, usage, accountability, network responsiveness and, of course, ease of use.

Services firms rely on ideas for survival. Therefore, it is important to foster interaction between groups of people and share ideas and expertise. This means the implementation of a number of cultural changes and new ways of doing things. In the world of knowledge management companies must recognize the primary need to manage tacit knowledge needs through leveraging their people's own insight and understanding. Sharing is not a skill learned overnight, and how to best create and tap into knowledge and expertise is an iterative exercise. Regarding continuous learning and cultural change, Jack Welch summed up some of the necessary components of a knowledge-based culture in his sixteenth annual letter to General Electric stockholders. He said that GE leaders must provide the executive support for a company to:

- Continuously learn from any source and rapidly convert this learning into action
- Reward the finding and sharing of ideas, even more than their origination
- Have the self-confidence to involve everyone and behave in a boundaryless fashion
- Create a clear, simple, reality-based vision
- See change as an opportunity, not a threat
- Have global brains, and build diverse and global teams

---

[8]Barbara Quint, "High Risk Online," *Searcher,* January, 1995, pp. 6–8.

The team-based information experience must be continuously reinforced through incentives and shared successes.

According to a recent study,[9] "people" problems constitute the majority of the reasons why knowledge management programs are not successful. Knowledge sharing doesn't happen because of: a resistant culture, top management's failure to signal the importance of the program, a lack of shared understanding of the business strategy or model, an organizational structure that doesn't support KM activities and, finally, lack of ownership of the problem. Therefore, it is critically important to address cultural and learning issues early in the knowledge management strategy.

Process

KM processes must support the business strategy. A knowledge policy and strategy must be created and advanced at an upper management level. To effectively and efficiently implement knowledge management efforts KM processes must be part of the mainline business processes that make an organization run. Contributing to a knowledge base can't be done in a random fashion. Therefore, a significant effort must be made to develop and manage the KM business processes as well as the knowledge and information. An actual knowledge organization might be built in a company or knowledge gathering and harvesting might be supported from within various departments' functions.

The value of a piece of "knowledge" must be assessed and properly "packaged" into a useful format. It is also unreasonable for companies to expect knowledge contribution to just magically happen. "To capitalize on the wealth of intelligence available in an organization, knowledge must be packaged in such a way that it is insightful, relevant, and useful. To support these efforts, budgets need to include allocations for content development either by staff or external consultants.[10]

Processes also need to be put in place for the development of knowledge "soft" skills, such as learning and sharing, as well as for the "hard" skills of actual creation of content, insight, and expertise; distribution, mapping, and measurement. There are a number of theories about the motivational factors necessary to make any new processes work. Incentive programs linked to performance evaluation, skills improvement, and overall measurement of success have been started at a number of companies to help the knowledge management process grow within the culture of a company. Ongoing participation in the KM program, and endorsement and use by a wide base of employees and management are necessary. Communications and use by senior management is critical. A stated clear purpose, description, and language about knowledge management helps employees understand KM needs and attributes, both in relation to the company and to their own personal growth.

The ultimate added value of knowledge management programs is indisputably linked to employee performance. The benefits derived from improved processes in knowledge management might include improved client and customer service, increased productivity and more timely deliverables, the creation of more knowledge capital, increased speed of information distribution and sharing, enhanced accessibility of knowledge, and increased extent of reuse. New knowledge products and processes (such as industry learning and best practices, information frameworks and context strategies, metrics and valuation) can contribute to business effectiveness, innovation and growth, and timely competitive response.

Examples of companies actively involved in KM projects include:

Buckman Labs — Using technology to enhance knowledge networking to better support customer contact.

Teltech — Building, networking, and managing teams of experts and tying their information together with other outside sources of information and research.

Price Waterhouse, Arthur Andersen — Codifying and harvesting best practices from their professionals and making those experiences available through tools and databases.

---

[9]Ernst & Young Center for Business Innovation and Business Intelligence, Survey of 431 organizations, 1997.
[10]Paul S. Myers and Richard W. Swanborg Jr., "Packaging Knowledge," *CIO Enterprise;* April 15, 1998, pp. 26–28.

Coopers & Lybrand — Creating a virtual library from internal and external sources, as well as an information content center; establishing an information value chain.

Sprint — Building tools to directly link staff to organizational objectives.

Hewlett Packard — Providing direct links between dealers and products to get better information more quickly to clients.

Xerox — Supporting collaborative work and knowledge-based productivity tools to facilitate knowledge sharing.

American Productivity & Quality Center — Researching team-based operating systems and innovative reward systems.

Hoffmann-LaRoche, Johnson & Johnson, Merck — Speeding the process of drug approval through better documentation and knowledge sharing systems.

Dow Chemical — Building tools to implement a corporate intellectual capital and asset management program.

Monsanto — Developing knowledge architectures and databases to help recognize opportunities.

Skandia — Developing asset measures help visualize internal knowledge and to support the growth of intellectual capital.

McKinsey, Price Waterhouse — Changing incentive and motivation practices to encourage employees to share ideas and experiences.

IBM, ICL — Building intellectual capital management tools and programs.

Hughes Space and Communications — Promoting collaboration and learning programs across project teams.

Canadian Imperial Bank of Commerce — Supporting intellectual capital and "collective memory" systems.

15.4 Application Case Study — Knowledge Management at Coopers & Lybrand L.L.P.

At one professional services firm in the United States, Coopers & Lybrand L.L.P.(C&L), the knowledge management program has been put into the overall context of a "Knowledge Services Value Chain," which includes all facets of service process delivery, e.g., leadership, operations and administration, marketing, and product development. At the heart of this Services Value Chain is knowledge. Knowledge use starts with acquisition, organization, and access and then progresses through analysis, application, and the processes surrounding sharing, creation, and innovation. The enablers of this Knowledge Value Chain are content, technology, learning, personal responsibility, culture, and measurement.

C&L started much of their information management restructuring in 1993. In 1994 a Chief Knowledge Officer (CKO) position was created by the firm's chairman; the CKO appointed was a senior partner and vice-chairman of the firm. Initially, libraries, research bases, and information centers were "reengineered." Many of the information professionals and analysts at C&L were incorporated into a national support unit — the Knowledge Strategies Group (KSG). As knowledge management initiatives have grown and evolved, C&L has addressed a wide variety of objectives, as well as new support processes, as part of the overall knowledge management programs.

Some of C&L's knowledge management goals and objectives include:

- Accelerate the timely delivery of valuable information to clients and employees
- Leverage the firm's size and expertise
- Effectively and efficiently organize collective knowledge information
- Enhance the sense of community through team learning and knowledge sharing
- Manage information and publications costs to improve communications and stimulate innovation and growth.

C&L's Knowledge Strategies Group initially focused their technology efforts by creating and taking responsibility for the firm's award-winning intranet, KnowledgeCurve^sm. KSG has responsibility for content and some aspects of the underlying intranet technology. C&L's intranet has now been successfully running for two years. As an initial KM investment, a conscious decision was made to make basic business sets of both internal and external information available through the intranet to all personnel throughout the firm.

The convenience and immediacy of a personal computer desktop was key to getting C&L's people access to the right information, at the right time, no matter where they were located or what time of day or night they needed the information. In reaching out to the end user population (many of whom were initially uncomfortable with the desktop access and use of information), programs and services were put in place to:

- Expand customer service through centralized information specialist services
- Create a content help desk, reachable via an "800" number
- Implement and continue a variety of usability testing, feedback, and usage measurement activities
- Form a "Power Users Group," which is used for testing and additional communications
- Expand other end user options

Building the knowledge, research, and specialist skills into focused and value-added support was critical. Discussion and feedback facilities were key components of knowledge management, as was providing performance incentives for experts to disburse insight and provide counsel and guidance. Careful efforts were made to avoid relapses into an information class system. A wide variety of job aids, training pieces, tutorials, and presentations were developed and distributed. In addition to the intranet, communications, training, and awareness building was supplemented with Lotus Notes. Now, in early 1998, global development and implementation of a global intranet are underway. Additionally, "target wizards," "subway maps" to information, information guides, and search tools are part of ongoing evaluations and beta testing. The first main "payoff" from this huge investment in knowledge assets has been large growth in terms of measurable application and use of information. Demonstrable bottom line dollar payback is still reasonably far off.

When C&L and Price Waterhouse's merger was first announced in late 1997, the first screen of the KnowledgeCurve^sm was used to provide an immediate interactive forum for merger news, including feedback that went back to executive management. The site on this intranet that consistently gets the most "hits" is C&L's virtual library for basic business information and news, the CyberLyb^sm. CyberLyb^sm provides all employees with ongoing access to a wide variety of external information. It provides the firm with high-quality external information tools and products, selected to provide anytime, anywhere answers to frequently asked questions that any staff member might have. This basic set of easy-to-use research sources includes links to best free Internet sites and access to fee-based sources, typically paid for on an "all-you-can-eat" basis by the firm. Content includes news; periodicals, journals, and magazines; financial, market, and economic spreadsheets; securities pricing; industry and equity research; company briefs and tear sheets; and SEC documents.

C&L's objectives in making a tool like CyberLyb^sm available to everyone in the firm are to:

- Focus users to look at pertinent, quality sources before they go on the Internet
- Provide easy-to-use access 24 hours a day, location and platform independent
- Replace many local print subscriptions and materials collections with desktop access to quality basic sources
- Leverage C&L's size and scope with "all-you-can-eat" or advantageous pricing to provide pertinent focused, timely information at considerable cost savings
- Help control and better measure information costs, quality, and usage

Lessons that were learned early in the process included dealing with some substantive technology issues, especially involving bandwidth and user authentication. They also encompassed many design problems, such as endorsing site speed instead of beauty and reworking the site for increased presentation simplicity. Other fundamental considerations involved key questions around quality rather than quantity and the delivery of substance. Finally, they embodied some of the most telling concerns and difficulties of implementing new knowledge processes — cultural changes, feedback from end users, and cross-functional use of information. Users had to be educated that the real knowledge edge came with understanding the why and how of a piece of information, in addition to the where and how to find it. Application and use of the information was what counted.

15.5 Research Issues and Future Directions

All types of business organizations are in the middle of a knowledge economy. The key to making knowledge work is to make information usable. Knowledge management programs and initiatives attempt to do that. Knowledge workers all over the world are charged with implementing new processes and doing new things. Each day brings more new technology, faster communications, and higher corporate hurdles to jump.

In facilitating information usage to grow into knowledge management, some of the biggest challenges ahead are:

- Improving usability, measurement, and feedback for a variety of users and specialists
- Continuing to build user familiarity and motivate training
- Establishing what McKinsey's Brook Manville calls the community of practice, through virtual community-building
- Constructing technology systems and tools that support knowledge management, as well as robust Internet access and bandwidth
- Visualizing a content architecture that works across organizations
- Continuing innovations in processes, products, and perceptions
- Successfully changing corporate cultures

There is no reason to expect that knowledge management will not also be a part of future programs and initiatives in business. It is equally probable, knowing the speed with which modern business drops one approach and embraces another, that KM and its terminology will be supplanted by the next iteration of business management "fads." The interesting aspect of most business fads, however, is that many of the precepts and principles of any management approach live on as fundamental parts of a business organization's culture and its processes. It is equally true that many learnings will continue to be incorporated into newer management approaches. This is part of the change cycle in industry innovation and learning and one in which knowledge management will continue to evolve. "Individuals and companies that can effectively harness the power of knowledge are more competitive, more efficient, more productive and more profitable. That is why the effective management of knowledge has become a strategic focus for so many companies."[11]

References

Working Knowledge: How Organizations Manage What They Know, Thomas H. Davenport and Laurence Prusak, Harvard Business School Press (1998).
The Fifth Discipline: The Art and Practice of the Learning Organization, Peter Senge, Doubleday (1990).
"Managing professional intellect," James B. Quinn, Philip Anderson, Sidney Finkelstein, *Harvard Business Review,* pp. 71-83 (March/April 1996).
"Librarians at the Gate," *Webmaster Magazine,* CIO, (September 1996).
The Post Capitalist Society, Peter F. Drucker, Butterworth-Heinemann (1993).
"Coming Soon: The CKO," Tom Davenport, *CIO* (May 9, 1996).

---

[11]Lorin David Kalisky and Timothy O'Brien, "The Knowledge Connection," *ContentWatch,* Vol. 1, no. 4, 1998.

The Knowledge Creating Company, Ikujiro Nonaka and Hirotaka Takeuchi, Oxford University Press (1995).

"The Balanced Scorecard — measures that drive performance," Robert S. Kaplan and David P. Norton, *Harvard Business Review,* (January/February 1992; see also October/November 1993 and January/February 1996).

"Getting Real About Brainpower," Thomas Stewart, *Fortune,* p. 97, (November 27, 1995).

"Improving knowledge work processes," T.H. Davenport, Sirkka L. Jarvenpan, and Michael C. Bens, *Sloan Management Review,* pp. 52-66 (Summer 1996).

Creating the Knowledge-Based Business, David J. Skyrme and Debra M. Amidon, Business Intelligence (1997).

Intellectual Capital: The New Wealth of Organizations, Thomas A. Stewart, Doubleday (1997).

The New Organizational Wealth: Managing & Measuring Knowledge-Based Assets, Karl Erik Sveiby, Berrett-Koehler (1997).

"Successful Knowledge Management Projects," Thomas H. Davenport, David W. DeLong, Michael C. Beers, *Sloan Management Review,* pp. 43-57 (Winter 1998).

Key Terms and Definitions

Balanced Scorecard — Measures that balance financial with nonfinancial criteria (customer perspective, internal business perspective, innovation and learning perspective) (from Robert S. Kaplan and David P. Norton).

Chief Knowledge Officer — In an organization, a position of responsibility that focuses on the promotion, creation, and facilitation of knowledge order to better meet the organization's mission, goals, and objectives.

Information Value Chain — The process that changes raw data to information, to knowledge, and, eventually, into wisdom.

Intellectual Capital — Intellectual material — knowledge, information, intellectual property, experience — that can be put to use to create wealth (Thomas A. Stewart).

Intranet — Using Internet protocols for an internal platform-independent system; an internet specifically for a company.

Knowledge Economy — An economy where knowledge is the critical factor of production. (Peter Drucker).

Knowledge Manager — An individual who is responsible for managing or facilitating the harvesting or sharing of knowledge.

Knowledge Mapping — A process to present a knowledge architecture or classification system.

Tacit Knowledge — Knowledge that usually exists only in people's heads and has not been written down. It is usually implicit and may be based upon experience and intuition.

Virtual Community — A community of practice, a group of people who are informally bound to one another by exposure to a common class of problem (Brook Manville).

16

Knowledge Management in Government

Kim Ann Zimmermann
Consultant

Take a quick trip to your town hall and get a whiff of the dusty old property records and you'll understand why improved knowledge management is essential to government at any level.

Government agencies — be they federal, state, or local entities — face huge challenges, often beyond those of the private sector. The U.S. Department of Labor has web-based expert systems to provide employment advisory advice to workers and small businesses (www.dol.gov/elaws).

Challenge #1. The sheer volume of records any government agency processes. For example, the Internal Revenue Service processed 120 million tax returns in 1997. Add to that the thousands and thousands of birth certificates, voting records, property deeds, and numerous other documents processed by government agencies and you've got an overwhelming dilemma.

Challenge #2. The necessity to keep those records accessible over a long period of time. Property records date back hundreds of years. The Internal Revenue Service can audit your tax records years after you file your taxes.

Challenge #3. Unlike the private sector, government agencies are mandated to provide public access to records. No one can walk into a hospital and demand to see the results of your latest blood test, but your neighbor has the right to walk into town hall and look up the records on your property. The challenge of meeting the mandate for public access is daunting in a paper-based environment.

Challenge #4. The need to keep documents secure. While it can be argued that other knowledge management users — such as those in the healthcare field or the legal profession — need to secure their documents as well, security is especially sensitive when dealing with government records. Pentagon records in the wrong hands could have serious national security consequences.

Challenge #5. Many of the documents handled by all government agencies must be handled in paper form, at least at the moment. While it would speed things tremendously if everyone filed their taxes on-line and renewed their driver's license over the Internet, it is very unlikely that will happen any time soon. Paper is here to stay for quite a while.

How are government agencies meeting these challenges? This chapter will take a look at how government agencies are meeting the knowledge management challenge in unique ways.

However, there are some issues common to all knowledge management installations that should be addressed up front.

The first is *workflow*. Workflow is essential for the expedient routing of documents to the right individuals in an agency as well as, at times, those outside the agency such as a lawyer requesting court documents or a title search company requesting property records.

Because these documents need to be reviewed by a multitude of people within and outside an agency, the flow of documents from the mailroom to the various individuals who need to review them is essential.

This is why many government agencies are looking to streamline and speed this process using sophisticated document capture systems that scan, recognize, and route documents with minimal intervention from operators.

Document imaging doesn't improve efficiency if operators are spending as much time processing the document images as they would have spent processing the paper documents.

The second issue for government application of knowledge management systems is *collaborative computing*. Because all of these various people are sharing these documents, there has to be some sensible way to control the sharing of these documents and regulate the way changes are made. By the nature of government, workers are spread out all over the state or even the nation to serve the people.

The Massachusetts Department of Mental Retardation, for example, has 350-plus caseworkers in 36 offices throughout the state serving 25,000 individuals with mental retardation. They are using a groupware system to share this information within the department as well as provide parents and guardians Internet access to some information about the plans for the person under their care. This system at the Massachusetts Department of Mental Retardation, as those in the following case studies, have greatly improved the flow of documents throughout a variety of government organizations

Here are some case studies that will hopefully inspire you and provide you with ideas as to how improving document management in your organization can improve knowledge management.

These case studies are divided into two areas: those that focus on the benefits of workflow and those that focus on the benefits of collaborative computing.

16.1   Workflow Applications

The Public Utility of Texas

The Public Utility Commission of Texas consists of 160 combined utilities with revenues of over $20 billion per year and a staff of 240 people. Brenda Jenkins, Executive Director, states, "Deregulation and competition is going to cause utilities to rethink their efficiency. How to do business better and how to make government smaller are growing, unavoidable questions."

"I look at the big picture," Jenkins declares., "I look at how to improve the process and the system. That's what people at my level do. I could care less about technology for technology's sake. Having the latest technology is highly suspect to legislative members, so you have to be sure you don't appear as if you just have an affinity for bells and whistles. I'm concerned with value, and with providing training, support, and instant response to our requests for help."

The repository of all legal documents relating to proposed and existing utility regulations is at Central Records of the Public Utility Commission of Texas. Central Records has by far the highest visibility of any portion of the PUC in the eyes of the public. While 85% of the PUC are professional staff, including lawyers, accountants, and engineers, most complaints were directed at the clerical functions of Central Records.

Jenkins says, "I wanted to upgrade information systems, so I brought in Internet connections. I wanted all of our staff to have, at minimum, access to all of the libraries on the Web. And for the same reasons, I upgraded our Central Records methods and systems. There was no way I could have succeeded in doing this without the support of the Board and our extremely qualified technical staff."

The current practice, prevalent in government offices all over the country, requires people who need information to go to certain offices and line up to gather copies of paper documents. The task is very time-consuming and expensive, especially for the clients who pay for all those people to stand in line.

In order to address these shortcomings in the clerical process, a decision was made to move to an electronic document filing and distribution system. Electronic access to records saves money for clients, vastly improving the responsiveness of the system by eliminating nonproductive time and cost elements. The greatest challenge to this task was the fact that 100% of this information was on paper.

InputAccel was chosen as the backbone of this knowledge management architecture that would enable the Texas PUC to move away from this antiquated document handling method.

The primary goals for the new system were to:

- Improve the document retrieval and distribution process and eliminate delays in document handling. The previous method typically caused long waits in line for paper documents to be manually retrieved from file cabinets and reproduced on copiers. Electronic document retrieval and high-speed printing of document images virtually eliminates the delays involved in the manual methods.
- Reduce overall costs to users of PUC Central Records.

Many users of the system were professionals and semiprofessionals who incurred high hourly charges while waiting for documents. By greatly reducing the time required to get documents, very large savings are realized by the organizations that were previously incurring these high costs.

To accomplish their goals of moving from paper to electronic document processing, they needed a document system with the following elements:

- High-volume and flexible document handling capacity. The ongoing volume of documents coming into the PUC Central Records department is very substantial, 3,000 to 10,000 pages per day, and only a scalable and robust network solution could meet these large requirements. InputAccel was chosen based on a proven capacity to handle these large volumes.
- Direct integration of document capture with the existing system The Texas PUC had an in-place Agency Information System (AIS) and the goal was to integrate the new document capture system with this system. InputAccel provided the means to integrate the previous paper-intensive process with the proven efficiency of the AIS database by delivering the means to index the documents and export the information and document images into the new system.
- Support rapid access for all users. After paper documents are captured and indexed through InputAccel, the file system provides virtually instantaneous access to 250 in-house users, and Internet access to the files is also available. Rather than waiting for documents to be laboriously copied, electronic images are rapidly output via laser printers.
- A component-based architecture, with flexibility for development and enhancement. InputAccel provides a flexible architecture so that the system can continuously grow to take advantage of future developments in hardware and software, and the PUC's changing needs.

Brenda Jenkins states her confidence in the InputAccel architecture of the system. "This project is critical to the future of other automation and efficiency improvements in local and regional government and organizations." The total cost of this project is $400,000, and the nominal fees charged to users should pay for the system in less than two years. Jenkins points out that "Once the fees have paid off the investment, all fees will be dropped and nobody will pay to enjoy the immense benefits of the electronic system."

Jenkins explains the national implications of such technological imperatives, saying, "This will be a prototype for all of Texas, and for utility commissions all across the country. Through the demonstration of our success, the efficiency and cost savings we produce will be emulated and employed nation wide.

"Our system is an operating example of how to do business better," Jenkins insists, "and how to make government smaller. In all of these processes, information in time is the essence. With today's technology we can deliver breakthroughs in both user satisfaction and system costs."

PUC recognized the need to capture paper documents and provide for robust document management capabilities as part of the overall automation of the Agency. Price Imaging Corp. was selected to provide system design and integration services. The PUC system includes:

- InputAccel, from Cornerstone Imaging, Inc. of San Jose, CA to automate, control, and manage all of the document capture processes. InputAccel software performs scanning, QC, image enhancement, optical character recognition, and data capture, automated indexing, forms processing, and image and data export to existing applications.
- Agency Information System — A database application developed by PUC staff using Sybase SQLAnywhere and PowerBuilder, running on Novell 4.1, that tracks the status of every matter that comes before the PUC.

"I wanted to accomplish several ideal goals," Brenda Jenkins says. "I wanted to improve technology, and I wanted to improve our processes. I wanted our entire organization and our contacts to get better connected, and I wanted to improve our public services." Pausing to reflect on these major improvements in the business functions of the PUC, Jenkins states: "Using best available resources, I've achieved all this."

The Texas PUC has achieved these ambitious business objectives by implementing best-of-breed solutions, which are tied together through Cornerstone InputAccel document capture technology. The investment in hardware and software will pay for itself through user fees initially, and will provide ongoing cost savings over the previous paper-based system.

"I intuitively knew it would work," Jenkins proclaims. The new system would have to work, because the Legislature demanded that the system pay for itself. "Other organizations who are considering the efficiency and automation measures we have implemented can be confident that it will work, because we have proven it," Jenkins says.

The key element that has enabled this successful evolution from paper to electronic documents is the robust and scalable functionality of Cornerstone InputAccel. In a carefully designed system that employs best-of-breed solutions for each step of the business process, InputAccel serves as the on-ramp for paper documents into the new electronic document environment at the Texas Public Utility Commission.

City of Boston

Boston taxpayers who dispute the city's property tax assessment can appeal by filing a tax abatement application with the City within 30 days of the original mailing. After the abatement application is filed, the Assessor has 90 days to render a decision regarding the application. If an application is not processed within the review period, the application is automatically denied by reason of inaction. The inability to process abatement applications on a timely basis results in lost revenue and credibility for the city of Boston.

The biggest problem confronting the Assessment Department was that 70% of the abatement applications were received within the last four days of the filing period. Although processing these applications under normal circumstances would take only a few days, during peak periods, the backlog would run into weeks. Such processing delays consumed a disproportionate amount of the 90-day review period.

The Boston officials realized that granting the assessors immediate access to scanned images of abatement-related documentation would expedite the process, making it far more efficient and less time-consuming. Recognizing the need to implement a high-quality, cost-effective knowledge management solution, the city determined a system had to meet the following criteria:

- Be scalable
- Perform full image processing (clean-up, deskew, OCR and barcode recognition)
- Index 70% of the documents without user intervention
- Be fully operational within thirty days

The new abatement application processing system begins with the daily batch scanning of 9000 abatement applications and supporting documentation pages. Key information is extracted via barcode and OCR and validated against data stored in the city's Sybase database.

If the document passes validation, the images are released to optical storage via a custom-developed release script that integrates with Ascent Storage. All documents that fail validation are sent to an exception queue and held for manual indexing. After the documents are manually indexed, they are then released to Ascent Storage using the same release script.

The benefits have been astounding:

- Image cleanup and highly accurate indexing features have enabled 70% of all documents to be processed without operator intervention;
- Ascent Capture's batch management features have made it easy for Assessors to reprioritize sensitive batches needing rapid processing;
- For the first time, the City of Boston is able to keep up with peak period workloads.

The city was able to realize significant cost savings by utilizing preexisting hardware.

Kansas Driver Control Bureau

The Driver Control Bureau of the Division of Vehicles for the State of Kansas maintains records for 1.7 million Kansas drivers. It is chartered with identifying problem drivers and suspending, restricting, or reinstating their driving privileges.

Another department in the Division of Vehicles, the Driver Review Section, is responsible for keeping records of drivers with medical and vision conditions that may affect the operation of a motor vehicle. Both of these departments used labor-intensive, paper-oriented systems to monitor and maintain records.

The Kansas Driver Control Bureau uses a mainframe application, the Kansas Driver License System (KDLS), to store information on individual drivers. However, the mass of official forms and supporting documentation required for license suspension or revocation was maintained on an antiquated, time-consuming, paper-based filing system. An extensive paper file folder existed for every Kansas driver whose license had been revoked or suspended. When drivers, their lawyers, or other officials called in for information, specially designated file clerks had to retrieve the paper files from a vast archive of folders.

Because of the paper-based system, the Bureau was not able to service the public in a timely manner. Telephone inquiries took an average of seven minutes to process and often required a "call-back" to allow time to locate the driver's folder. In January 1992, the Bureau was receiving an average of 633 calls per day.

However, computer monitoring of the phone system showed an average of 2,170 attempted calls. The Bureau was unable to provide customers with adequate service with the existing system.

Meanwhile, the state's Driver Review Section was also experiencing operational problems. Chartered with reviewing cases for applicants with medical or vision problems, this department issues, denies, or restricts licenses based on a driver's physical ability. Heavy paper volumes and lean staffing led to poor customer service. At one time, there was a 14-month backlog of cases.

In 1992, the Kansas Department of Revenue launched a project to improve the efficiency of the Driver Control Bureau and the Driver Review Section. In a report to the Department of Revenue, the Internal Audit Bureau cited numerous recommendations that would improve the workflow and efficiency of the Driver Control Bureau.

However, the key recommendation stated that document-image and WorkFlo software was the long-term solution to the many problems identified. Based on this recommendation, the Director of Vehicles obtained funding to implement a document-imaging and workflow solution.

Betty McBride, the Director of Vehicles for the State of Kansas, explains the choice of FileNET: "We also wanted to redo our business process. Along with the optical imaging software, FileNET had the WorkFlo solution to allow for this. Their WorkFlo software was the clincher for us."

During the procurement process, FileNET differentiated itself from the other vendors in two ways: the WorkFlo solution and the strength of FileNET customer references. The opportunity to visit FileNET customers who were in full production was a major factor in making the FileNET choice. As Marti Gonzales, the project manager, explains:

"FileNET's expertise was proven again and again at each customer site that we visited. We saw systems working at full capacity and received positive recommendations from each company. You can't ask for more than that."

With an objective to maximize productivity and enhance customer service, the document-imaging project was intended to:

- Reduce processing time
- Improve internal workflow
- Provide faster service to customers
- Provide advanced technology worker tools
- Provide better supervisory control tools to manage work
- Priorities, monitor productivity, and apply standards
- Improve worker morale by eliminating or reducing backlogs and recognizing quality performance

The implementation of FileNET document-imaging and WorkFlo software has allowed Kansas to reengineer and streamline processes at the Driver Control Bureau and the Driver Review Section.

Now incoming documents are scanned, indexed, and routed to Bureau personnel for processing. During processing, electronic images of original documents and driver information stored in the main-frame database can be viewed simultaneously using multiple windows. When a telephone inquiry is received, Bureau personnel can compare records and answer questions at their workstation. This means customers receive immediate response to telephone inquiries.

Often a law enforcement agency or court has an inquiry about a driver's record with an immediate need for supporting documentation. In this case, the capability to electronically fax documents using FileNET's WorkFlo/Fax software eliminates delays.

The Bureau uses FileNET's FolderView software to organize and manage electronic folders. Customized icons are set as tabs to allow fast access to specific documents. Annotations and folder notes provide a convenient way to summarize case transactions. In addition, the FileNET WorkFlo software allows Bureau personnel to easily browse through the documents in a driver folder; prepare and file a form letter; route a document to another worker or supervisor; fax or print a document.

Intelligent character recognition (ICR) software is used in conjunction with imaging and WorkFlo processing to automatically index and route driver's license applications.

The system currently has 2.5 million records on-line and 33 users who process more than 3,000 documents per day. Since FileNET's document-imaging and WorkFlo software was installed, there is no backlog and the Bureau is current with daily workloads.

The Division of Vehicles also implemented a document-imaging and WorkFlo system to support the processing of 4,000 new driver's license applications per day. Once scanned, the applications are auto-matically indexed using ICR (intelligent character recognition) software and filed onto optical disks, with exceptions routed to the appropriate processing clerk. WorkFlo scripts and ICR determine the appropriate route based on check-box information, hand-entered onto the form. Driver photos and signatures are also captured and stored on the system for on-line access, which is useful for other state agencies such as law enforcement.

The Kansas Division of Vehicles has realized tangible and intangible benefits as a result of the FileNET document-imaging and WorkFlo system. Management reports produced by the system provide informa-tion to allow the State to continuously improve the process. These management reports include infor-mation on work produced, work in-process, individual processing performance, and accuracy.

Some of the benefits of FileNET WorkFlo and imaging for the State of Kansas are:

- Increased productivity: Reduced number of manual steps in Driver Control from 27 to 9; elimi-nated 30 hours per week looking for lost files; eliminated 20 hours per week of clerical effort; reduced average length of telephone inquiry from 7 to 3 minutes; eliminated 11 full-time positions for annual savings of $227,000.
- Reduced storage: Freed up 1,624 square feet of office and warehouse space.

- Improved customer service and document security: Immediate, concurrent access to driver's files with no lost or misplaced folders. Management reports on work statistics allow for better control of work.
- Reduced training time: Users are guided through each step of a transaction, so new employees are productive more quickly.

16.2 Collaborative Computing Applications

Maricopa County, Arizona

The Office of the Maricopa County Recorder has responsibility for maintaining a perpetual and comprehensive set of public records and making these records easily accessible for viewing by any member of the public. The Elections Department of the County Recorder's Office conducts elections for the county and its 24 cities and towns. The county is the sixth largest voter registration district in the United States and the department verifies and maintains voter registration for Maricopa's 1.3 million voters.

In Arizona, the Office of the County Recorder is the only office that is mandated to record, index, and preserve documents as a permanent public record. The records maintained by the Office affect the real estate, personal, and collateral interests of all its citizens.

In the 1980s, the population of the county expanded dramatically, leading to a large increase in recording transactions. The County Recorder recognized that an alternative to the paper-based, manual system of document recording was required to keep up with the growing volumes. The goal was to provide more cost-effective and better service to the public, including on-line access to documents at remote locations.

The Maricopa County Recorder needed a solution that could deliver on-line document images quickly to the public at remote locations, replace the manual cash drawer management of recording fees, and help to manage the process of recording transactions. The process of signature verification for voter registration could also benefit from automated storage and retrieval of original voter documents, especially at election time.

The county began evaluation of document management systems including microform-based computer-assisted retrieval (CAR) systems. They quickly determined that only workflow and document-imaging technology could meet all the storage and retrieval requirements as well as provide management of the recording process.

In September 1991, a FileNET WorkFlo and document-imaging solution became operational with customized, integrated front-end software, called the recorder's document information system.

By reducing the need for valuable staff resources to find and deliver needed documents, the county has seen enormous public service benefits.

Every day between 3000 and 8000 new documents are recorded. The software assigns a unique identifying number and prints a label for the first page of the document with barcode, date, and time. Since original documents are not kept by the county, return instructions and mailing labels are automatically generated. The document is scanned, indexed, and stored on optical disk in a FileNET OSAR (optical storage and retrieval) library using WorkFlo software to guide the operator through the process.

The recorder's document information system handles the accounting functions, such as fee computation and reconciliation reports, on an IBM AS/400 minicomputer. The WorkFlo software integrates AS/400 screens and displays the option at the clerk's PC workstation.

"The time saved in our recording and accounting process has been phenomenal. The Office of the County Recorder has gone from a "cigar box" mentality to complete automation of all of our accounting functions," said Helen Purcell, Maricopa County Recorder.

User-friendly, on-line search capability by grantor/grantee, recording number, recording date, or document type is available in county maintained public search facilities in metro Phoenix and Mesa. On-line subscription services to the index information and document images are also available.

Public reaction to the system has been extremely positive and each day more than 200 customers use the search facilities. Public PC workstations have large color monitors with color-coded function keys and graphical interface. A customer can find a document and print it from the optical disk in a matter of seconds, a process that used to take 20 minutes or more.

In July 1997, the County Recorder expanded its service by giving Internet access to its grantor/grantee index and allowing the public to view document images of the recorded documents with FileNET WebSeries software. WebSeries software allows Internet users to easily view electronic information stored in a FileNET IMS server using a plug-in that turns a standard Internet browser into a document-imaging viewer. Through WebSeries software, people all over the U.S. have the ability to access public documents from Maricopa County using the Internet. "Every day we have between 350 and 450 persons who access documents in this way. Offering immediate, on-line, and remote access has been a great time saver for the Recorder's Office and has allowed us to enhance our public service," said Barbara Frerichs, project leader for the Office of the County Recorder.

By law the County Recorder is required to record daily, by 5 p.m., all documents presented for recording, but the high volume of documents often made that goal unattainable. Before WorkFlo and document-imaging, the manual process was labor-intensive and time-consuming. A backlog of several days work was not uncommon. Now the recording process is completed that day. The Elections Department performs signature verification for election petitions and all early voter (absentee) ballots.

During the biannual election years this can be an onerous task requiring added temporary staffing. With original voter registration affidavits stored on-line, a clerk can verify a batch of 10 signatures in a matter of minutes instead of an hour. Today the County Recorder has 120 FileNET image-enabled PC workstations. Plans include the addition of more satellite offices for expanded, on-line access to documents and Internet access. Other enhancements under consideration are the addition of OCR (optical character recognition) and full-text retrieval technology to reduce data entry and allow for expanded searches of the County Recorder's information.

With the workflow and document-imaging solution, Maricopa County has:

- Provided a return on investment to the county in approximately 18 months
- Allowed the county to double its recording transaction volumes over the past five years with no increase in staff.
- Improved service to the public through Internet access and walk-in centers for immediate, on-line access to public records at two area locations.
- Enabled sharing of recorder's information with other county offices, including the assessor, treasurer, and county transportation department.
- Improved control and management of recording transactions, allowing the County Recorder to be immediately responsive to the public.
- Reduced staff time spent on signature verification for election petitions and early voter ballots.
- Automated the process of recording documents, reduced administrative costs, allowed parallel processing of documents, and compressed processing time using WorkFlo software.

"Our Recorder's document information system with FileNET WorkFlo and document-imaging has set the standard for a state-of-the-art recorder's office. From one workstation, multiple recording tasks can be accomplished in minutes and customer service has been greatly enhanced by giving immediate access to documents. We have found a better way for county government to serve the public with FileNET," said Helen Purcell, Maricopa County Recorder.

The City of Maastricht Department of Social Services and Welfare

The Department of Social Services and Welfare of the City of Maastricht, located in The Netherlands, has implemented a unique Kofax Ascent software solution that facilitates the scanning and automatic processing of the monthly incoming periodic income reports (PIRs). The new Document Information

System (DIS), was implemented by Square, a Dutch systems integrator and Ascent Certified Reseller. It is the first step in the process of creating a fully digitized client file.

The processing of PIRs creates a very labor-intensive work effort for the 200 employees of the department. It begins when the Social Services Department mails PIR forms out to its clients, who then fill in their forms and return them for processing at the end of each month. The moment these forms — bearing name, address, and a bar code — are sent to clients by mail, their unemployment benefits become blocked. Every month 7000 PIRs (80% of which include a supplement) are sent in to the Social Services Department within a period of two to seven days. Because payment of clients' unemployment benefits depends on the processing of the PIRs in the shortest amount of time possible, an intensive effort by Social Services Department employees to expedite the PIRs begins.

Evaluating these forms was done manually, i.e., the information on the form was read, and it was deblocked and paid — only if there was no reason to alter the benefit. Thus, it became extremely important for clients to fill out their monthly forms clearly and without mistakes, so processing would not be held up. The time needed to complete the process became extremely critical.

With the "old" way of processing, throughput was too low. First, there was the "human" aspect: clients often gave incorrect or vague information on their monthly forms, increasing the risk of employees making mistakes while processing and evaluating these forms.

Second, information on paper has certain limitations. Clients' PIRs weren't always immediately available, nor could they be simultaneously consulted at different working desks. And paper can get lost.

In 1995, realizing that the old system could no longer meet the ever increasing demand for speed and service, the Maastricht Department of Social Services and Welfare formed a project group and started to look for alternatives.

The members of the project group selected system integrator and document management specialist, Square b.v., to analyze and implement a solution. After a careful needs analysis was conducted, Square decided to implement a solution based on Kofax's Ascent Capture software. Ascent software was chosen because it met both the out-of-the box and integration requirements needed by the Social Services and Welfare Department.

Square added a unique OMR (optical mark recognition) functionality to the Ascent Capture software to handle the processing of PIRs. OMR technology now allows filled-in check boxes on the PIRs to be automatically recognized. It doesn't matter whether these boxes are completely colored black, or whether they are merely ticked; they are recognized.

The advantages of using the document capture solution for automatic processing of PIRs have been innumerable:

- Time-consuming manual processing of forms has been completely automated.
- Simultaneous access to documents is easily accomplished.
- Long searches of paper files have been eliminated.
- Automated processing of PIRs has dramatically reduced human error.
- The compression of time to process PIRs has resulted in significant cost savings.
- Additional cost savings have also been realized at various other departments of the Social Service Department.

The U.S. Defense Technical Center

With more than 2 million documents on file and more being added daily, the U.S. Defense Technical Information Center (DTIC) needed to radically change its storage and retrieval system to keep up with the workload.

The DTIC, headquartered in Ft. Belvoir, Va. and with offices around the country, is the central facility for the Department of Defense in the acquisition, storage, retrieval, and dissemination of scientific and technical information. The DTIC also archives major research reports from various branches of science, including chemistry, electronics, and human factors engineering.

The system, which uses viewer technology from TMSSequoia and OCR technology from Xerox Imaging Systems, can scan 10,000 pages per day.

The center has reduced the turnaround time on reports from five days to one and enables the center to output more than 11 million printed pages per year.

State of Nebraska

The State of Nebraska had a major problem with farmers taking out multiple loans using the same agricultural commodity as collateral.

Once a farmer received a loan from one bank, he would apply for an additional loan at another bank, using the same commodity as collateral. The second or third bank would be unaware of the first loan, because the loan information was not stored or available electronically. Microfiche was used to distribute and access the data, and due to the length of time needed to create the fiche, the information was typically out of date.

The State of Nebraska recognized the need for an electronic solution for its banking problem, one that would allow the banks to instantly store, access, and share lien information relating to multiple loans.

This new system would provide information on multiple loans and enable the banks to tell what liens were outstanding on any loan at any given instant in time.

All lien information was physically stored and distributed on paper or microfiche. In addition, the U.S. Department of Agriculture maintained a centralized database (with 350 subscribers) to track loan information for 18 states. Although the loans were being recorded, the information was not being tracked or distributed quickly enough to prevent multiple loans from being made against the same collateral.

The State of Nebraska also knew that microfiche was quickly becoming obsolete and that information on microfiche could not easily be added to the central databases.

The State of Nebraska UCC Division, and DATASTOR Inc., located in Lincoln, Nebraska, installed OTG Software's Report Distribution System (RDS) for electronic dispersal of loan account information. The goal was to provide the statutory liens and farm commodities information on CD-ROMs.

RDS provides the capability to take information in the form of scanned image documents, computer output report data, and any type of object-oriented data (Word documents, spreadsheets, etc.), and format it for writing to CD-ROM. RDS, through any commercially available CD-ROM authoring system, builds an index to the data for retrieval purposes and writes the data and the index to the CD using any commercially available CD-ROM authoring system. In addition, RDS includes OTG's ApplicationExtender viewing software on the CD.

Acknowledgments

Thanks to all of the users and vendors who shared their success stories:

Cornerstone Imaging, San Jose, CA
800.562.2552 or 408-435-8900

FileNET Corporation, Costa Mesa, CA
800-345-3638

Kofax Image Products, Irvine, CA
714-727-1733

OTG Software, Bethesda, MD
800-324-4222

TMSSequoia, Stillwater, OK
800-944-7654

Index

DATE DUE

FEB 1 7 2003

JAN 1 0 2004

NOV 2 8 2004

Demco, Inc. 38-293